Nonmetalliferous Stratabound Ore Fields

EVOLUTION OF ORE FIELDS SERIES

Wilfred Walker, Series Editor

ORE FIELDS AND CONTINENTAL WEATHERING/
Jean-Claude Samama
SILICEOUS SEDIMENTARY ROCK-HOSTED ORES
AND PETROLEUM/*James R. Hein*
EVOLUTION OF CHROMIUM ORE FIELDS/*Clive W. Stowe*
NONMETALLIFEROUS STRATABOUND ORE FIELDS/
Milka K. de Brodtkorb

NONMETALLIFEROUS STRATABOUND ORE FIELDS

Edited by

MILKA K. de BRODTKORB

Consejo Nacional de Investigaciones
Científicas y Técnicas,
Universidad Nacional de La Plata
Argentina

VNR SPRINGER SCIENCE+BUSINESS MEDIA, LLC

Originally published by Van Nostrand Reinhold in 1989
Softcover reprint of the hardcover 1st edition 1989

Library of Congress Catalog Card Number 88-17199

ISBN 978-1-4684-6556-3 ISBN 978-1-4684-6554-9 (eBook)
DOI 10.1007/978-1-4684-6554-9

16 15 14 13 12 11 10 9 8 7 6 5 4 3 2 1

Library of Congress Cataloging-in-Publication Data
Nonmetalliferous stratabound ore fields/edited by Milka K. de Brodtkorb.
p. cm. — (Evolution of ore fields series)
Bibliography: p.
Includes index.

1. Ore deposits. 2. Rocks, Sedimentary. 3. Barite.
4. Celestite. I. Brodtkorb, Milka K. de. II. Series.
QE390.N66 1989 88-17199
549′.753—dc19 CIP

CONTENTS

SERIES EDITOR'S FOREWORD

The series editor's foreword provides the opportunity to give the rationale for a series on the evolution of ore fields. In brief, it meets a need that I, an explorationist, perceived as I became involved in a multidisciplinary exploration program in the late 1960s. We were looking for mines while prospecting at the ore-field scale. The practicalities demand that we know more about the ore field and that we do not just study individual deposits, the analogy being the three blind men who attempted to describe the elephant as they felt the trunk, ear, or leg.

Two considerations in identifying ore fields are the different perception as one changes scale from orebody to ore field and the problem of stratigraphic classification and nomenclature in terms appropriate to metallogenesis.

Two workers, Brock (1972) and Carey (1976), have been particularly concerned with the different appreciation brought about by a change in scale. Brock, for example, noted the contrast between a scale model of a large orebody and a microscopic section of the ore. These workers considered scale changes quantitatively and noted the changes from outcrop to regional, continental, and global scales. As one can commonly walk out the surface equivalent of an ore deposit, I think of the deposit as outcrop scale, whereas the ore field is at the regional scale. If, then, we move up a scale and study ore fields, we may reach new perceptions. We may then be able to prospect for new ore fields, not merely new deposits. Such is the aim of this series of volumes. Those of us in the business of finding ore might take a lesson from our oil geologist brethren, but have need to consider not only basin analysis but the controls of ore-field positioning related to regional deformation and metamorphism. I have asked the authors and editors of this series to consider such matters.

A major problem for the regional explorationist is the absence of an accepted stratigraphic classification and nomenclature for the Precambrian. John Cowie, chairman of the International Commission on Stratigraphy, has perceptively noted that in today's dynamic phase of development in stratigraphy, recurrent cyclic events that can be shown to be global and isochronous can be uniquely valuable in elucidating Earth history, and that cyclicity is still being sought in the modern search for the "pulse of the earth" (Cowie, 1985).

By plotting histograms of radiometric dates against time, continent by continent, Gastil (1960) has in fact shown that major thermal events—the granite-forming events—are global and episodic. The thermal events are the hallmarks of mobile belt formation. This is true even though in detail they relate to the component tectonic cycles that comprise mobile belts.

In Canada, Archean greenstones and granites are followed by Proterozoic greenstones and granites, just as Sederholm (1932) found in Scandinavia, where Proterozoic greenstones and granites are followed by Paleozoic greenstones and granites. Herein I found a means of subdividing the entire geological record (Table 1): there is a continuum from Precambrian through Phanerozoic (Walker, 1972, 1980). We now know that the pretectonic greenstones (with immature sediments) and syntectonic granites were followed by posttectonic mature platform sediments in every era. Commonly, the posttectonic sediments were ascribed to the following period, for example, the Archean Keewatin greenstones and Kenoran granites followed by the Proterozoic Huronian sediments (Hutchinson et al., 1971), and this is a problem in stratigraphic classification and nomenclature. As Joe Wargo said, "Nature made the rocks but man made the classification."

The Tectonic Subcommission of the Commission of the Geological Map of the World (Khain, 1972) provides a formal role in leading us to use the pattern of crustal evolution as a means of classifying rock sequences from the beginning of the geological record to the present (Table 2). It is noteworthy that the broad groupings—Archean, Proterozoic, Precambrian, and Phanerozoic—are absent from the legend.

The importance of the pattern of crustal evolution (and the pulse of the Earth) to the exploration geologist comes when the work of Bilibin (1968) is introduced. Bilibin taught that ore formation is closely linked to the development of Phanerozoic mobile belts. My role was to show that as Precambrian mobile belts followed the same evolutionary path as Phanerozoic ones, so too, should ore formation.

As Bilibin showed, ore formation is related to the tectonic stages of development in the one or more tectonic cycles that comprise a mobile belt. Thus a common evolutionary path for metal is introduction in pretectonic vulcanism and contemporary deposition in the volcano-sedimentary environment; modification by deformation and metamorphism during syntectonic orogeny; and erosion and redeposition in the posttectonic environment. Many ore fields are subsequently reworked when they are crossed by a younger mobile belt, so posttectonic and, indeed, epicratonic platform sediments may or may not be modified by deformation and metamorphism.

Table 1 Table of Eras

Era	*Age in m.y. (approx.)*
Alpine	200–present
Paleozoic	600–200
Katangan	900–600
Grenvillian	1,600–900
Karelian	2,200–1,600
Superior	3,000–2,200
Swazian	3,500–3,000
Godthaabian	4,000–3,500
Enderbian	4,400–4,000

Source: From Walker (1982).
Note: The Table of Eras has approximate ages. It is not a chronostratigraphic column in which time takes precedence.

Table 2 Legend of the Tectonic Map of the World

Areas of Continental Crust	*Consolidated before*
Mesozoic and Cenozoic	present
Paleozoic	200–190 m.y.
Katangan-Baikalian-Cadomian	650–570 m.y.
Grenville-Dalslandian-Satpuran	1,100–900 m.y.
Gothian-Elsonian	1,500–1,350 m.y.
Karelian-Hudsonian-Mayombian-Eburnian	2,000–1,700 m.y.
Kenoran-Belomorian-Rhodesian-Guiana	2,600 m.y.
Transvaalian	3,000 m.y.

Source: From Khain (1972).

I have asked each author or editor to consider crustal evolution as a sequence of developing mobile belts and related platforms. In working on his volume on *Continental Weathering,* Jean-Claude Samama noted first the difficulty in transposing from the classical stratigraphy, then the excitement as the pieces came into place. In the present volume, Ramos and de Brodtkorb (Chap. 14) have compiled the data on barite, celestite, magnesite, and fluorspar on a time scale (Fig. 14-9), while in their section on barite, with their approval, I have transposed the nomenclature to that in my Table of Eras (Table 1). In the next edition of this volume, I, for one, would like to see the authors of the case history chapters develop the theme initiated by Ramos and de Brodtkorb in this final chapter. We look forward, also, to input from areas and authors not represented here.

WILFRED WALKER

REFERENCES

Bilibin, Y. A. 1968. *Metallogenic Provinces and Metallogenic Epochs,* trans. E. A. Alexandrov. Flushing, N.Y.: Queens College Press.

Brock, B. B. 1972. *A Global Approach to Geology.* Capetown: Balkema.

Carey, S. W. 1976. *The Expanding Earth,* Developments in Geotectonics, No. 10. Amsterdam: Elsevier.

Cowie, J. W. 1985. Guest Editorial, Stratigraphy and the International Commission. *Episodes* **8**(2):86.

Gastil, R. G. 1960. The distribution of mineral dates in time and space, *Am. Jour. Sci.* **258**(1):1.

Hutchinson, R. W., R. H. Ridler, and G. G. Suffel. 1971. Metallogenic relationships in the Abitibi Belt, Canada, a model for Archean metallogeny, *Canadian Institute of Mining and Metallurgy Bulletin 64,* p. 48.

Khain, V. 1972. *Legend, Tectonic Map of the World.* Tectonic Subcommission, Commission of the Geological Map of the World.

Sederholm, J. J. 1932. Geology of Fennoscandia. *Commission Geologique Finlande Bulletin 98,* 30p.

Publishers Note: We are sorry to inform our readers that Dr. Wilfred Walker died during the production of this volume.

Walker, W. 1972. Mantle cells and mineralization, *Soc. Min. Engineers, AIME, Trans.* **252**:314.

Walker, W. 1980. The development of mobile belts since 4 b.y. and related ore paragenes. *IAGOD* **5**(1):185.

Walker, W. 1982. Evolving earth, Pacific opening, and related mineral belts. Society of Mining Engineers, AIME Preprint number 82–319.

PREFACE

This book is the result of the modern knowledge from research devoted to stratabound and stratiform nonmetalliferous ore fields of barite, celestite, magnesite, and fluorite.

The metalliferous stratabound ore fields have been grouped and described in several specific publications, whereas less emphasis has been placed on the nonmetalliferous ores in the international geologic literature. In editing this volume I have requested the cooperation of a qualified team of experts in order to narrow the gap.

The studies of processes and products of recent sedimentary environments are important in interpreting present-day controls of ore accumulation and have therefore contributed to a better understanding of the past sedimentary ore deposits.

In the evolution of a stratiform or stratabound deposit we must take into account the origins of cations and anions that may be provided separately to the system, during the formation of the deposit, or in a later diagenetic process. An important step toward the understanding of these processes was the work of M. Erhard (1956) on the biorhexistasis and the more recent studies by French researchers such as J. C. Samama on weathering, which is the result of atmospheric, hydrospheric, and biospheric processes over the continental surface. On the other hand, and no less important, were the investigations on recent environments, mainly in the sub-, intra-, and supratidal settings, conducted in the sabkhas of the Persian Gulf by D. J. J. Kinsman and others in the 1960s.

A better understanding of submarine volcanism was an important conceptual step, as was the detailed knowledge of the submarine hydrothermal activity known from the ocean rift systems, which show outstanding synsedimentary mineral concentrations. This produced new insight into the genesis and interpretation of different types of ore deposits.

The term *stratabound,* although widely used today, is comparatively new in the literature. In 1970 P. Nicolini distinguished two morphologic groups of stratiform deposits, the stratiform vrais and the péneconcordants. The *stratiform vrais* group was restricted to the ores where mineralization is disseminated throughout the host rock, having the mineral concentration in volume similar to that of the host rock. The *péneconcordant* group comprises ores where mineralization is somewhat discontinuous and more or less discordant with the volume of the host rock.

R. L. Stanton (1972) indicated that *stratiform ores* occur as layers concordantly disposed with respect to the stratification of the enclosing sedimentary materials. In general, they are sharply confined stratigraphically and may occur on one or a very few preferred sedimentary horizons. The *stratabound ores* also habitually favor specific horizons within local sedimentary sequences and show marked preferences for certain rock types. They are found within specific sedimentary units as pore-space and breccia fillings, cave and solution cavity fillings, crosscutting veins, and minor layers. On the other hand, for J. W. Gabelman (1976), a *stratiform deposit* has real stratigraphic layers of varying but generally high purity and of great uniform continuity. Conversely, a stratabound ore is a mineral deposit confined to a single stratigraphic unit and could be either a stratiform or a randomly oriented body contained within a single stratigraphic unit. The term *stratiform* was also used for bodies in a pseudostratigraphic position of certain metamorphic rocks or ultrabasic magmatites.

Other related terms are *concordant,* meaning an ore whose external limits coincide with the outline of the host rock structure; *manto,* a Spanish word with the meaning of a tabular ore body, mostly in horizontal position and confined to a determined stratigraphic horizon. *Ore bed* is used by G. C. Amstutz to denote an ore deposit of wide extent in sedimentary rocks.

The mineral deposits confined to a specific stratigraphic interval can result from a variety of geologic processes. But the terms *stratiform* and *stratabound* in the past had had either a geometric or a genetic connotation. Stratiform ores were restricted for some authors to the bedded or stratified ores. Before 1950 the stratabound deposits had been frequently considered as epigenetic replacement ores, but later, based on their concordant outline, they have been considered mainly as syngenetic. An early to late diagenetic origin with eventual epigenetic remobilization is the most common interpretation today for such deposits.

Sulfate evaporite minerals, until recently, were not considered as deposited before 800–600 Ma ago,* but in recent years sulfate evaporite deposits have been found in the Amadeus basin, central Australia. They consist of a warm-water shallow marine sea sequence of stromatolitic dolomite and limestones, microfossiliferous chert, and evaporites, 900 Ma old. Also in the McArthur group, Northern Territory, Australia, is a major sulfate evaporite deposit, as old as 1,600–1,400 Ma, to which huge stratiform Pb–Ag deposits are related. They are among the world's largest deposits of pre-Paleozoic evaporites. Smaller evaporite bodies ranging in age from 2400 to 600 m.y. are known from northern Australia, Canada, and South Africa.

In these deposits, pseudomorphs after halite and gypsum were reported in the 1960s. The most frequent evaporite relicts are carbonate pseudomorphs after euhedral and subeuhedral gypsum and anhydrite crystals, chert pseudomorphs after anhydrite nodules and halite casts.

The present knowledge goes as far back as 3.0 to 3.5 Ga for *stratabound barite deposits* known from the Swaziland supergroup of South Africa, the East Pilbara Block of western Australia, and the Peninsular Gneissic complex of India.

The chapters in this book incorporate previous investigations with recent research data. The modern study of geometric forms together with geochemical and isotopic data helps genetic interpretation and is, in this way, a tool for facies characterizations and new ore findings.

*The traditional abbreviation for million years, m.y., has recently been replaced by Ma for milliannum. Likewise, thousand million years is now being expressed as Ga for giga-annum. The reader will find both the traditional and modern expressions in this book as the usages are still in transition.

The chapters are grouped by themes and regional studies. Chapters 1 through 11 are devoted to celestite and barite. After the introduction on the geochemistry of Sr and Ba, regional studies are described. Worldwide celestite ore fields with their strong behavioral characteristics are mentioned in one chapter. The barite deposits can occur in different environments, such as littoral, distal platform, and deep oceanic, and also in karst fillings. The origin of the barium cations are different in the diverse types and can result from the interaction of sea water with released barium or from the introduction through ocean floor hydrothermal events. Regional descriptions follow, subdivided into big districts or countries so we can find the old known mines of central Europe, the karst fillings of Sardinia, the Kuroko deposits of Japan, and the different ore types of South America and North America. Barite nodules of continental or marine origin are described in another chapter.

Chapter 12 deals with magnesite ores. They are divided into cryptocrystalline magnesite in or affiliated with ultramafics, crystalline magnesite in sedimentary or metamorphic rocks, saline to hypersaline and/or evaporite magnesites, and magnesites and related Mg-bearing minerals in intramontane freshwater basins. The crystalline magnesite deposits are widespread and at present are the most important commercially, but the occurrences in Cenozoic freshwater basins have been overlooked because of insignificant aspects. However, the findings are rapidly increasing.

The next chapter, 13, describes the principal types of fluorite deposits: the ores connected to lacustrine environments, those related at the base of transgressive series, and those lodged within carbonate series.

Finally, there is a summary, Chapter 14, in which the tectonic setting and the time-bound characteristics of the described ore fields are discussed.

Most of my interest in stratabound deposits is the result of several years of learning, studying, and discussing with my former teachers, Professor Maucher, Professor Ramdohr, and Dr. Stipanicic. Active interaction with my colleagues, Professor Schneider, Professor Amstutz, Professor Arribas, and the reading of papers by Professor Routhier, Professor Bernard, Professor Samama and other colleagues have directly or indirectly contributed to my present understanding of the various problems of stratabound genesis.

ACKNOWLEDGMENTS

The results achieved in this book are possible through the encouragement of Wilfred Walker, who gave me the opportunity to prepare the present volume in the Evolution of Ore Fields Series. The skill and expertise of outstanding authors of the various chapters have contributed immensely to the success of this book.

To all of them, my gratitude, as well as to the publisher for the fine preparation and editing of the manuscripts.

MILKA K. de BRODTKORB

REFERENCES

Erhard, M. 1956. *La genèse des sols en tant que phénomène géologique.* Paris: Masson et Cie.

Gabelman, J. W. 1976. Classification of strata-bound ore deposits. In *Handbook of Strata-bound and Stratiform Ore Deposits,* Vol. 1, ed. K. H. Wolf, pp. 79–110. Amsterdam: Elsevier.

Nicolini, P. 1970. *Gîtologie des concentrations minérales stratiformes.* Montreuil: Gauthier-Villars.

Stanton, R. L. 1922. *Ore Petrology.* New York: McGraw-Hill.

Contributors

Silvia Ametrano
Facultad de Ciencias Naturales y Museo, Paseo del Bosque, 1900 La Plata, Argentina

G. Christian Amstutz
Mineralogisch-Petrographisches Institut de Universität, Im Neuenheimer Feld 236, D-6900 Heidelberg, Federal Republic of Germany

Mario Barbieri
Dipartimento di Scienze della Terra, Univesità degli studi di Roma "La Sapienza," P. le Aldo Moro 5, 00185 Roma, Italy

Milka K. de Brodtkorb
Instituto de Recursos Minerales, Universidad Nacional de La Plata, Calle 47 No. 522, 1900 La Plata, Argentina

Jacques P. Cassedanne
Instituto de Géociencias, Universidad Federal do Rio de Janeiro, Cidade Universitária, Ilha do Fundão, 21910 Rio de Janeiro, Brasil

Sandra H. B. Clark
U.S. Geological Survey, National Center, MS 954, Reston, Virginia 22092, USA

Yves A. Fuchs
Laboratoire de Geochimie et Minéralogie Expérimentale et Appliquée, Tour 25-26, 2 ème Et., Université Pierre et Marie Curie, 4 Place Jussieu, 75232 Paris Cedex 05, France

Horacio Magliola-Mundet
Facultad de Ciencias Exactas y Naturales, Universidad Nacional de Córdoba, Avda Velez Sarsfield 299, 5000 Córdoba, Argentina

Katsumi Marumo
Geological Survey of Japan, 1-1-3 Higashi, Yatabe, Ibaraki, 305 Japan

Giorgio Padalino
Istituto de Giacimenti Minerari, Geofisica e Scienze Geologiche, Università di Cagliari, Cagliari, Sardegna, Italy

Forrest G. Poole
U.S. Geological Survey, Denver Federal Center, MS 905, Denver, Colorado 80225, USA

Victor A. Ramos
Facultad de Ciencias Exactas, Físicas y Naturales, Universidad de Buenos Aires, Pabellón 2, 1428 Nuñez, Argentina

Isidoro B. A. Schalamuk
Instituto de Recursos Minerales, Universidad Nacional de La Plata, Calle 47 No. 522, 1900 La Plata, Argentina

Sandro Tocco
Istituto di Giacimenti Minerari, Geofisica e Scienze Geologiche, Università de Cagliari, Cagliari, Sardegna, Italy

Maurizio Violo
Dipartimento di Ingegneria Chimica, dei Materie Prime e Metallurgia, Universita degli studi di Roma "La Sapienza," Via Eudossiana 18, 00184 Roma, Italy

Wolfgang Wetzenstein
Am Milchbornsberg 4 b, 5060 Bergisch Gladbach 1, Federal Republic of Germany

Richard A. Zimmermann
Mineralogische-Petrographisches Institut der Universität, Im Neuenheimer Feld 236, D-6900 Heidelberg, Federal Republic of Germany

Nonmetalliferous Stratabound Ore Fields

1

Strontium Isotopes as Natural Tracers in Minerogenetic Processes

Mario Barbieri

It is well known that strontium can enter the lattice of minerals as calcite, fluorite, gypsum, anhydrite, and barite. Therefore, the geochemistry of the strontium in these minerals can be a suitable, though indirect, method for studying the genesis of such minerals and thus for investigating the origin of the ore deposits in which they occur. In fact, the strontium isotope composition has been largely used for studying the genesis of ore deposits since the late 1970s (Barbieri et al., 1977, 1982, 1984; Kesler and Jones, 1981; Brodtkorb et al., 1986).

Four naturally occurring strontium isotopes show the following approximate percentage abundances: $^{84}Sr = 0.55$; $^{86}Sr = 9.75$; $^{87}Sr = 6.96$; and $^{88}Sr = 82.74$. The ^{84}Sr, ^{86}Sr, and ^{88}Sr isotopes are constant in nature. They are neither radioactive nor the decay products of any natural radioactive isotope. The ^{87}Sr isotope is also stable, but its abundance shows some small variations due to the variable increments of ^{87}Sr derived from the radioactive β decay of ^{87}Rb. The variations of ^{87}Sr are small due to both the low Rb abundance in the Earth's materials and the extremely long ^{87}Rb half-life (50.10^9 years).

The Rb–Sr chronometer is very useful in determining geologic ages, and the strontium isotope composition, expressed as $^{87}Sr/^{86}Sr$, is an important tracer of various geologic processes. In particular, the strontium isotope composition measured in calcium-carbonates and Ca, Sr, and Ba sulfate minerals, which are known to contain no valuable amounts of rubidium, should be the original one when they first crystallized. This evidence allows us to determine the source of strontium in these latter minerals by simply comparing their strontium isotope composition with that of the surrounding rocks.

Barite, celestite, and Ca-sulfates can occur as stratabound minerals within the sedimentary rocks of any age. These minerals likely have been formed in equilibrium with sea water and/or originated, in the case of celestite and barite, during early or late diagenetic processes. In the first occurrence, celestite and barite would show a strontium isotope composition close to that of the original sea water. In the second occurrence, the $^{87}Sr/^{86}Sr$ ratios of the minerals are related to that of the strontium-, barium-, and rubidium-bearing sedimentary rocks.

At any one time strontium is isotopically well mixed in ocean water, and this homogeneity is related to the ratio of the residence time of strontium in the oceans, t, and to the mixed time of the oceans, tm. For strontium, $t/tm = 4000$, so that a high degree of homogeneity is expected. Replicate analyses of modern sea water indicate a worldwide strontium isotope ratio close to 0.709198 ± 0.000020 (De Paolo and Ingram, 1985). However, during the geologic evolution of the Earth, the strontium isotope ratios in the oceans modified because of changes in the relative inputs of strontium to the oceans from different sources, each with its own strontium isotopic value. The main fluxes are represented by submarine hydrothermal activity, continental fresh waters, and submarine dissolution of ancient sedimentary carbonates. A reliable record of the strontium isotope composition can be used to understand the role of the various sources in controlling present and past sea water chemistry.

There is increasing evidence that a close relationship exists between the strontium isotope composition of sea water and geologic time. Wickman (1948) proposed that the decay of ^{87}Rb in the Earth's crust would lead to an increase of about 30 percent in the $^{87}Sr/^{86}Sr$ ratio of sea water in the last 3350 m.y. Gast (1955) and Hedge and Walthall (1963) found much smaller increases than those originally predicted by Wickman. According to Hedge and Walthall, this discrepancy is due to an overestimation of the Rb/Sr ratio of the Earth's crust. Koepnick et al. (1985) suggest that the $^{87}Sr/^{86}Sr$ ratio of sea water has increased only about 1.4 percent throughout geologic time.

Peterman et al. (1970) determined $^{87}Sr/^{86}Sr$ ratios of aragonitic shells from paleontologically dated time intervals within the Phanerozoic. They documented both an increased and decreased strontium isotope ratio with geologic time (see Fig. 1–1). These early findings have generally been supported by Dasch and Biscaye (1971; Fig. 1–2) and Veizer and Compston (1974; Fig. 1–3). The latter authors also argued about the reasons why the marine carbonates must have the $^{87}Sr/^{86}Sr$ ratio of the sea water at the time of their formation and the role of the diagenetic processes that generally produce the increase of this ratio or leave it unaltered.

Since 1974, additional papers on $^{87}Sr/^{86}Sr$ ratios determined by marine carbonate, evaporite, and phosphate samples of known ages have been published (see Tremba et al., 1975; Brass, 1976; Faure et al., 1965; Kovach, 1980; Burke et al., 1982; De Paolo and Ingram, 1985; Koepnick et al., 1985; Fig. 1–4).

All these data permit definition of the curve of the sea water $^{87}Sr/^{86}Sr$ ratios versus geologic time (Late Cambrian to Holocene), as shown graphically in Figure 1–5. This curve confirms the general shape of that previously defined by Peterman et al. (1970) and Veizer and Compston (1974).

Strontium isotope ratios, close to that of modern sea water (0.709198 ± 0.000020), during Late Cambrian to Early Ordovician time started to fluctuate from Ordovician time, with the lowest ratios (close to 0.7068) in Jurassic and Late Permian time. These temporal variations of the strontium

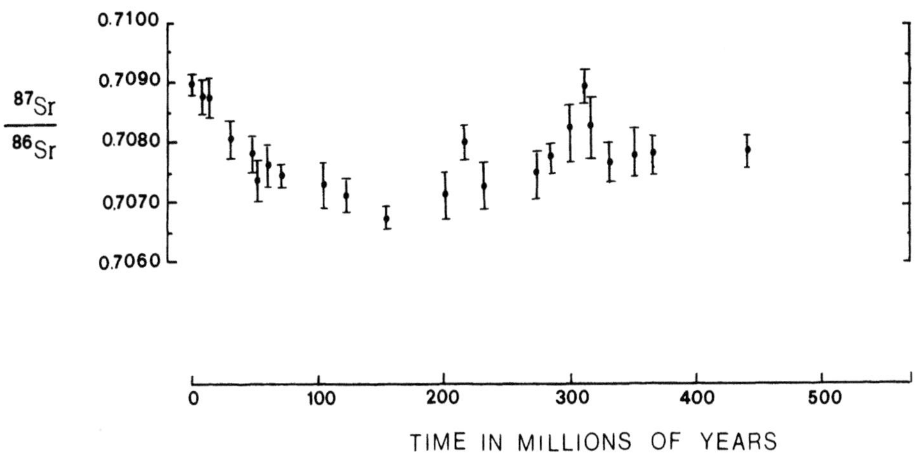

Figure 1-1. $^{87}Sr/^{86}Sr$ ratios versus time for aragonitic fossils. Vertical line gives range of analytical uncertainties (after Peterman et al., 1970).

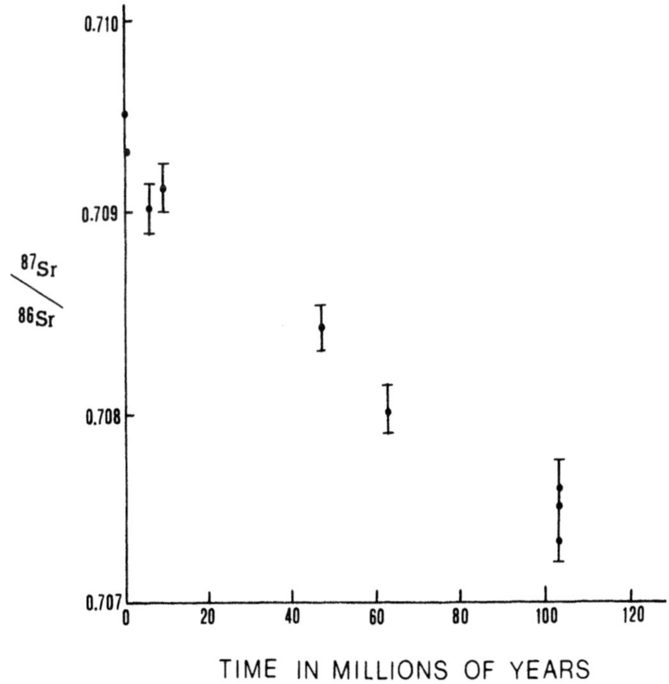

Figure 1-2. $^{87}Sr/^{86}Sr$ ratios versus time for pelagic foraminifera from Cretaceous to Recent (after Dasch and Biscaye, 1971).

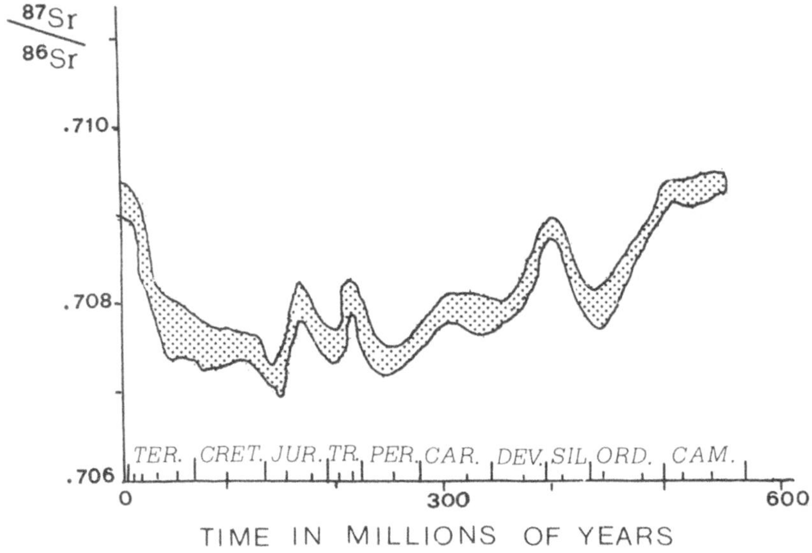

Figure 1-3. $^{87}Sr/^{86}Sr$ ratios versus time from Cambrian to Recent (after Veizer and Compston, 1974).

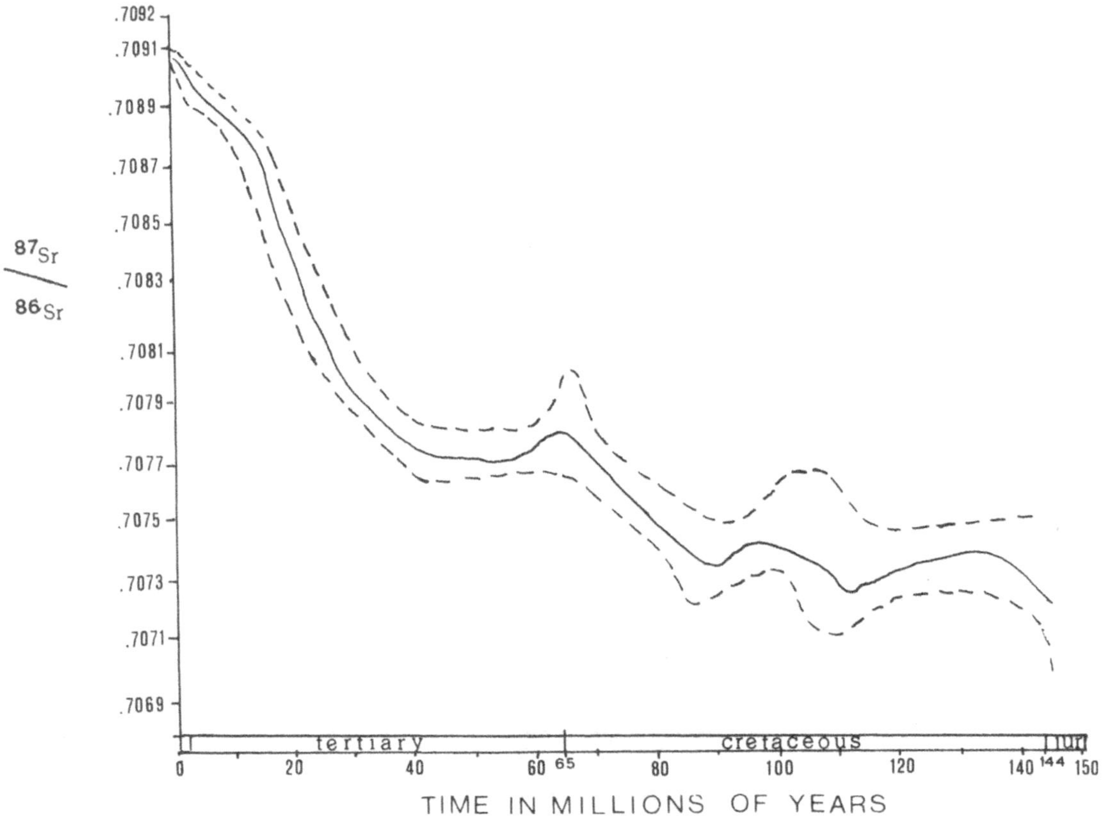

Figure 1-4. $^{87}Sr/^{86}Sr$ ratios versus time for the Cenozoic, Cretaceous, and Late Jurassic (after Koepnick et al., 1985).

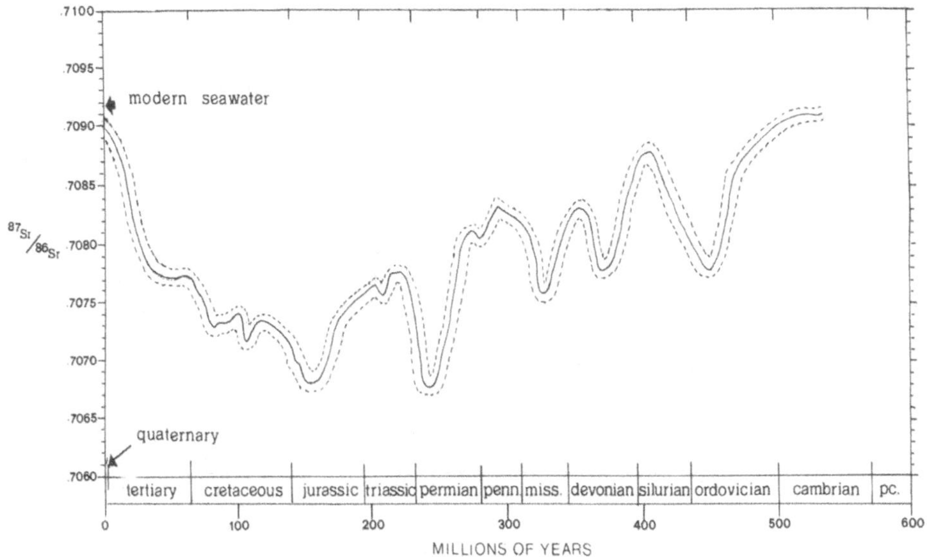

Figure 1-5. $^{87}Sr/^{86}Sr$ isotope ratios in sea water from Cambrian to Present; dotted line represents the range of analytical values (data from Burke et al., 1982; De Paolo and Ingram, 1985).

isotope ratios of sea water are related to the changing fluxes of strontium from isotopically distinct sources within the crust. High ratios would indicate prevailing old sialic rocks as the ^{87}Sr radiogenic source. On the contrary, low strontium isotopic ratios would suggest a major role of mafic volcanic and intrusive rocks from active margins and ocean basin ($^{87}Sr/^{86}Sr$ average values close to 0.704; Faure and Powell, 1972). The shape of the $^{87}Sr/^{86}Sr$ curve seems to be controlled significantly by the geologic evolution of the Earth through the Phanerozoic time, as, for example, the systematic rising ratio during the Tertiary.

According to Burke et al. (1982), the actual circumstances leading to these fluctuations likely involve a great number of factors, such as variations in the lithologic composition of the crust involved in the weathering processes, landscape of the continents, changes in climate, extent of magmatic activity, and nature of the volcanic materials. Many of these factors may be related to global tectonic processes, thus the temporal $^{87}Sr/^{86}Sr$ variations in sea water may be evaluated in terms of tectonic implications.

The curve of sea water $^{87}Sr/^{86}Sr$ ratios versus geologic time provides a useful basis for dating marine carbonates, evaporites, and phosphate rocks but only when these materials show high strontium concentrations and a very low insoluble residue.

Moreover, strontium isotope geochemistry is widely used to test the possible sources for this element and the relationship between Sr-bearing minerals and the associated surrounding rocks.

Wood and Shaw (1976) measured $^{87}Sr/^{86}Sr$ ratios in celestite samples from the Yate area, near Bristol, England. In this area, celestite occurs as massive and geodetic nodules, beds, or veins within the Keuper marl, in the Coal Measures and Silurian rocks, which are overlain, unconformably, by the Keuper marl. According to Wood and Shaw, the high $^{87}Sr/^{86}Sr$ values of celestite indicate that this mineral did not originate by primary precipitation from normal sea water or solutions directly derived from sea water. The most likely source of the Sr would be via the dolomitization of the Carboniferous limestone releasing the element into solution. Moreover, the strontium isotope composition of celestite indicates that this mineral formed during several episodes, each one with a slightly different ground water composition.

Brodtkorb et al. (1982) studied the celestite–barite deposits of the Early Cretaceous from the Extra Andean region of Neuquén (Argentina). The ores are located in a Subandean Mesozoic belt and show a conspicuous stratabound character related to evaporitic facies. The study of the geologic evolution of the sedimentary sequence allows prediction of the temporal and spatial setting of the evaporitic facies. Stratigraphy is the main geologic control for the celestite–barite ores, and their distribution depends on the structure of the region. Based on the geometry and texture of the mineralization, three types of ores have been recognized: (1) celestite beds syngenetic with the evaporitic facies; (2) epigenetic stratabound ore formed by in situ mobilization of celestite with stalactite growth and cavity filling; and (3) epigenetic vein type ores formed by mobilized Ba-enriched solutions. Lithofacies and strontium isotope ratios, close to the values reported in the literature for the Lower Cretaceous sea water, in addition to the geologic evidence confirm an evaporitic origin for the celestite beds.

Barbieri and Masi (1984) analyzed the only important deposit of celestite in the entire Italian territory, located in the northwestern part of the Tolfa Mountains (Latium, Central Italy). The mineralization shows a concretionary structure, the concretions being made up of alternating layers of barite, celestite, and rare calcite. Ore is host in a stratigraphic unconformity between the Pliocene gypsum-bearing sediments and the Tertiary flysch. The Tolfa Mountains were affected by major volcanic activity of Plio-Pleistocene age whose products of acidic composition outcrop mostly in the central part of the area. Likely connected with the now extinct volcanism, travertine-depositing thermal springs occur throughout the Tolfa Mountains, mostly on the flanks of this mountain system. Strontium isotope composition of celestite–barite samples, very similar and close to 0.70820, indicates that the bulk of Sr of the minerals was provided by the same source. This latter is likely represented by the Upper Triassic marine evaporites, which are present in the local stratigraphic series.

References

Barbieri, M., and U. Masi. 1984. Sr geochemical evidence on the origin of celestite-barite deposit at Pian dell'Organo in the Tolfa Mountains area (Latium, Central Italy). *Min. Petrogr. Acta* **28**:33–37.

Barbieri, M., U. Masi, and L. Tolomeo. 1977. Geochemical evidence on the origin of the epithermal fluorite deposit at Monte delle Fate near Cerveteri (Latium, Central Italy). *Mineralium Deposita* **12**:393–398.

Barbieri, M., U. Masi, and L. Tolomeo. 1982. The strontium geochemistry in the epithermal barite deposits from the Apuan Alps (northern Tuscany, Italy). *Chem. Geology* **35**:351–356.

Barbieri, M., U. Masi, and L. Tolomeo. 1984. Strontium geochemical evidence for the origin of the barite deposits from Sardinia (Italy). *Econ. Geology* **79:**1360–1365.

Brass, G. W. 1976. The variation of the marine $^{87}Sr/^{86}Sr$ ratio during Phanerozoic time: Interpretation using a flux model. *Geochim. Cosmochim. Acta* **40:**721–730.

Brodtkorb, M. K. de, V. Ramos, M. Barbieri, and S. Ametrano. 1982. The evaporitic celestite-barite deposits of Neuquén, Argentina. *Mineralium Deposita* **17:**423–436.

Brodtkorb, M. K. de, I. B. A. Schalamuk, M. Barbieri, H. Puchelt, S. Ametrano, R. Fernandez, and R. Etcheverry. 1986. The nodular barite deposits of Canota, Argentina. *Quadrennial IAGOD Symposium, 7th, August 1986, Proc.* (in press).

Burke, W. H., R. E. Denison, E. A. Hetherington, R. B. Koepnick, H. F. Nelson, and J. B. Otto. 1982. Variation of $^{87}Sr/^{86}Sr$ throughout Phanerozoic time. *Geology* **10:**516–519.

Dasch, E. J., and P. E. Biscaye. 1971. Isotopic composition in Cretaceous-to-Recent pelagic foraminifera. *Earth and Planetary Sci. Letters* **11:**201–204.

De Paolo, D. J., and B. L. Ingram. 1985. High resolution stratigraphy with strontium isotopes. *Science* **227:**938–941.

Faure, G., and J. L. Powell. 1972. *Strontium Isotope Geology.* Berlin: Springer-Verlag, 188p.

Faure, G., P. M. Hurley, and J. L. Powell. 1965. The isotopic composition of strontium in surface water from the North Atlantic ocean. *Geochim. Cosmochim. Acta* **29:**209–220.

Gast, P. W. 1955, Abundance of ^{87}Sr during geologic time. *Geol. Soc. America Bull.* **66:**1449–1454.

Hedge, C. E., and F. G. Walthall. 1963. Radiogenic strontium-87 as an index of geological processes. *Science* **140:**1214–1217.

Kesler, S. E., and L. M. Jones. 1981. Sulfur and strontium isotopic geochemistry of celestite, barite and gypsum from the Mesozoic basin of Northeastern Mexico. *Chem. Geology* **31:**211–224.

Koepnick, R. B., W. H. Burke, R. E. Denison, E. A. Hetherington, H. F. Nelson, J. B. Otto, and L. E. Waite. 1985. Construction of the sea water $^{87}Sr/^{86}Sr$ curve for the Cenozoic and Cretaceous: Supporting data. *Chem. Geology* **58:**55–81.

Kovach, J. 1980. Variations in the strontium isotopic composition of sea water during Palaeozoic time determined by analysis of econodonta. *Geol. Soc. America Abstracts with Programs* **12:**465.

Peterman, Z. E., C. E. Hedge, and H. A. Tourtelot. 1970. Isotopic composition of sea water throughout Phanerozoic time. *Geochim. Cosmochim. Acta* **34:**105–120.

Tremba, E. L., G. Faure, G. C. Katsikatsos, and C. H. Summerson. 1975. Strontium-isotope composition in the Tethys sea, Euboea, Greece. *Chem. Geology* **16:**109–120.

Veizer, J. and W. Compston. 1974. $^{87}Sr/^{86}Sr$ composition of sea water during the Phanerozoic. *Geochim. Cosmochim. Acta* **38:**1461–1484.

Wickman, F. W. 1948. Isotope ratios: A clue to the age of certain marine sediments. *Jour. Geology* **56:**61–66.

Wood, D. W., and H. F. Shaw. 1976. The geochemistry of celestite from the Yate area near Bristol (U.K.). *Chem. Geology* **17:**179–193.

2

Geochemistry of Barium

Mario Barbieri

Among the most common divalent positive ions, barium has the largest ionic radius. The crystal chemistry character of Ba leads us to believe that it may replace (or may be replaced by) large ions such as Pb^{2+}, Sr^{2+}, and K^{+}. Less common is replacement of the smaller Ca^{2+}. Various substitutional schemes have been proposed for the Ba–K ions (Shmakin, 1984; Wentlandt, 1977). In igneous rocks of the Earth's crust, Ba usually does not form minerals of its own, thus it is distributed among a number of silicate minerals, mainly feldspars and micas, where it substitutes for potassium.

During the first stages of magmatic differentiation, Ba is enriched in the liquid phase. With progressing crystallization, barium is incorporated essentially in K-feldspar and micas; residual melt then results in depletion of this element. Pegmatites resulting from magmatic differentiation generally contain less barium in comparison with their embedding wall rocks of magmatic or metamorphic origin. This is most apparent in the main Ba-bearing minerals such as feldspars and micas (Heier and Taylor, 1959). It is likely that magmatic-hydrothermal fluids do not originally contain significant amounts of Ba; this element is probably derived by leaching suitable wall rocks (Engelhardt, 1936). Another mechanism working in sedimentary environments is represented by dissolution of barite from the sediments by bacterial reduction in a suitable diagenetic environment (Puchelt, 1967). Ba-bearing solutions may also be produced by a reaction that may give off barium because structures (such as micas and feldspars) that had incorporated this element become unstable.

The possibility of Ba transport by hydrous gas phases was first demonstrated by the experiments of Strübel (1967). Naboko (1945) and Minguzzi

(1948) determined traces of Ba in fumarole sublimates of the Klyucherskoy volcano (U.S.S.R.) and in fumarole products of Vesuvius (Italy), respectively. During weathering and alteration of the rocks, Ba shows a geochemical behavior that depends on a great number of factors such as type of clay minerals, which form during weathering of primary minerals, amount and type of organic matter present, and sulfur or sulfate content.

Barite is the least soluble sulfate mineral and the most abundant barium mineral in the Earth's crust. According to Strübel (1967), $BaSO_4$ solubility in the hydrothermal range up to 600°C and 1.084 bars is 9.61 ± 1.95 mg/kg H_2O. The presence of electrolytes in solution considerably increases the barite solubility.

Weathering of igneous rocks bearing feldspars yields Ba ions that are adsorbed by clay minerals, hydroxides, and organic substances. Therefore, these phases control the amount of Ba content in natural waters. Oceanic water contains up to 78 ppb, and the estimated mean is about 20 ppb (Turekian and Johnson, 1966). The Ba concentration in sea water is not uniform and, according to Chow and Goldberg (1960), steadily increases with depth due to the large amount of biological activity. In fact, at the surface Ba is adsorbed on the organic matter and is then transported downward by the organic debris.

Because of the Ba supplied by rivers, the ocean water should be almost saturated in $BaSO_4$. If we assume a stream discharge of $3.6 \cdot 10^4$ km^3/year, a Ba concentration of 45 ppb (Durum and Haffty, 1963), an ocean mass of 1,372 $\cdot 10^6$ km^3, and an average Ba content of 20 μg/1 (Turekian and Johnson, 1966), it is possible to calculate a residence time of Ba in the sea water of 17.10^3 years. Assuming a sedimentation rate of 0.05 g SiO_2/year for diatomaceous oozes with an average barium content of 6000 ppm and a mean ocean depth of 5000 m, Turekian and Johnson (1966) calculate a residence time of only 33 years. This is a very low value and, because Ba increases with depth, indicates an additional source of Ba, likely from sea-floor volcanism (Canadian-American Seamount Expedition, 1985).

The Ba content of common recent and fossil sediments and sedimentary rock types is reported by Puchelt (1967) and Marchig et al. (1985). Deep-sea clay shows Ba contents ranging from 200 ppm to 6700 ppm, whereas deep-sea carbonate materials show contents ranging from 210 ppm to 5000 ppm and deep-sea siliceous muds from 700 ppm to 10,400 ppm. Manganese nodules, occurring on wide areas of the sea bottom, contain up to 20,000 ppm Ba, either as adsorbed element or as $BaSO_4$ (Arrhenius, 1963). Consolidated sediments show a variable Ba amount; in particular, quartz-bearing sediments show low values, whereas feldspar-bearing sediments are enriched. Among silica-rich sediments, chert generally exhibits very high Ba concentrations. Puchelt (1967) and Vinogradov (1956) measured average Ba values from 250 ppm to 800 ppm, and Leutwein (1951), measuring shales from Germany found an average of 750 ppm with values ranging from 50 ppm to 5000 ppm. Black shales generally show Ba contents higher than common shales, suggesting a relationship between Ba and the organic matter. Some shales show the original content of Ba, whereas others show a gain or loss of barium. Diagenetic processes can produce a Ba redistribution in the sediments (Marchig et al., 1985). Carbonate rocks generally show very variable (from 1 ppm to 10,000 ppm) Ba content (Puchelt, 1967). High Ba values in carbonate rocks are mainly due to detrital clay minerals, redistribution and precipitation of $BaSO_4$ during diagenetic evolution of sedimentary rocks, or Ba incorporation in carbonate minerals. The latter process is quantitatively of minor extent. During sea water evaporation, Ba is precipitated as barite in the form of finely disseminated minerals within the calcium carbonates (Puchelt, 1967; Gandin et al., 1973). Diagenetic and secondary alteration of carbonate evaporitic rocks can yield very high barite concentrations (Gandin et al., 1973). The organic matter generally shows a variable, but very high, Ba content. Marine plants exhibit Ba contents ranging from 10 ppm to 30 ppm.

The most common barium ore is barite. This mineral is deposited by fluids with high oxidation potential, where sulfur is present as $SO_4^{=}$. Such conditions occur at the Earth's surface or underground, where sulfate-bearing solutions mix with Ba-bearing solutions. Puchelt (1967) discussed barite formation and the possibility of Ba transport in natural solutions. The Earth's most important barite occurrences are: (1) synsedimentary barite; (2) barite originated by $CaSO_4$ metasomatism; (3) karst deposits; and (4) veins.

Barite from different occurrences contains variable, but generally high, Sr content. According to Church (1970), marine barites host up to 3.4 percent Sr, and Starke (1964) measured an $SrSO_4$ content of 12 percent in vein barite.

Since barite does not incorporate Rb in its structure, the strontium isotope ratios measured in strontium-bearing barites represent the original unmodified values. Thus $^{87}Sr/^{86}Sr$ ratios can be used for discussing barite genesis and the source of Sr; this helps to explain the genesis of some ore deposits.

Goldberg et al. (1969) determined U, Th, and Sr content and Sr isotope ratios of several barites. The isotopic data, along with Th and U content, help to distinguish between marine and continental origins of the mineral.

Farrell et al. (1978) measured Sr isotope ratios between 0.70677 and 0.70820 in barites from Kuroko deposits (Japan). These data are explained as the result of a reaction of Miocene sea water with Miocene volcanics of $^{87}Sr/^{86}Sr$ ratios ranging from 0.703 to 0.705.

Kesler and Jones (1981) measured values between 0.7077 and 0.7094 in barites from Northern Mexico, thus ruling out an origin of Sr from evaporites and suggesting instead its provenance from "basement-derived arkose in the Cretaceous basin."

Kesler et al. (1983) studied the strontium isotope geochemistry of the widespread fluorite deposits of Coahuila, Mexico. In these deposits, fluorite occurs as limestone replacement at the contact with Tertiary rhyolites and microsyenites, as well as mantos and veins that are not clearly associated with any known igneous body. The Sr isotope ratios permit us to delimit the possible sources of this element in the ore-forming solutions, as well as the mechanisms of ore deposition.

Synsedimentary barites from west central Argentina (Barbieri et al., 1981) show Sr isotope ratios close to the published values for Early Cretaceous sea water, suggesting a marine origin.

Barbieri et al. (1982) studied the strontium geochemistry in the epithermal barite deposits from the Apuan Alps (northern Tuscany, Italy) that occur at the contact between dolomitic rocks of Upper Triassic age and phyllites of Palaeozoic age. No typically evaporitic minerals, such as gypsum and anhydrite, are recorded in that area. The importance of these minerals as a possible source for Sr-bearing brines is well known (Barbieri et al., 1976). Moreover, the epithermal mineralization is referred to a plutonic intrusion beneath the

Apuan Alps (Carmignani et al., 1976), which could not merely be a "hot spot." Thus the provenance of Sr could not be inferred a priori, making the Apuan Alps barites of some interest. The Sr content and Sr isotope ratios measured in barites collected from three different deposits are very similar (Sr = 5, 500 ppm and $^{87}Sr/^{86}Sr$ = 0.7101), suggesting a provenance of this element from the same source. In particular, Sr could be derived by (1) direct precipitation from sea water; (2) precipitation from waters leaching marine limestone; (3) deposition from late-magmatic fluids of the inferred granitic intrusion beneath the Apuan Alps; or (4) from the phyllites of the Verrucano formation. This latter possibility is the most likely one. In fact, the Verrucano phyllites show $^{87}Sr/^{86}Sr$ ratios very close to the Apuan barites. Moreover, the Verrucano phyllites are likely a possible source for barium, which shows content up to 7000 ppm in some horizons.

Barbieri and Masi (1984) measured Sr content and Sr isotope ratios of barite samples from Tennessee and Nevada mines (U.S.A.). Barites from Tennessee were collected from the Lost Creek mine, which is located in Union County (Fagan, 1978). The most important feature of this mine is a nearly circular breccia body about 40 m in diameter, apparently involving both the Mascot dolomite and the Dot formation. Barite fills formerly open spaces between breccia blocks and/or forms a coating on brecciated rock fragments. The mineralizing solutions have completely dolomitized the breccia mass, whose matrix is made up of fine-to-medium crystalline argillaceous dolomite. The barite-filled features decrease in spacing with distance from the central core. This pattern probably is related to the origin of the breccia body itself, which was likely due to collapse of the Dot formation into the Mascot dolomite, induced by solution stopping the rising mineralizing solutions. The age of the mineralization is unknown, however, it cannot be younger than the Middle Ordovician, as the basal Dot formation bed collapsed shortly after deposition or complete lithification. The Lost Creek barite shows an Sr content ranging from 14,500 ppm to 19,600 ppm and a virtually identical Sr isotope composition, thus indicating that the bulk of Sr was provided by the same source. This latter was likely marine in origin, the Sr isotope ratios being very close to that of Ordovician sea water. Thus Sr may derive either from Ordovician sea water directly involved in the mineralizing process or from dissolution of marine carbonate of the same age. The source of $SO_4^=$ and Ba is unknown.

The Mountain Springs mine is located on the west flank of the Shoshone Range, some 40 km south of Battle Mountains, Nevada (U.S.A.). Barite occurs both as massive beds and as interlaminations with chert and shales. It is of a dark gray color, and fresh fractures yield a fetid odor. The primary gangue mineral is silica. Both barite and surrounding country rocks are similar to those of the Grey Stone mine, Argenta mine, and others in the Shoshone Range. The age of the mineralization is unknown (Valcarce, 1978). The Mountain Springs barite Sr content ranges from 2600 ppm to 5200 ppm with an average value of 4120 ppm, that is, distinctly lower than that of the Lost Creek barites. This probably indicates a poorer Sr environment or perhaps lower temperatures (see Barbieri et al., 1984*a*). Barites show quite different Sr isotope ratios, suggesting that Sr was provided by more than one source. In particular, the Sr isotope data rule out that the bulk of Sr was supplied by sea water or marine carbonate of any age, which is known to have lower ratios. A likely source of Sr might be the shales associated with the barites. These rocks can display a wide range of Sr isotope ratios, due to their Rb and Sr contents. Sulfate ions were likely provided by decay and oxidation of the organic matter associated with the sediments. Ba perhaps derived from the shales, as suggested by studies of similar deposits from Argentina (Brodtkorb et al., 1986).

Barbieri et al. (1984*b*) studied the barite deposits from Sardinia (Italy), which can be distinguished as karst or hydrothermal types. The karst deposits are all located in the southwestern corner of the island. The age of the mineralization, which occurs within Middle Cambrian marine carbonate formations, is difficult to establish, but it is likely that most of the karst deposits formed after the Hercynian orogenesis (Padalino et al., 1980). Geologic evidence shows that the karst deposits derived from a marine precursor such as the Cambrian stratiform barite, which occurs interbedded with layers of dolostone in a rhythmic evaporitic sequence. The evaporitic rocks formed by a mixture of fresh water leaching continental rocks with marine water supplying $SO_4^=$ in a sabkhalike environment (Gandin et al., 1973). The hydrothermal vein deposits are epithermal and spread throughout Sardinia. The host rocks, whose age ranges from the Cambrian to the Permian, can be classified as marine carbonates, pelitic schists, gneiss and high-grade metamorphic rocks, and granitic rocks. The mineral assemblage of the hydrothermal deposits is made up of galena, sphalerite, fluorite, barite, quartz, and calcite. The age of the mineralization is not known in detail for many deposits; however, most of them are related to the Hercynian orogenesis, during which there were emplacements of a huge batholith. Remobilization processes likely occurred in many deposits in connection with the major geologic events that affected the island during the Triassic, Eocene, and Alpine orogenesis (Cocozza et al., 1974). Barite from Sardinian deposits show an Sr content range similar to that reported in the literature for barite from other deposits throughout the world (see Tufar, 1980).

The Sr isotope compositions of the Sardinian barites are quite different. Stratiform barites show an average ratio of 0.70867 ± 0.00003, which is similar to that of Cambrian sea water, suggesting that the bulk of Sr was provided by sea water. This is in agreement with the occurrence of barite, which was deposited in a sabkhalike environment (Gandin et al., 1973). Karst barites show an average Sr isotope composition of 0.70947 ± 0.00001, which is slightly higher than that of stratiform barites. This points out that the karst barites derived from recycling of both stratiform barites and other ^{87}Sr radiogenic-enriched sources. Hydrothermal barites show $^{87}Sr/^{86}Sr$ ratios ranging from 0.70990 to 0.71837. These values are distinctly higher than those of both karst and marine barites, ruling out the latter as parental materials. Despite the hydrothermal occurrence and the relationship between the ores and the Hercynian magmatism, this latter cannot, in general, be a major source of Sr as their initial ratios range from 0.7085 to 0.7107. On the contrary, some Ordovician–Silurian marine shales, which represent the host rocks of most mineralization, exhibit Sr isotope ratios that fall in the range of the hydrothermal barites.

Brodtkorb et al. (1986) described the geochemical features of stratiform barites from Canota, Argentina. In this district, barite occurs as crystals, nodules, and lenses, all of which are associated with primary bedding of the Ordovician shales of the Empozada formation in Mendoza Province, southern sector of the Precordillera. The stratigraphy and composition of the Empozada formation suggest a reducing, low-energy marine platform environment. Of the three forms of barite present, the nodular one is the most characteristic and frequent. The nodules are generally ellipsoids, with the long axis between 15 cm and 20 cm in length, and they are linked to form chains of a few decimeters to several meters in length. The barite is dark gray to black in color, and its density ranges from 3.8 to 4.14. It yields a fetid smell when hit. Barite nodules show Sr isotope ratios very close to 0.70885. If this ratio is plotted in the diagram in Burke et al. (1982), it falls in a field that would indicate a marine origin for the Sr contained in the barite and, more precisely, that it derived from Upper Ordovician sea water. The isotopic values of barite crystals occurring in the layers are significantly higher than the ratios of nodules and of Ordovician shales. On the basis of Sr geochemistry, REE distribution, and sulfur isotopes, it is suggested that the barite crystals formed at the expense of the Ba host in the shales. As these latter also contain Rb, it is reasonable to assume that solutions deposited a radiogenic ^{87}Sr-rich barite.

REFERENCES

Arrhenius, G. 1963. Barium. In *Handbook of Geochemistry,* II, 4, ed. K. H. Wedepohl, 87p. Berlin, Heidelberg, New York: Springer-Verlag.

Barbieri, M., and U. Masi. 1984. Sr geochemical evidence on the origin of celestite-barite deposit at Pian dell'Organo in the Tolfa Mountains area (Latium, Central Italy). *Min. Petrogr. Acta* **28:**33–37.

Barbieri, M., U. Masi, and L. Tolomeo. 1976. Distribuzione dello stronzio nei gessi e nelle anidriti delle formazioni evaporitiche dell'Italia Centrale. *Soc. Ital. Min. Petrol. Rend.* **32:**551–560.

Barbieri, M., M. K. de Brodtkorb, S. Ametrano, and V. Ramos. 1981. Datos isotopicos de ^{87}Sr/^{86}Sr relacionados a los yacimientos de celestina y baritina de la Formacion Huitrin, provincia del Neuquén. *VIII Cong. Geol. Arg.* **11:**787–796.

Barbieri, M., U. Masi, and L. Tolomeo. 1982. Strontium geochemistry in the epithermal deposits from the Apuan Alps (Northern Tuscany, Italy). *Chem. Geology* **35:**351–356.

Barbieri, M., B. De Vivo, V. Perrone, and E. Turco. 1984*a*. Strontium geochemistry of the San Donato barite mineralization (Calabria, Italy). *Chem. Geology* **45:**279–288.

Barbieri, M., U. Masi, and L. Tolomeo. 1984*b*. Strontium geochemical evidence for the origin of the barite deposits from Sardinia (Italy). *Econ. Geology* **79:**1360–1365.

Brodtkorb, M. K. de, I. B. A. Schalamuk, M. Barbieri, H. Puchelt, S. Ametrano, R. Fernandez, and R. Etcheverry. 1986. The nodular barite deposits of Canota, Argentina. *Quadrennial IAGOD Symposium, 7th, August 1986, Proc.* (in press).

Burke, W. H., R. E. Denison, E. A. Hetherington, R. B. Koepnick, H. F. Nelson, and J. B. Otto. 1982. Variation of ^{87}Sr/^{86}Sr throughout Phanerozoic time. *Geology* **10:**516–519.

Carmignani, L., G. Dessau, and G. Duchi. 1976. I giacimenti a barite, pirite ed ossidi ferro delle Alpi Apuane. Studio minerogenetico e strutturale. Nuove osservazioni sui giacimenti polimetallici. *Soc. Geol. Ital. Boll.* **95:**1009–1061.

Chow, T. J. and E. D. Goldberg. 1960. On the marine geochemistry of barium. *Geochim. Cosmochim. Acta* **20:**192–198.

Church, T. M. 1970. Marine barite. Ph.D. dissertation. San Diego: University of California.

Cocozza, T., A. Jacobacci, R. Nardi, and I. Salvadori. 1974. Schema stratigrafico del Massiccio Sardo-Corso e minerogenesi della Sardegna. *Soc. Geol. Italiana Mem.* **13:**85–186.

Durum, W. H., and J. Haffty. 1963. Implications of the minor element content of some major streams of the world. *Geochim. Cosmochim. Acta* **27:**1–11.

Engelhardt, W. von. 1936. Die Geochemie des Bariums. *Chem. Erde* **10:**187–246.

Fagan, J. M. 1978. Geology of the Lost Creek Barite Mine. In *Guidebook for Eastern Field Excursion A-2,* IAGOD V, 40.

Farrell, C. W., H. D. Holland, and U. Petersen. 1978. The isotopic composition of strontium in barites and anhydrites from Kuroko deposits. *Min. Geology* **28:**281–291.

Gandin, A., G. Padalino, S. Tocco, and M. Violo. 1973. Un esempio di deposizione stratiforme di barite nella "dolomia rigata" del Cambrico della Sardegna sudoccidentale. Tentativo di correlazione tra l'ambiente di sedimentazione e la precipitazione del solfato di bario (nota I). *Soc. Geol. Ital. Boll.* **92:**329–354.

Goldberg, E. D., B. L. K. Somayajulu, J. Galloway, I. R. Kaplan, and G. Faure. 1969 Differences between barites of marine and continental origins. *Geochim. Cosmochim. Acta* **33:**287–289.

Heier, K. S., and S. R. Taylor. 1959. Distribution of Ca, Sr and Ba in southern Norvegian pre-Cambrian alkali feldspars. *Geochim. Cosmochim. Acta* **17:**286–304.

Kesler, S. E., and L. M. Jones. 1981. Sulfur and strontium geochemistry of celestite, barite and gypsum from the Mesozoic basin of northern Mexico. *Chem. Geology* **31:**211–224.

Kesler, S. E., J. Ruiz, and L. M. Jones. 1983. Strontium isotopic geochemistry of fluorite mineralization (Coahuila, Mexico). *Isot. Geosci.* **1:**65–75.

Leutwein, F. 1951. Geochemische Untersuchungen an Alaun-und Kiesel-schiefern Thüringens. *Arch. Lagestättenforsch.* **82:**1–27.

Marchig, V., P. Moeller, H. Baker, and P. Dulski. 1985. Foraminiferal ooze from the Galapagos rift area—hydrothermal impact and diagenetic mobilization of the elements. *Marine Geology* **62:**85–104.

Minguzzi, C. 1948. Spectrography of some products of Vesuvius fumaroles (abstract). *Soc. Mineral. Ital. Rend.* **5:**60.

Naboko, S. I. 1945. The sublimates of the Klyuchevskoy volcano. *Acad. Sci. U.S.S.R. Bull.*, Geol. Ser. 1, 50–58.

Padalino, G., S. Tocco, and M. Viola. 1980. Different genetic environments may produce similar occurrences of stratiform barite: Examples from Sardinia (Italy). *Quadrennial IAGOD Symposium, 5th, August 1978, Proc.*, pp. 459–466.

Puchelt, H. 1967. Zur Geochemie des Bariums im exogenen Zyklus, *Sitzungsber. Heidelb. Akad. Wiss. Math. Nat. Kl.* **4:**85–205.

Shmakin, B. M. 1984. Causes and consequences of high contents of barium in sheet muscovite and phlogopite. In *27th International Geological Congress Proceedings,* vol. 15, *Nonmetallic Mineral Ores.*

Starke, R. 1964. Die Strontiumgehalte der Baryte. *Freiberger Forschungsh. C.,* **150:**1–86.

Strübel, G. 1967. Zur Kenntnis und genetischen Bedeutung des Systems $BaSO_4$–NaCl–H_2O. *Neues Jahrb. Mineral. Monatsh.* **1:**223–233.

Tufar, W. 1980. Ore mineralization from Eastern Alps, Austria, as stratabound syngenetic formations of pre-Alpine age. *Quadrennial IAGOD Symposium 5th, August 1978, Proc.,* pp. 513–544.

Turekian, K. K., and D. J. Johnson. 1966. The barium distribution in sea water. *Geochim. Cosmochim. Acta* **30:**1153–1174.

Valcarce, J. S. 1978. The Mountain Spring Barite Mine, Lauder Country, Nevada. In *Guidebook to Mineral Deposits of the Central Great Basin,* Nevada Bureau of Mines and Geology Report 32, pp. 49–50.

Vinogradov, A. P. 1956. Barium. In *Handbook of Geochemistry,* II, 4, ed. K. H. Wedepohl, 87p. Berlin, Heidelberg, New York: Springer-Verlag.

Wentlandt, R. F. 1977. Barium-phogopite from Haystack Butte, Highwood Mountains, Montana. *Carnegie Institution of Washington Yearbook* **76:**534–540.

3

CELESTITE: WORLDWIDE CLASSICAL ORE FIELDS

Milka K. de Brodtkorb

Some of the most important celestite ore fields in the world are described in this chapter. Deposits in England, Mexico, Spain, the German Federal Republic, Iran, Canada, Algeria, and others are mentioned here; those in Argentina are described in Chapter 4.

The geological similarities of these deposits are striking: They are located only in a certain facies of the evaporite sequence, always related to limestones and gypsum, having been obliterated in many cases by diagenesis and later dissolution and precipitation. The deposits have been found in terranes ranging from Carboniferous to Quaternary. They are different from stratabound barite deposits because they are found in a great variety of host rocks and can be of different origins.

Data on the geology, sedimentology, chemistry, and isotope chemistry of the different celestite ore fields are fairly irregular; for this reason, different studies on the formation of celestite, undertaken in recent decades, are mentioned for the reader's better understanding before describing the ore fields.

In 1956, when the Hemmelte West oil field was drilled in the northwest of the German Federal Republic, the sedimentological characteristics of the drill cores were analyzed. The study of anhydrite showed a higher specific weight than expected, due to the presence of celestite. By the time study of all the holes from the oil well was concluded, an important not-outcropping celestite orebody, more than 1200 m deep known as Hemmelte West had been found in the upper Malm. The ore reserves are calculated at several million tons, in beds up to 4.5 m thick. In order to explain the genesis of this type of deposit, Müller (1962) studied the geochemical behavior of strontium in sea water, mainly in the evaporation stage, and the formation of these sedimentary deposits in saline basins. Strontium is normally associated with calcium in magmatic rocks as well as in sedimentary rocks because it can substitute for the calcium in the crystal structure of minerals due to the similar ionic radii. Their chemical properties are also similar because of their positions in the periodic table. Strontium does not form independent minerals in the magmatic cycle, whereas deposits containing Sr without Ba are not known, so far, in the hydrothermal sequence. The Ba:Sr ratio ranges from 10:1 to 100:1 (Gundlach, 1959). The Sr content of hydrothermal veins varies between 0.1 percent and 0.7 percent. The solubility of $SrSO_4$ is higher than that of $BaSO_4$, such that barite precipitates first and takes the small amount of Sr present in the solution. When suitable conditions for the crystallization of celestite are reached, most of the barium has already precipitated. However, if considerable quantities of Sr are still available, celestite would form only exceptionally because it requires a higher activity of the $SO_4^{=}$ ion. Thus, the formation of strontianite ($SrCO_3$), with a lower solubility than $SrSO_4$, is more feasible and it would be more frequent than celestite in hydrothermal veins.

In the exogenous cycle, celestite is the most important mineral of strontium and precipitates mainly in the evaporite environment.

According to Mason (1960), the chemical composition of sea water, with a chlorinity of 19 percent and a salinity of 35 percent, is:

Ion	*Percent in Sea Water*
Cl	18.900
Na	10.556
SO_4	2.649
Mg	1.272
Ca	0.400
K	0.380
HCO_3	0.140
Br	0.065
H_3BO_3	0.026
Sr	0.008
F	0.001

From these 8 g/t of Sr (Müller, 1962), about 62.58 percent is available for the formation of celestite and 37.42 percent is included in other evaporite minerals.

Analytical data on several samples from various localities show that gypsum or anhydrite associated with evaporites usually has an average content of strontium of 0.20 percent to 0.25 percent. Strontium's ability to accommodate in solid solution is higher for anhydrite, up to 0.74 percent (Dean and Tung, 1974), than for gypsum, up to 0.10 percent (Noll, 1934). Therefore, in the transformation from anhydrite to gypsum, Sr can be released to form celestite.

To understand the evaporite formation of celestite (Müller, 1962), one can take into account the fact that in sea water from the present oceans, $SrSO_4$ is undersaturated by a factor of 5 (8 g Sr/t correspond to 1.677 mg $SrSO_4$/100 ml of sea water, with an $SrSO_4$ solubility of 8.25 mg).

The precipitation of celestite from average present sea water begins in the boundary between the Ca-carbonate and the Ca-sulfate precipitation when the concentration of sea water reaches approximately one-fourth of its original volume, not before it reaches one-third and not after it reaches one-fifth (Fig. 3-1). According to Posnjak (1940, in Müller, 1962), gypsum begins to precipitate with a concentration of 1/3.35 times the original volume. Comparatively, halite begins to precipitate when sea water is evaporated to approximately one-twelfth of its original volume.

The beginning of precipitation of celestite in sea water at different initial concentrations (between 2 g/t and 40 g/t) is shown in Figure 3-2. Müller (1962) suggested that the amounts of Sr taken up by the carbonates is minimal (calcite was considered as the mineral that formed in the primary chemical precipitation), and the saturation of $SrSO_4$ is reached when there is an enrichment of 40 g Sr/t. When the point of precipitation of $CaSO_4$ is reached (i.e., at a concentration of sea water of 1/3.35) most of the Sr precipitates as celestite, whereas with 4 g Sr/t, celestite begins to precipitate when most of the Ca-sulfate is already formed.

Unfortunately, this interesting paper by Müller is in German, a language not often consulted; thus, his detailed work is not always taken into account when the genesis of similar deposits is analyzed.

Whereas Müller's work (1962) deals mainly with the chemical conditions for the precipitation of celestite, in the 1960s attention was paid to the detailed study of the coastal areas with their subtidal, intratidal, and supratidal facies and the coastal and continental sabkhas. These sedimentologic studies helped explain the formation and precipitation of the carbonate and evaporite facies in which the celestite deposits are found.

Müller (1962) mentions that aragonite may contain several percent of Sr

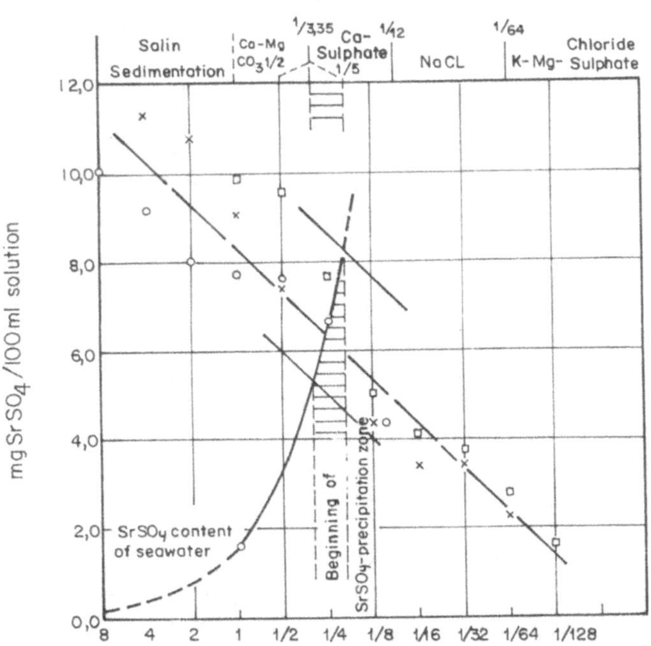

Figure 3-1. $SrSO_4$ solubility for different dilutions and concentrations of sea water at 25°C (after Müller, 1962).

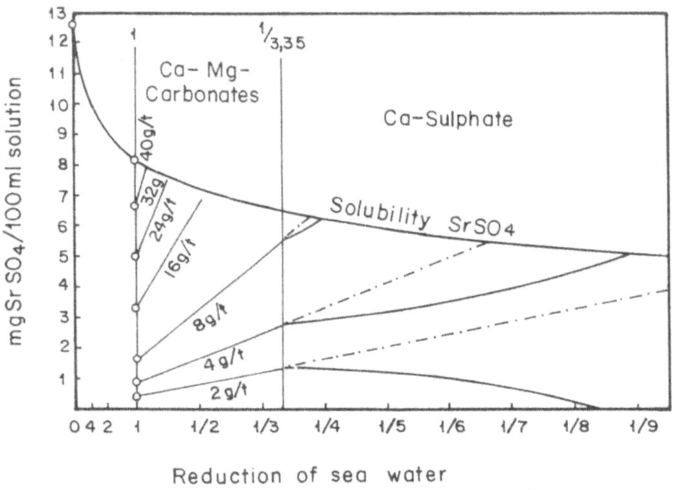

Figure 3-2. Solubility curve of celestite for different initial Sr ratios of sea water (after Müller, 1962).

and, in the diagenetic processes in which aragonite transforms into calcite, Sr is released and enriches the formation waters circulating through the sediments. If these are enriched in $SO_4^=$, then celestite may precipitate.

Bausch (1968) mentions that the different capacities of calcite and aragonite to accommodate Sr in the crystal structure are the main reason for the differences in Sr content between recent and fossil limestones. He also gives importance to the Sr content of clay minerals.

Kinsman (1969) believes that sabkha diagenesis involves the interstitial emplacement of evaporite minerals within the host sediments, but brine pans may accumulate more soluble evaporites because of the greater evaporation losses. According to Kinsman, celestite is a common early diagenetic mineral of the coastal sabkhas and is abundant in areas of intense dolomitization. It forms during the replacement of aragonite (7000–8000 ppm Sr) by dolomite (600–700 ppm Sr). In recent environments, for instance, celestite occurs in sabkhas from the Persian Gulf (Evans and Shearman, 1964), in Baja California, and in shallow lagoons in southeast Australia (Skinner, 1963).

Butler (1973) investigated the complexity of Sr behavior during diagenesis of sedimentary gypsum and anhydrite.

West (1973) indicates that the presence of celestite and length-slow chalcedony provides evidences of the former presence of evaporites. West points out that celestite is usually formed by the selective removal of strontium ions from ground water by reaction with gypsum or anhydrite. Celestite beds of appreciable thickness, which are associated with either underlying or overlying calcium sulfate layers, may be formed by replacement of these.

In 1974, Veizer and Demovic studied the distribution of Sr in Mesozoic carbonate rocks from the central western Carpathians and concluded that they are facially controlled, with high concentrations in hypersaline deep sea rock types and low Sr concentrations in littoral limestones of organogenic and organodetrital types.

In 1976, Kranz investigated the Sr content of the different facies of the

Oberen Wettersteinkalk (middle Trias) in the eastern Alps. His studies comprised the reef, margin, lagoonal, and basin facies and the influence of diagenesis, salinity, clay content, and dolomitization.

In the Lower Paleozoic sediments of the Oslo, Norway, region, the formation of celestite is known in an evaporite depositional environment (Olaussen, 1981). According to Olaussen, celestite is found in a high intertidal to subtidal facies in the form of anhedral or prismatic crystals infilling pores. He presents three models to explain such formation: (1) the early diagenetic dolomitization of aragonite mud releases Sr that may react with the brine to form celestite; (2) Sr enrichment in interstitial water by the process of dolomitization of limestones; and (3) in the latter case, the Sr released may precipitate forming celestite as a late stage authigenic mineral in geodes and veins.

From the above-mentioned study and the work of other researchers, the distribution of Sr in a fossil limestone complex can be interpreted genetically. Variations in the Sr content of these limestones are influenced mainly by diagenesis, dolomitization, and the salinity of the basin.

Kendall (1979) mentions that only recently have evaporites been considered as sediments rather than as chemical precipitates.

The S and Sr isotope studies undertaken in the last few years are also important (Barbieri et al., 1982, 1984*a,* 1984*b*; Kesler and Jones, 1981), and they aid in the interpretation of the origin of these elements.

HEMMELTE WEST DEPOSIT, GERMAN FEDERAL REPUBLIC

The structure at Hemmelte West consists of Jurassic, Lower Cretaceous, Tertiary, and Quaternary sediments folded in soft anticlines cut in blocks by faults. At the end of the Jurassic, a basin formed in northwestern Germany that was subdivided in several lowlands separated by shoals, with only a narrow seaway connection to the open sea (Fig. 3–3). A more than 300 m thick saline sequence with a prevalence of Ca carbonates and sulfates in the margins and a thick NaCl deposition in the center precipitated in this basin (Müller, 1962). A facies of oolitic calcareous and in part dolomitic rocks developed in the Upper Kimmeridge. Following deposition of a 0.40 m thick argillaceous-marly bed, the deposition of the celestite facies begins in the Upper Malm. Dolomitic limestones with minor celestite, frequently inserted as thin layers up to 3 mm thick, appear first. The celestite content increases until it forms 70 percent of the rock. Then anhydrite begins to precipitate, changing the conditions for both minerals, reflecting the constant changes in the salinity of sea water. At Hemmelte West, the different lenses are formed from bottom to top by carbonates, carbonate rocks and celestite, celestinite, celestite alternating with anhydrite, and anhydrite.

The precipitation of calcium sulfate at the different sub-basins began at different times, in such a way that, while anhydrite was still precipitating in some of them, halite was precipitating in others. There was an important contribution of pelitic material from the continent, changing in short vertical distances. At the end of the Malm, the basins dried up rapidly without giving place to the precipitation of the Mg and K chlorides sequence.

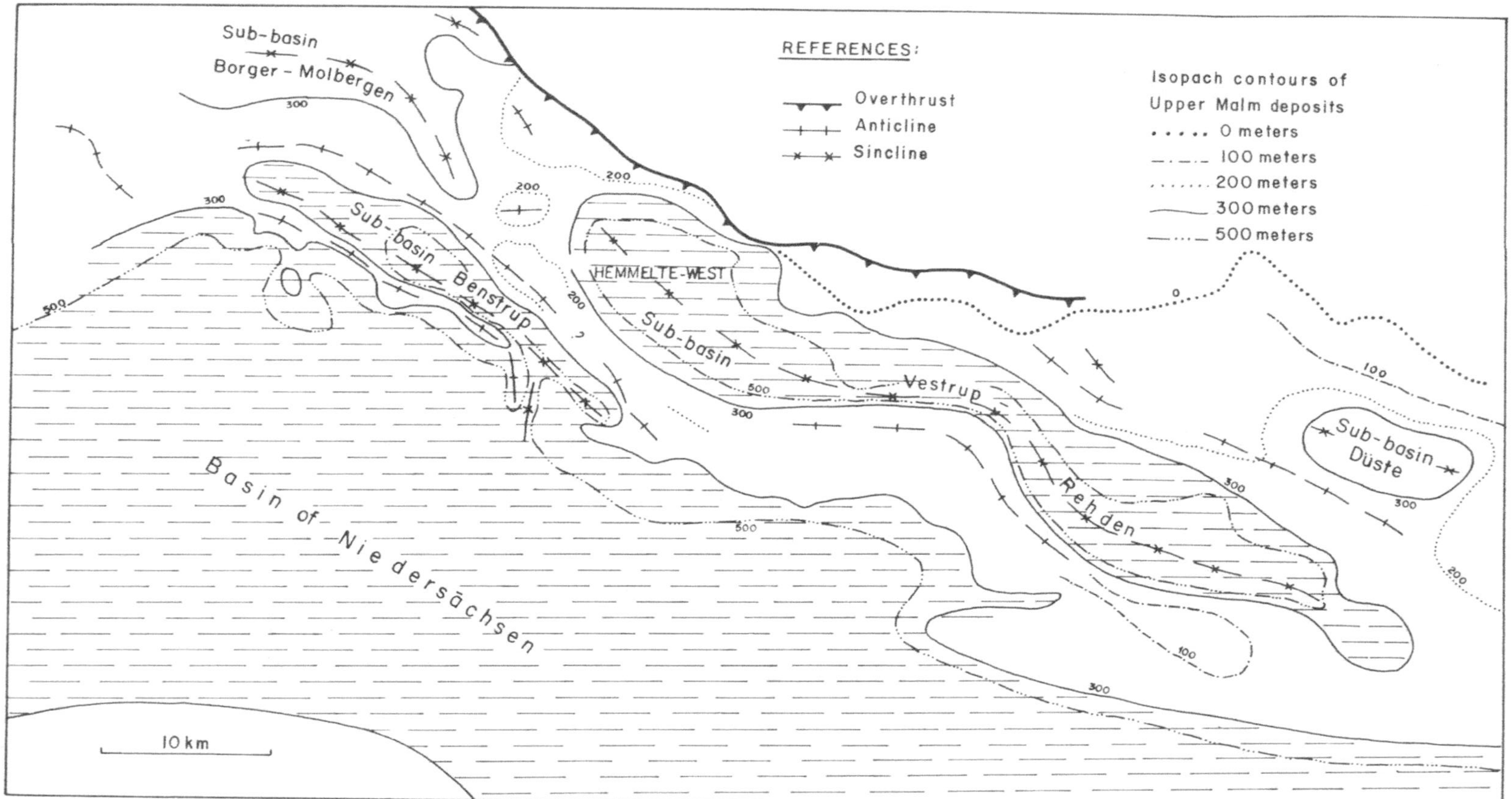

Figure 3–3. Isopach map of the Upper Malm in the Hemmelte West area with indication of basins and sub-basins of evaporite deposits (after Müller, 1962).

The celestinite textures are similar to those of anhydrite; Müller (1962) described schlierig-cloudy, phacoidal, and varved textures, all typical of shallow water formations. Celestite has undergone only a few transformations. The most frequent one is the replacement of celestite by calcite in isolated crystals or in groups of crystals present as microcrystal lime aggregates, which can completely fill the celestite crystals. The replacement of celestite by anhydrite can only be observed in isolated crystals, which keep their optical orientation.

From the textures, the mineral paragenesis, and its stratabound position, Müller (1962) suggested that the genesis of celestite at Hemmelte West is synsedimentary, formed during the evaporation of sea water in the Upper Malm.

CELESTITE ORE FIELD OF THE BRISTOL AREA

Celestite from the Keuper marl and the Coal Measures has been mined during the last 100 years at various localities in the Bristol area (Fig. 3–4). One of the best-known deposits is at Yate; it is very big and has been mined by means of shallow open pits.

The celestite deposits of the Bristol area are located close to the top of the Triassic Keuper marl. Following the marine transgression of the Carboniferous, which deposited vast thicknesses of limestone, the area rose above sea level in the Triassic and became dry land with a hot arid climate. Wood and Shaw (1976) suggest that the Keuper marl deposited in desert basins of inland drainage, with a basement consisting mainly of Carboniferous rocks in which deposits of saliniferous sediments formed. According to Wood and Shaw, celestite deposits in which celestite is associated with gypsum, and halite is the main evaporite mineral, are also known in inland sabkhas from the Bristol Dry Lake of California and in large celestite deposits in Libya.

The Triassic of the region comprises desert sandstones and evaporite minerals: gypsum and anhydrite formed mainly in the Midlands of England, halite in the Cheshire Basin, potash in Yorkshire, and celestite in the Bristol area (Anonymous, 1974). In England the Keuper marl contains two major horizons of gypsum and several horizons of celestite, but only the one related to the Severnside Evaporite Bed is important (Nickless et al., 1976). This unit outcrops for a distance of over 50 km. Between Bitton and Wapley the horizon is named "Celestite Bed" or "Celestite horizon" (Nickless et al., 1976). Celestite has also been reported from the overlying Tea Green marl.

The celestite beds in the Bristol area are found in alternation with clay, dolomite, and evaporite horizons. Three main types of mineralization are (Thomas, 1973; Nickless et al., 1976):

1. Stratiform deposits, bedded or more often horizons bearing nodular and disseminated celestite, thin stringers and veins, within the Keuper marl, associated with the Severnside Evaporite Bed, typically 10–15 m beneath the top of the Keuper, that is, the bottom of the Tea Green marl.
2. Nodular and disseminated celestite at the unconformity between the

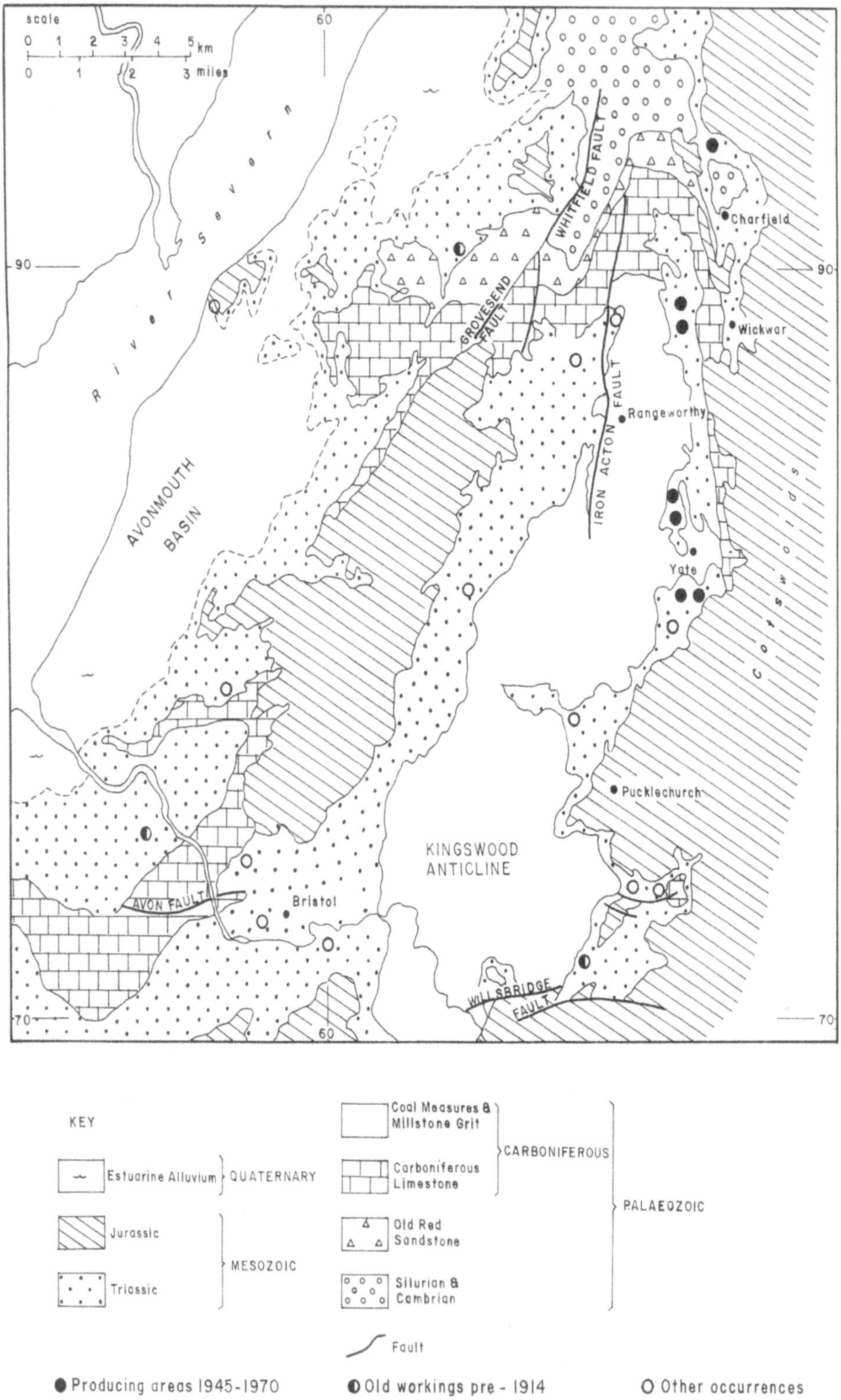

Figure 3–4. The area northeast of Bristol and location of celestite deposits (data from Thomas, 1973; Nickless et al., 1976).

Keuper marl and the underlying Paleozoic rocks (mainly the Carboniferous Coal Measures).

3. Veins, sheets, and infillings of celestite in Paleozoic rocks below the Keuper marl and, in some instances, forming the matrix in a breccia of Paleozoic material. According to Thomas (1973), the type 1 ore is primary or early diagenetic, and for the type 2 ore, the origin may possibly be primary in places but is probably secondary, redistributed, having similarities to the supratidal evaporites of the Persian Gulf.

According to Nickless et al. (1975, 1976), even though minor amounts of celestite occur as a primary mineral, the celestite from the Severnside area is thought to have formed after gypsum and/or anhydrite. Examination of thin sections from the deposits north of Bristol shows that the replacement of calcium by strontium in either gypsum or anhydrite took place during early diagenesis, the Sr probably released from aragonite. Late diagenetic replacement of both anhydrite and celestite by calcite and quartz has been observed.

An article in *Industrial Minerals* (Anonymous, 1975) postulates rather, that the longstanding theory that celestite is a product of sea water evaporation is now largely discarded. In this paper it is thought that the celestite was formed in warm saline lagoons within desert basins, and the strontium itself is thought to have been released into solution during the dolomitization of the underlying Carboniferous limestone.

Wood and Shaw (1976) arrive at similar conclusions. The occurrence of relict anhydrite in celestite crystals made some researchers think that most of the celestite is secondary after anhydrite. Evidence for this process exists; however, the apparent absence of widespread anhydrite relicts within the celestite would indicate that most celestite precipitated primarily. Wood and Shaw (1976) proposed that celestite most likely precipitated in arid desert basins of inland drainage. The percolating ground waters must have been enriched in Sr and the probable source of the large amounts of Sr could be the dolomitization of the Carboniferous limestones, which released Sr into solution.

The values of $^{87}Sr/^{86}Sr$ ratios for celestite found by these authors are higher (approximately 0.7095–0.7110) than the values for the Triassic sea water mentioned by Peterman et al. (1970) and Veizer and Compston (1974). These values are also higher than those for Carboniferous limestones but lower than the ratios for the Keuper marl sediments.

Celestite has an average Ba content of 1 percent with a range of 0.3 percent to 2 percent. The enrichment of Sr relative to Ca is 10^3–10^4 times greater than that of natural waters and the Sr/Ba ratio of the celestite is 1500 times greater than that of sea water. It is improbable that celestite precipitated directly from the Triassic sea water, although one cannot totally exclude, on the Sr isotope evidence, the occurrence of secondary processes.

If Sr is considered to have been released during dolomitization, the isotope data would indicate that Sr derived from this source must have been modified by exchange processes when the ground water passed through the Keuper marl sediments before being included into the celestite. From the geochemical evidence and the isotope data it can be concluded that there were several different episodes of celestite formation: primary, secondary after anhydrite, and in-vein occurrences (Wood and Shaw, 1976).

MEXICAN CELESTITE AND BARITE ORE FIELDS

The existence of celestite deposits in Mexico has been known for a long time, but mining in the San Agustin mine of Coahuila and in the Torreon region only began in 1971. The celestite deposits of northeastern Mexico are found in a thick Mesozoic sedimentary sequence that was deposited on a Precambrian(?)–Paleozoic basement (Kesler and Jones, 1981). In early Mesozoic times the Coahuila and Tamaulipas peninsulas were surrounded by gulfs that joined a sea located to the south (Fig. 3–5). The early Neocomian marine transgressions covered the Tamaulipas Peninsula. In the Upper Neocomian and Aptian, a continuous marine transgression formed a major reef system, which extends from Monterey to Texas with a back-reef evaporite deposition. According to Kesler and Jones (1981), in early Aurora times (middle Albian) this reef system was submerged and a new system of reef and back-reef evaporites formed, surrounding an island located to the south of the Coahuila Peninsula. The back-reef evaporites were covered by pelagic lime mudstones in late Albian times.

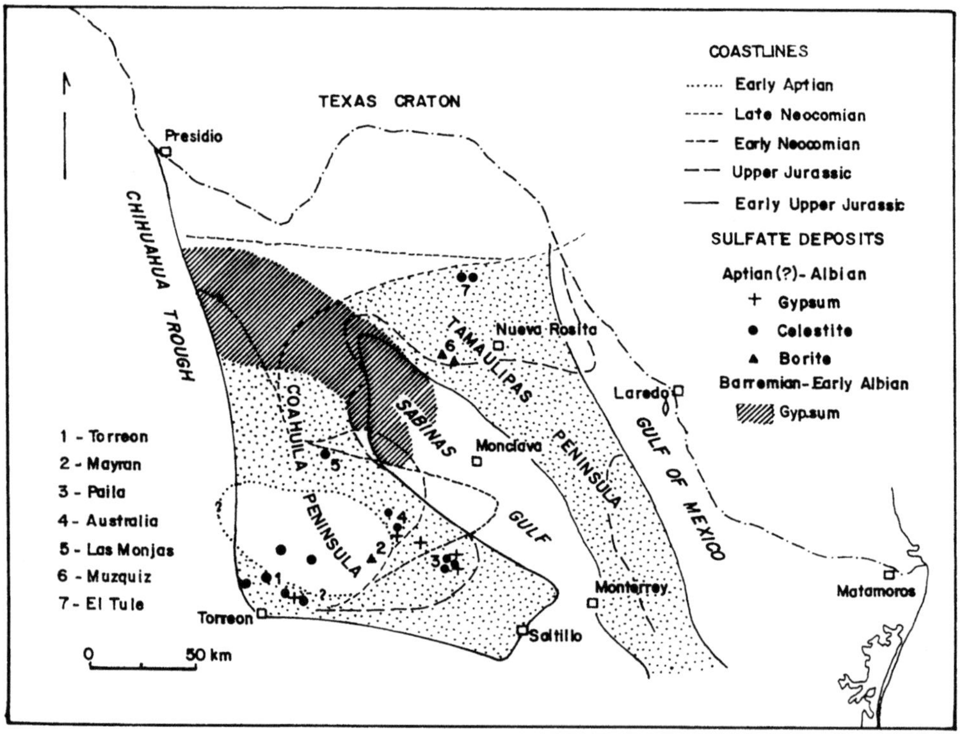

Figure 3–5. Paleogeographic map of the Mesozoic sedimentary basin of northeastern Mexico, showing distribution of gypsum, celestite, and barite (after Kesler and Jones, 1981).

Although evaporite gypsum is widespread in different formations of the Mesozoic sedimentary basin, the celestite- and barite-bearing beds have a much more restricted distribution and are related to the Albian Aurora formation.

The celestite and barite stratabound layers are up to 4 m thick and up to 1 km long. Each layer is confined to a single stratigraphic horizon and some of them may be followed for 20 km (Kesler and Jones, 1981). The Paila area is the only one presenting celestite together with barite beds. In a stratigraphic interval of approximately 100 m up to 10 layers of celestite were found. The celestite beds are present in a northeastern trending zone located to the northwest of the barite layers. Kesler and Jones (1981) recognized that there are not enough data to explain this distribution.

In the Mayran and Muzquiz districts only barite layers are present, whereas in Torreon, Las Monjas, and El Tule only celestite layers are present (Fig. 3–5). In the Australia, Paila, and Torreon areas bedded gypsum is found in another stratigraphic horizon, a few meters below the celestite layers. There is also overlapping fluorite mineralization, in veins and layers, related to an Oligocene igneous activity in the Paila and El Tule areas (Kesler, 1977, in Kesler and Jones, 1981). Regarding the genesis of these deposits, Salas (1973) and Rickmann, (1977), (both in Kesler and Jones, 1981) think that the celestite beds from the Paila region formed by replacement of limestone. In their S and Sr isotope study (Fig. 3–6) Kesler and Jones (1981) concluded that:

- The celestite and barite deposits are grouped in two different geological systems: (1) the deposits around the Coahuila island, and (2) the deposits of El

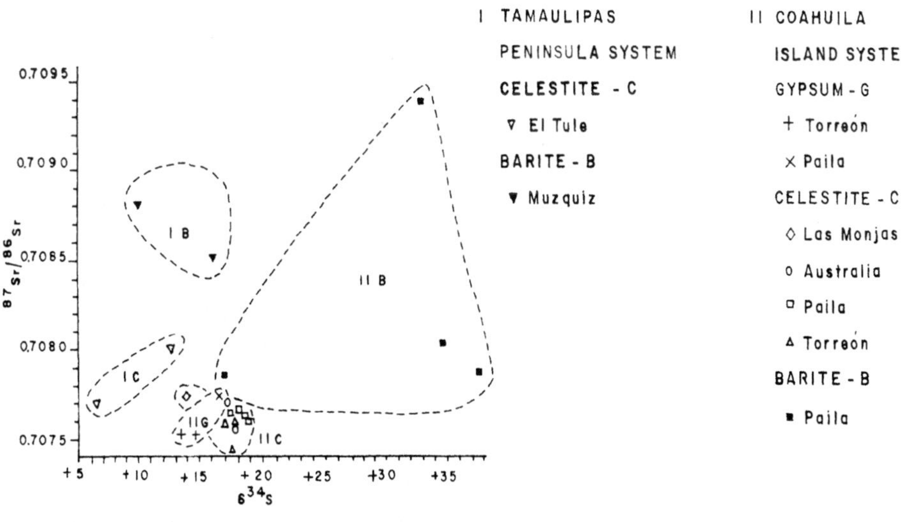

Figure 3–6. Variations of $^{87}Sr/^{86}Sr$ versus $\delta^{34}S$ for gypsum, celestite, and barite from early Albian carbonate rocks in northeastern Mexico (after Kesler and Jones, 1981).

Tule–Muzquiz on the northern Tamaulipas high. Kesler and Jones suggest that these systems had a different geochemical evolution. (There is more information on the Coahuila area deposits such as the isotopic composition of the sedimentary gypsum.)

- The sulfur isotope composition of gypsum ranges from 14 percent to 17 percent and agrees with the values indicated for Lower Cretaceous marine sulfate by Claypole et al. (1980). The sulfur isotope values of celestite from the Paila, Torreon, and Australia regions range from 16.3 percent to 18.2 percent, indicating that the celestite has homogeneous and similar $\delta\,^{34}S$ compositions. They are 1–2 percent higher than the values for gypsum from the same area. This difference probably resulted either from inorganic fractionation between celestite and dissolved sulfate derived from local gypsum or from limited enrichment of gypsum sulfate by bacterial reduction. The sulfur isotope values of barites are quite different and seem to require a process such as bacterial sulfate reduction.

- The Sr isotope values of gypsum from Paila and Torreon and of limestones from northeastern Mexico agree with the values of Peterman et al. (1970) and Veizer and Compston (1974) for Aptian–Albian limestones. The Sr isotope data on celestite from the Paila, Torreon, and Australia areas show strong similarities. The $^{87}Sr/^{86}Sr$ compositions of gypsum and celestite are similar, thus the Sr could have been derived directly from limestones or gypsum. However, the Sr isotope values of barites show a wide range. Values such as 0.7094 are higher than those proposed by Veizer and Compston (1974) for Mesozoic carbonates (0.7085); for this reason, a different source must be invoked to explain the unusual radiogenic Sr in barites. It is postulated that it could come from the arkoses present in the Precambrian(?)–Paleozoic basement of the Coahuila Peninsula.

- The isotope values of sulfur from Muzquiz and El Tule are different and inhomogeneous. Even though more isotope data on the district are needed, sulfate from the Tamaulipas system could have been derived by partial oxidation of reduced sulfur coming from coal, which is widespread, and crude oil, found in trace amounts, in this area.

- Fluid inclusions in the celestite contain single-phase, very low temperature liquids.

- The estimated Sr content of the abundant carbonate sediments range from 3×10^{15} kg to 75×10^{15} kg of Sr, thus only 10 percent of the Sr was necessary to form the deposits existing in this region and the remainder was diluted in the Cretaceous sea.

- In conclusion, the limestone-hosted stratiform beds of celestite and barite appear to have formed by replacement, with the Sr derived from the diagenetic release of the carbonate rocks. The Sr of the celestite seems to be derived from local limestones. The more radiogenic Sr of the barites, on the other hand, appears to have come mainly from basement-derived arkoses. Sulfur was derived from evaporites in one district and from a lighter source in the other. At least two local solutions existed and mixed locally during the formation of celestite and barite beds.

CELESTITE ORE FIELD OF THE GRANADA BASIN, SPAIN

Two different types of evaporite formations, which developed in two different tectonosedimentary units, are known in the Granada Basin (Dabrio et al., 1982). They are separated by a generalized sedimentary break: the lower evaporites (with gypsum, celestite, and halite), which were found in a marine environment, and the upper evaporites (detrital gypsum derived from the lower gypsum) deposited in a lacustrine environment.

In the Upper Tortonian–Lowermost Messinian (Martin et al., 1984), from open marine conditions, the basin becomes restricted and the lower evaporites were deposited. Following turbidite sandstones and marine shales, a facies developed that laterally changes from conglomerates, sandstones, and shales to stromatolites, with celestite, gypsum, and halite, at the center of the basin (Figs. 3–7 and 3–8). Then, due to karst, dissolution cavities formed in the gypsum beds that were filled by celestite pebbles and other older materials.

Over these, in Messinian–Pliocene times, were deposited shales and lacustrine turbidites, fluvial sediments, and the upper evaporite unit with gypsiferous turbidites, followed by more lacustrine sediments. Several continental conglomerates were deposited in the Upper Pliocene–Quaternary.

Celestite outcrops exist in two parallel east–west belts (Fig. 3–7). In the shorter northeastern belt are found the Montevives and La Atalaya deposits, whereas in the longer (about 10 km) southwestern Escuzar belt, there are only celestite exposures. Montevives is the only deposit mined.

In Montevives, the deposit consists of a carbonate lens partially replaced by celestite. It is about 40 m thick and outcrops for approximately 1 km, dipping 20°–50° toward the northwest. Gypsum has been found only in drill holes. According to Martin et al. (1984), the carbonate body is well stratified in layers 20–50 cm thick and contains horizons of different detrital clays. The layers are finely laminated with alternations of clear and dark laminae of 0.5–1 mm that correspond to algal mat structures. Both types of laminae contain calcified filaments of blue-green algae that possibly developed in fresh water.

In La Atalaya, poor lateral continuity is observed and only some old mining galleries exist. Celestite has partially replaced the laminated stromatolitic carbonate. The layers are only 10–30 cm thick and are interbedded in the upper part of the selenitic gypsum bed.

Two types of celestite occurrences outcrop in the Escuzar belt. One is the product of partial replacement of stromatolitic carbonates, and the other is the karst-filled Miocene breccia with celestite pebbles (dolinas with diameters of 1–5 m and depths of 2–3 m of minor economic importance). The celestite layers that partially replace the stromatolitic bed are 20–50 cm thick and dip 20°–30° northward. This bed, approximately 20 m thick, occurs marginally and progrades into selenite gypsum of the lower evaporite unit. In this case, the predominant structure is also laminated, although where the stromatolites, replaced by celestite, interfinger with gypsum, some spongy pustular textures are present. Buchbinder (1981, in Martin et al., 1984) recognized that the pustular stromatolite structures are the result of calcification of mats of coccoid unicellular algae that tolerate a salinity of 20–25 percent.

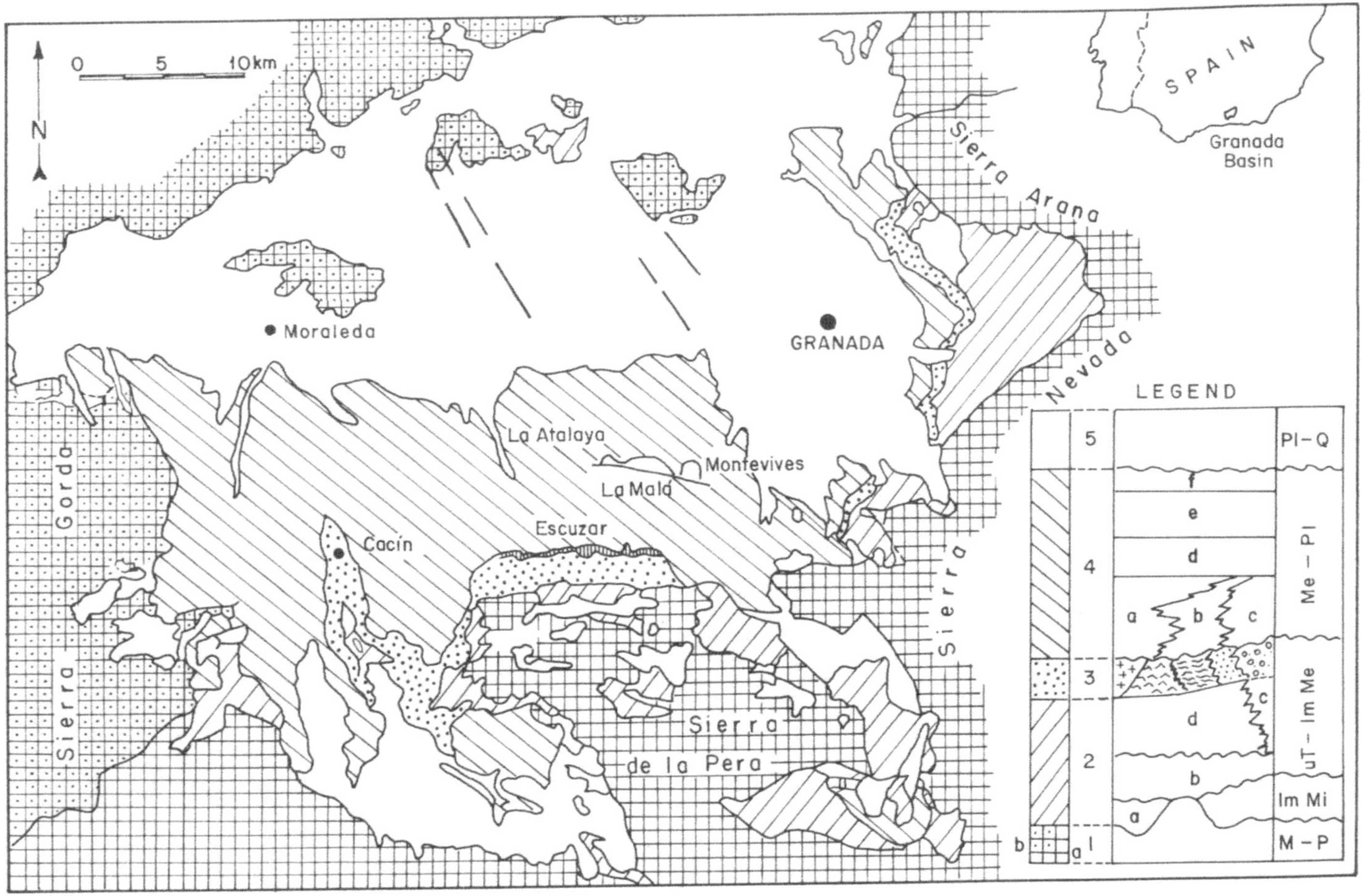

Figure 3–7. Geologic map and schematic stratigraphic section of the Granada basin, showing distribution of the main celestite outcrops (from Martin et al., 1984).

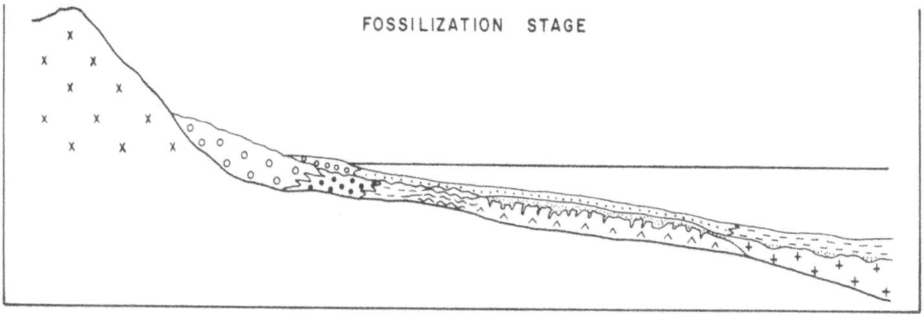

Figure 3–8. Origin and evolution of strontium deposits of the Granada Basin. Fossilization stage of the primary (replacement type) and secondary (karstic type) celestite mineralization, and renewal of the sedimentation in lacustrine conditions (after Martin et al., 1984).

1. Basement rocks; 2. Conglomerates; 3. Sands; 4. Lutites; 5. Stromatolites; 6. Selenite gypsum; 7. Halite; 8. Karst-filled breccia; 9. Coral reefs

The celestite content of the ore-bearing bed from Montevives is 70–80 percent, whereas that in the Escuzar belt is 50–55 percent and that in the karst cavities is only 15–20 percent.

The genesis of celestite from the Granada Basin was studied by Martin et al. (1984), whose detailed observations are summarized as follows:

- The stromatolites are found in a coastal belt limited to the east by the tectonically active Sierra Nevada. This was intensely eroded under semiarid

---

Figure 3–7. Legend inset.

1a. Nevado-Filábride and Alpujárride basements
1b. Sub-betic basement
2a. Conglomerates, sandstones, clays, limestones, and evaporites
2b. Conglomerates, sandstones, and limestones
2c. Delta fan conglomerates
2d. Turbiditic sandstones and marine lutites with planktonic forams
3. Mineralized interval
4a. Lacustrine lutites of Cacín
4b. Lacustrine turbidites of La Malá
4c. Fluviatile sediments
4d. Gypsiferous turbidites
4e. Lutites, sandstones, and lignites
4f. Lacustrine limestones
5. Most recent conglomeratic formations
M–P : Mesozoic–Paleozoic
lm Mi: Lower-Middle Miocene
uT-lm Me: Upper Tortonian–Lowermost Messinian
Me-Pl: Messinian–Pliocene
Pl-Q: Pliocene–Quaternary

climatic conditions and, at the base, alluvial fans were deposited forming a fringe around these highlands.

- Erosion of the Permo-Triassic dolostones, limestones, marbles, and micaschist allowed Sr to be released.
- The infiltration and/or run-off water coming from these highlands formed a fresh water lens, when approaching the coastal zone, that allowed the development of stromatolites lateral to the gypsum.

  The deposition of celestite and the partial replacement of stromatolites took place at the same time through the mixing of Sr-rich salt water with deeper sulfate-rich salt water. The porosity of the stromatolites and the close contact of both types of water favored the process of deposition of celestite and the replacement of stromatolites. They are of synsedimentary or early diagenetic origin.
- At the same time that these carbonates formed in a peripheral position, the basin was occupied by a hypersaline water body from which gypsum was precipitated. During basin desiccation, a layer of halite, up to 20 m thick, developed at the center of the basin.
- Following desiccation, the gypsum was weathered and the cavities were filled with breccias with abundant celestite pebbles, forming the second celestite horizon. Lacustrine materials were deposited on top of the erosion surface and the karst-filling breccia.

  At present, abundant celestite pebbles can be observed in the slopes and redeposition of strontianite and celestite is taking place along fractures.

The occurrence of celestite in the Granada Basin has been known for a long time, but extensive mining only began in the 1950s. In the last few years, the rate of extraction has been 50,000 tons/year with a grade of 95 percent.

LOCH LOMOND ORE FIELD, NOVA SCOTIA, CANADA

Celestite was found in the Loch Lomond area for the first time in 1962. The Loch Lomond deposits are located near the village of Enon on the island of Cape Breton, Nova Scotia. The three properties in the area are the Enon Lake (where most mining took place), the Mac Rae, and Armac prospects. The celestite deposits are found in the Loch Lomond Basin, a structural depression filled with Carboniferous sedimentary rocks (Forgeron, 1984).

The Mississippian Lower and Upper Windsor groups were deposited on a pre-Carboniferous basement. They contain several stratiform bodies within a series of layers composed of limestones, gypsum, anhydrite, shales, and siltstones. These layers were deposited in a near-shore marine environment. According to Forgeron (1984), the Lower Windsor group is divided into five formations (Table 3–1):

Table 3-1. Classification of Loch Lomond Celestite Showings

Age	*Group*	*Formation*	*Rock Type*	*Strontium Showings*
Pennsylvanian			sandstone, shale, conglomerate, siltstone, coal	0
Mississippian	Upper Windsor		limestone, dolomite, gypsum, siltstone, shale	0
	Lower Windsor	Uist	conglomerate, sandstone, gypsum, limestone	2
		Loch Lomond Evaporite	gypsum, anhydrite, conglomerate, sandstone, limestone	7 (4 major)
		B_2 Limestone	limestone, gypsum	3 (2 major)
		Enon	siltstone, gypsum, conglomerate	2 (1 major)
		B_1 Limestone	limestone, sandstone	
		Basal Conglomerate	conglomerate, sandstone	0
Ordovician (probable)	Horton		quartzite, shale	0
			granodiorite, rhyolite, andesite, granite, quartz veins	0

Source: Based on data from Forgeron, 1984.

The B_1 Limestone contains only minor celestite. The upper member of the Enon formations, a mottled calcareous siltstone up to 4 m thick, contains celestite in the so-called 15 manto, whose base is the middle stromatolitic-dolomitic member. The stromatolitic columns are separated by shales, which are sometimes replaced by celestite. Down dip, the siltstones grade into celestite, which grades into gypsum; the celestite is concentrated near the base of the member.

The B_2 Limestone is 5–8 m thick and occurs throughout the Loch Lomond Basin. At Lake Enon, the middle limestone facies has alternating limestone and shale layers that grade into celestite of the 54 manto and then into gypsum and anhydrite. The upper facies is a highly fossiliferous limestone with a vuggy texture, and celestite or gypsum fills the holes.

Following this is the Loch Lomond formation. At Enon Lake it is predominantly evaporitic, but in the Mac Rae property, the formation is mostly clastic. At Enon Lake, the uppermost member contains significant celestite; it belongs to the 75 and 125 mantos and replaces the original sandstone and siltstone. In the Mac Rae deposit, celestite replaces limestone in the lower members, whereas, in the upper members, it replaces a silty limestone conglomerate. This member contains the highest grades (60–70 percent $SrSO_4$) of this ore field, and the celestite is medium to coarse grained.

The Uist formation may hold important celestite reserves associated with a silty, sandy limestone, but more exploration is needed in the area.

Forgeron (1984) believes that the primary deposition of celestite at Loch Lomond took place from evaporation of sea water in extensive mud or silt sabkhas that were bounded inland by hot, desert conditions. Evaporation concentrated solutions, causing them to precipitate out in the usual solubility sequence of calcium carbonate, celestite, gypsum/anhydrite, and, finally, the salts of sodium, potassium, and magnesium.

According to Forgeron, the deposits known as 15 and 7 mantos and zone 2 and zone 3 formed in this way. Conversely, those deposits hosted by biostrome facies, such as the 54 and 54 West mantos, possibly formed later. Strontium ions leached from preexisting rocks or from the overlying gypsum could percolate through the porous B_2 Limestone and precipitate when finding sulfate-rich waters. Forgeron (1984) believes that paleo shorelines and offshore islands, with their own shoreline environments, influenced celestite deposition.

In 1975, the Loch Lomond deposit was expected to produce 19,000 tons of strontium products.

CELESTITE DEPOSITS IN IRAN

In 1970, Schiebel undertook a geological investigation of the celestite deposit of Nalehjir, Iran, about 200 km southeast of Tehran. The celestite deposit is situated in the northwestern part of a northeast–southwest trending ridge that rises to about 1550 m above sea level. This mountain ridge is formed by an alternation of a marine series of limestones, marls, clayey shales, breccias, salt clays, and cherts of the Lower Miocene. Taking into account the occurrence of the stratiform celestite deposits in a marine marl–gypsum–clay–limestone series over a distance of about 4 km, as well as the lateral transition with rocks of a marine saline facies, Schiebel (1978) concluded that the strontium seam must be classified as a deposit formed by an evaporite process of enrichment. The celestite horizon shows sedimentary textures, in particular stratification. During the diagenetic process, recrystallization took place. One to two percent $BaSO_4$ was found at the top of the celestite layers. Under a microscope, celestite is seen as hypidiomorfic needlelike grains in spatial-granular aggregates with a loose grain texture.

The reserves reach 470,000 tons (Schiebel, 1978), with an average thickness of 2.5 m and a maximum thickness of about 4 m. The total proven reserves, down to the level of the main valley (based on a seam thickness exceeding 1 m), are about 2 million tons of celestite ore, thus being one of the largest deposits in the world.

The average $SrSO_4$ content is 91.48 percent; 1.75 percent $BaSO_4$ and 3.29 percent $CaCO_3$.

CELESTITE IN TUNISIA AND EASTERN ALGERIA

During Late Cretaceous and up to Miocene times, an epineritic platform remained emerged in central Tunisia and Eastern Algeria (Constantinois). In the Cretaceous there were small islands, but in the Eocene–Paleocene the island of Kasserine-Tebessa formed, extending 80 km in the east–west direction and 60 km in the north–south direction (Chabou-Mostefai et al., 1978).

A shallow water environment surrounds the island. In the south, an evaporite basin (Gefsa Basin), located between the island and the mainland, is in continual regression. In contrast, in the northeast several changes of facies represent quick oscillations of the coastline. Karstic structures, calcretes, and silcretes formed in the island. Different mineral occurrences were found at different stratigraphic levels of the sequence (Chabou-Mostefai et al., 1978). Thus, layers of celestite are found in rocks from the Paleocene–Lower Eocene, phosphorites in the Lower and Middle Eocene, oolitic iron in the Upper Eocene, galena in evaporites from the Upper Eocene, and within the island, karstic fillings with sphalerite, which formed during the Miocene emergence.

Chabou-Mostefai et al. (1978) and Fuchs (1979) list certain characteristics of these deposits, which can be summarized as follows:

- In the southern margin (Fig. 3–9), the celestite beds are found in two stratigraphic horizons of the Lower and Upper Montien, related to the Lumachellic beds. In the different localities either one horizon or the other is present, although in some cases both can be found together.
- Celestite layers and nodules occur in the lower stratigraphic horizon; the thickness of the layers can range from a few centimeters to several meters.
- In the north and east, a horizon related to the Upper Montien is always present. It is made up of beds, several decimeters thick with geodes, which are overlain by beds with an important biogenetic activity.
- In Kass el Ain, the celestite bed with geodes is associated with silcretes, and close to Loridge, celestite is found together with anhydrite beside an

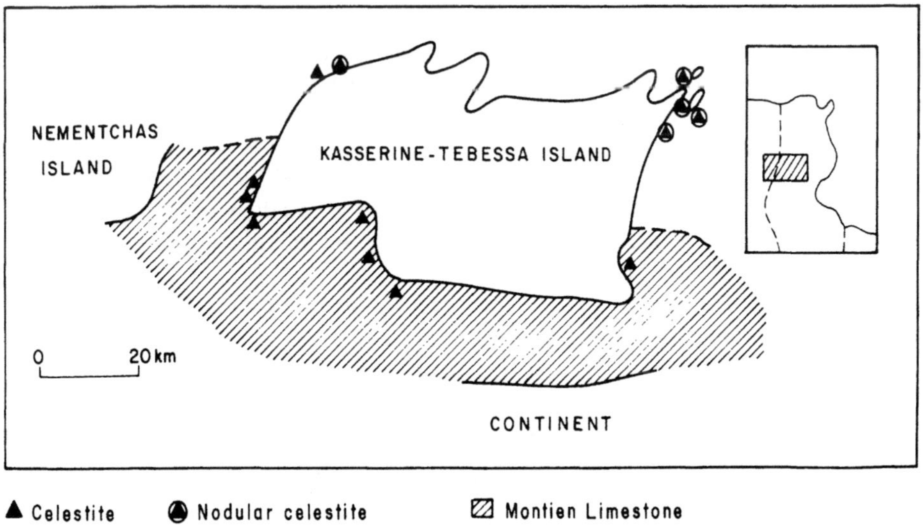

Figure 3–9. Paleogeographic sketch map showing the Eocene–Paleocene Kasserine-Tebessa Island in central Tunisia and eastern Algeria (after Fuchs, 1979).

emerged area with strong silicification. In addition to the silcrete, smectite was found, which would also indicate a high SiO_2 content in the basin.

Chabou-Mostefai et al., (1978) suggest that celestite is located in a narrow belt at the boundary of an emerged area, in which an arid to semiarid climate predominated, and in coastal evaporite lagoons, according to the following scheme: silcrete–celestite–gypsum–neritic sediments.

In the opinion of Chabou-Mostefai et al. (1978), the source of strontium could have been the erosion and dissolution of several celestite–barite veins from the Cretaceous formations present in Kasserine-Tebessa Island. However, Fuchs (1979) only refers to leaching of rocks present in the island.

The solubility of $SrSO_4$ is higher than that of $BaSO_4$, and the Sr leached could have circulated in surficial waters till it found $SO_4^=$-rich waters in the evaporite facies of the coastal belt.

According to Chabou-Mostefai et al. (1978), celestite is a good paleogeographic and climatic indicator for arid and semiarid regions. They emphasize that celestite showings are present at the same stratigraphic level, and in similar paleogeographic conditions, in different places of the epineritic platform that surrounds the north part of the African plate (Libya, Palestine, Syria).

Celestite in Italy

A major occurrence of celestite and barite in the Tolfa Mountains constitutes the most important celestite deposit in Italy (Barbieri and Masi, 1984). This deposit is located at Pian dell'Organo, north of Civitavecchia.

The sequence begins with evaporites of Upper Triassic age, overlain by marine carbonates ranging in age from Lias to Cretaceous–Paleocene, indicating an evolution through time from a shelf to a pelagic environment. The overlying terranes are flysch formations, some hundred meters in thickness, locally represented by calcareous facies of Upper Cretaceous–Paleocene age. The Lower Pliocene is represented by gypsum-bearing clay sediments. The deposit is composed of an almost vertical flat body of variable thickness of about 10 m and stretches as far as 500 m; it has a concretionary structure formed by barite, celestite, and some calcite. In Barbieri and Masi's opinion (1984) and on the evidence of Sr-isotopes, strontium was supplied by the Upper Triassic evaporites, whereas $SO_4^=$ ions may have been originated by leaching and percolation through the Pliocene gypsum-bearing sediments, coming up to the surface along a major fault.

The importance of this deposit has recently increased since the production has been exported.

Celestite in the United States of America

In California, during World Wars I and II, bedded deposits of celestite were mined sporadically, but all the known deposits are uneconomic in today's market. The deposits are located in Cenozoic lacustrine sediments in which concretionary layers of strontium minerals are associated with gypsum and occasionally halite.

ECONOMIC ASPECTS OF STRONTIUM MINERALS

Commercially, celestite is the only significant strontium mineral. Strontium is manufactured as carbonate to be used in glass for television tube face plates, in ferrites for magnets, and as nitrates for use in pyrotechnics because of the brilliant red color. Minor amounts are also used in the manufacture of electrolytic zinc, grease, soap, ceramics, alloys, and pharmaceuticals.

The principal producers are Great Britain, Mexico, Spain, and Canada. Small quantities of celestite have also been mined from time to time in Argentina, Italy, Pakistan, Morocco, and Turkey. The principal consumers are the United States (about 75 percent of total production), Japan, and Europe. The United States imports all its celestite since U.S. production stopped in 1959.

The Bristol area produced between 50 percent and 70 percent of world production from 1875 to approximately 1970. In 1960, it produced 90 percent of the world output of strontium minerals (approximately 10,000 tons). Production has increased in both Mexico and Spain since then. In 1969, the United States imported 80 percent of its strontium from Mexico and Great Britain equally. Mexico's share of the U.S. importation of strontium increased annually until 1973, when Mexico supplied 81 percent and Great Britain only 12 percent. Canadian deposits were mined in the 1970s, with 19,000 tons being produced in 1975.

By 1975, Great Britain's reserves in existing mines were becoming exhausted. Spanish output increased, and in the early 1980s, Montevives contributed 50,000 tons/year with about 95 percent strontium content.

According to *Mineral Yearbook* (U.S. Department of the Interior, 1984), production by different countries was as follows: Mexico, 44,000 tons; Turkey, 38,000 tons; Spain, 37,000 tons; Great Britain, 13,200 tons; Algeria, 6000 tons; Iran, 5000 tons; Italy, 3500 tons; and Argentina, 660 tons.

REFERENCES

Anonymous. 1974. U.K. celestite operations to close. *Industrial Minerals,* August, pp. 27–33.

Anonymous. 1975. Yate celestite. Rebirth of Bristol Mineral Co. *Industrial Minerals,* July, pp. 24–29.

Barbieri, M., and U. Masi. 1984. Sr geochemical evidence on the origin of celestite-barite deposits at Pian dell'Organo in the Tolfa Mountains area (Latium, central Italy). *Min. Petrogr. Acta* **28**:33–37.

Barbieri, M., U. Masi, L. Tolomeo. 1982. Strontium geochemistry in the epithermal barite deposits from the Apuan Alps (northern Tuscany, Italy). *Chem. Geology* **35**:351–356.

Barbieri, M., U. Masi, L. Tolomeo. 1984*a*. Strontium geochemical evidence for the origin of the barite deposits from Sardinia, Italy. *Econ. Geology* **79**:1360–1365.

Barbieri, M., B. De Vivo, V. Perrone, and E. Turco. 1984*b*. Strontium geochemistry of the San Donato Unit barite mineralization (Calabria, Italy). *Chem. Geology* **45**:279–288.

Bausch, W. M. 1968. Outlines of distribution of strontium in marine limestones. In *Recent Development in Carbonate Sedimentology in Central Europe,* eds. G. Müller and G. Friedman, pp. 106–115. Berlin: Springer-Verlag.

Butler, G. P. 1973. Strontium geochemistry of modern and ancient calcium sulphate minerals. In *The Persian Gulf,* ed. B. H. Purser, pp. 423–452. Berlin: Springer-Verlag.

Chabou-Mostefai, S., J. J. Devolve, Y. Fuchs, G. Merrant, and M. Riviere. 1978. Sur les niveaux a celestite de Tunisie centrales et du Sud-Constantinous. *Sci. Terre* **22**(3):291–300.

Claypole, G. E., W. T. Holser, I. R. Kaplan, H. Sakai, and I. Zak. 1980. The age curves of sulfur and oxygen isotopes in marine sulfates and their mutual interpretation. *Chem. Geology* **28**:199–260.

Dabrio, C. J., J. M. Martin, and A. Megias. 1982. Signification sédimentaire des évaporites de la depression de Grenade (Espagne). *Soc. Geol. Fr. Bull.* **24**(4):705–710.

Dean, W. E., and A. L. Tung. 1974. Trace and minor elements in anhydrite and halite, Supai Fm. (Permian), east-central Arizona. In *4th Symposium on Salt,* ed. A. H. Coogan, pp. 287–301. North Ohio Geological Society.

Evans, G., and D. J. Shearman. 1964. Recent celestine from the sediments of the Trucial Coast of the Persian Gulf. *Nature* **202**:343–386.

Forgeron, S. V. 1984. Loch Lomond celestite properties. In *The Geology of Industrial Minerals in Canada* (Special Volume 29), eds. G. R. Guillet and W. Martin, pp. 19–21. Canadian Institute of Mining and Metallurgy.

Fuchs, Y. 1978. Uber Erzlagerstatten im Tertiar von Tunesien und Constantinois. *Geol. Verh.* **3**:311–319.

Gundlach, H. 1959. Untersuchungen zur Geochemie des Strontiums auf hydrothermalen Lagerstätten. *Geol. Jahrb.* **76**:637–712.

Kendall, A. C. 1979. Facies models 13. Continental and supratidal (sabkha) evaporites. In *Facies Models,* ed. R. G. Walker, pp. 145–157. Geoscience Canada.

Kesler, S. E., and L. M. Jones. 1981. Sulfur and strontium isotopic geochemistry of celestite, barite and gypsum from the Mesozoic basins of northeastern Mexico. *Chem. Geology* **31**:211–224.

Kinsman, D. J. J. 1969. Modes of formation, sedimentary association and diagnostic features of shallow-water and supratidal evaporites. *Am. Assoc. Petrol. Geol. Bull.* **53**:830–840.

Kranz, J. R. 1976. Strontium-ein Fazies-Diagenese-Indikator in Oberen Wettersteinkalk (Mittel Trias) der ostalpen. *Geol. Rundschau* **65**(2):593–615.

Martin, J. M., M. Ortega Huerta, and J. Torres Ruiz. 1984. Genesis and evolution of strontium deposits of the Grenada Basin (southeastern Spain): Evidence of diagenetic replacement of a stromatolite belt. *Sed. Geology* **39**:281–298.

Mason, B., ed. 1960. *Principios de Geoquimica.* Barcelona: Omega, 333p.

Müller, G. 1962. Zur Geochemie des Strontiums in ozeanen. Evaporiten unter besonderer Berucksichtigung der sedimentaren Coelestin-lagerstatte von Hemmelte-West. *Geol. Beiheft* **35**:1–90.

Nickless, E. F. P., S. J. Booth, and P. N. Mosley. 1975. Celestite deposits of the Bristol area. *Inst. Min. and Metal. Trans.* **84B**:62–64.

Nickless, E. F. P., S. J. Booth, and P. N. Mosley. 1976. The celestite resources of the area northeast of Bristol. *Institute of Geological Sciences Mineral Assessment Report N25,* 83p.

Noll, W. 1934. Geochemie des Strontiums, mit Bemerkungen zur Geochemie des Bariums. *Chem. Erde* **8**:507–600.

Olaussen, S. 1981. Formation of celestite in the Wenlock, Oslo Region, Norway: Evidence for evaporitic depositional environments. *Jour. Sed. Petrology* **51**:37–46.

Peterman, U. E., C. E. Hedge, and H. A. Tourtelot. 1970. Isotopic composition of strontium in sea water throughout Phanerozoic time. *Geochim. Cosmochim. Acta* **34:**105–120.

Schiebel, W. 1978. New strontium deposits in Iran. *Industrial Minerals,* September 1978, pp. 54–57.

Skinner, H. C. W. 1963. Precipitation of calcian dolomites and magnesian calcites in Southeast Australia. *Am. Jour. Sci.* **261:**449.

Thomas, I. A. 1973. *Celestite Mineral Dossier N°6.* London: Her Majesty's Stationery Office, 26p.

U.S. Department of the Interior. 1984. *Minerals Yearbook: Metals and Minerals,* vol. 1. Washington, D.C.: Government Printing Office, 1034p.

Veizer, J., and W. Compston. 1974. $^{87}Sr/^{86}Sr$ composition of sea water during the Phanerozoic. *Geochim. Cosmochim. Acta* **38:**1461–1484.

Veizer, J., and R. Demovic. 1974. Strontium as a tool in facies analysis. *Jour. Sed. Petrology* **44:**93–115.

West, I. M. 1973. Vanished evaporites—significance of strontium minerals. *Jour. Sed. Petrology* **43:**278–279.

Wood, M. W., and H. F. Shaw. 1976. The geochemistry of celestites from the Yate area near Bristol (U.K.). *Chem. Geology* **17:**179–193.

4

Barite and Celestite Stratabound Ore Fields in Argentina

M. K. de Brodtkorb, I. B. A. Schalamuk, and S. Ametrano

Most barite stratabound deposits and all celestite deposits known to date are found in the Mesozoic marine basin located in central western Argentina, in the provinces of Mendoza and Neuquén. Two marine sedimentary cycles are recognized in this basin: The first, of Jurassic age, is formed by two subcycles, whereas the second corresponds to the Tito-Neocomian. Although both barite and celestite deposits occur in the Jurassic cycle, only celestite deposits are found in the second cycle. However, the Mamina barite deposit in Chile is located in the upper member of the Nantoco formation, which belongs to the Neocomian marine sequence (Amstutz et al., 1985). Cura Mallin and Mallin Quemado in the province of Neuquén are the most important barite deposits from an economic point of view (Fig. 4–1).

Several authors have been interested in the barite and celestite deposits occurring in this area: Angelelli (1941), Oliveri et al. (1964), Brodtkorb et al. (1975, 1982, 1985), Lyons et al. (1978), Bengochea et al. (1984), Barbieri et al. (1981, 1984), Llambías and Malvicini (1978).

Isotope analysis to determine the $^{87}Sr/^{86}Sr$ ratios were performed to support geologic evidence that would indicate a sedimentary syndiagenetic origin for these deposits.

Other barite stratabound deposits in Argentina are La Helvecia (province of La Rioja) and Canota (province of Mendoza), both in Ordovician rocks. La Helvecia contains sulfides together with barite, whereas the Canota ore field is characterized by nodular textures and is described in Chapter 11.

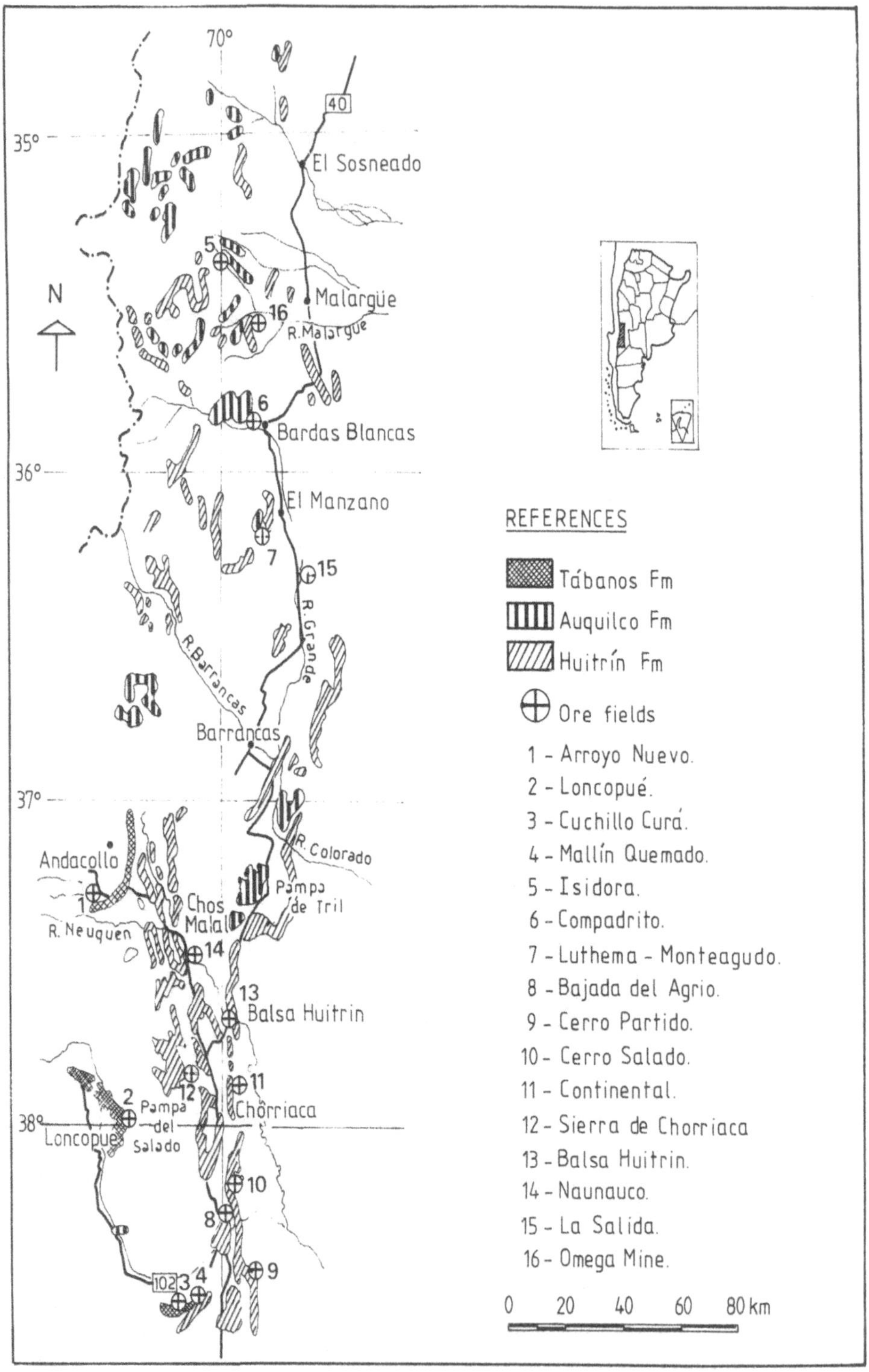

Figure 4–1. Location of celestite and barite ore fields in Mendoza and Neuquén provinces.

BARITE AND CELESTITE ORE FIELDS OF THE MESOZOIC BASIN

GEOLOGIC SETTING

The study area comprises a back-arc basin located to the east of the Argentine Central Andes, in the extra-Andean region of Neuquén and southern Mendoza. This back-arc basin developed over a relatively stable platform that is conformed by a basement composed of Precambrian to Lower Paleozoic metamorphic rocks, Carboniferous sedimentites, plutonic rocks of different ages, ranging from Precambrian to Permo-Triassic, and volcanic rocks from the Choiyoi group (Pre-cuyean "basement"). The geologic province known as the Neuquén–Mendoza basin is an area of accumulation of Jurassic–Cretaceous marine, continental, and transitional sediments that unconformably overlie the basement on a regional scale.

Groeber (1946) and later Stipanicic (1969), among others, divided this sedimentary sequence into three cycles: "Jurassic" (Lias–Lower Kimeridgian), Andean (Upper Kimeridgian–Albian), and Riograndic (Senomanian–Paleocene?), separated by unconformities or by evident changes in the sedimentary regime. Oil drilling in the east and several exposures in the occidental regions contributed to the knowledge of this sequence in Neuquén as well as in southern Mendoza (Fig. 4-2).

The initial phase of basin filling corresponds to the "Jurassic" cycle, and two sequences of transgression and regression known as the Cuyean and Lotenian–Chacayan have been recognized. The sea encroached from the west and northwest and formed an embayment with a large extension to the east. In the Neuquén sector, the Cuyean Cycle (Lias–Lower Callovian) is characterized by a long period of subsidence in which up to 2500 m of sediments were deposited. Deposition started with continental sediments filling the depressions developed on the surface of Triassic volcanic rocks. The filling consisted of conglomerates, sandstones, and tuffs. Interbedded with these sediments, or lying directly over the Triassic volcanic rocks, is the marine sequence in the occidental and central parts of the basin. The Cuyo group is divided into the Los Molles, Lajas, and Tábano formations. The Los Molles formation consists of turbidites made up of black shales and mudstones and gray calcipelites. The Lajas formation is formed by fine- to coarse-grained sandstones interbedded with mudstones as well as calcareous and sandy shales. According to Digregorio (1978), the Los Molles formation corresponds to an intermediate to distal marine environment with restricted circulation, and the Lajas formation to a littoral to sublittoral environment with moderate agitation. The Cuyean subcycle finishes with the evaporites (gypsum and dolomite) of the Tábanos formation. Based on their geographic location and their facies, these evaporite deposits were defined by Gulisano et al. (1984) as corresponding to the center of the basin.

The second subcycle, Lotenian–Chacayan, constitutes a new transgression–regression. The Lotena formation corresponds to the initial phase and is formed by littoral and continental sandstones. A new marine transgression

	SYSTEM	CYCLE	SUBCYCLE		NEUQUEN	MENDOZA ROCK UNITS	DIASTROPHIC PHASES
CRETACEOUS	UPPER	RIOGRANDIC	NEUQUENIAN	NEUQUEN G.	Río Colorado Fm Río Neuquén Fm Río Limay Hm	Malargüe Fm	
							Intercretacic
	LOWER	ANDIC	RAYOSIAN	RAYOSO G.	C.de la Zorra Fm Ranquiles Fm Huitrín Fm	Diamante Fm- Salas Fm	
			MENDOCIAN	MENDOZA GROUP	Agrio Fm Mulichinco Fm Quintuco Fm Vaca Muerta Fm	Mendoza Fm Cienagüitas Member Chachao Member Vaca Muerta M.	
							Intermalmic
JURASSIC	UPPER	"JURASSIC"	CHACAYAN — LOTENIAN		Tordillo Fm Auquilco Fm La Manga Fm Lotena Fm	Tordillo Fm Auquilco Fm Calizas del Calabozo Fm Puchenque Fm	
							Intercalovian
	MIDDLE		CUYEAN	CUYO GROUP	Tábanos Fm Las Lajas Fm	Tábanos Fm El Codo Fm Puesto Araya Fm	
	LOWER				Los Molles Fm	EI Freno Fm	
							Interliassic
PRECUYEAN							

Figure 4-2. Mesozoic marine sedimentary cycles, rock units, and diastrophic phases in the Neuquén-Mendoza basin.

started in the middle Callovian and finished with the deposition of the lower part of the La Manga formation. The high-energy carbonate deposits of the upper part of La Manga suggest a regressive trend with a basinwide restriction grading transitionally into the evaporites of the Auquilco formation (also known as "Yeso principal"). This formation extends in a more or less continuous form for over 77,000 km^2 (Lambert, 1956). The Auquilco formation is mainly made up of white gypsum with laminated and nodular textures. It is often interbedded with limestones and green and red silts. It corresponds to a shallow water deposit that grades into supralittoral sabkhas (Digregorio, 1978). The upper part of the Auquilco and La Manga formations define the regressive Chacayan Cycle. It follows the Tordillo formations, representing a period of basin filling with detrital material. The lower part of this unit consists of red sandstones and mudstones, whereas the upper part is made up of green sandstones and conglomerates (Stipanicic, 1969). Digregorio and Uliana (1980) suggest that the lower part is related to the accumulation of the Auquilco formation whereas the upper part is transitional to the marine sediments of the Vaca Muerta formation, both being separated by the Araucana diastrophic phase.

The Andean Cycle (Mendocian and Rayosian) begins with a new basin subsidence characterized by several transgressions and regressions. The Mendoza group reaches a thickness of 2400 m and the sequence is composed of the transgressive Vaca Muerta formation followed by the regressive Quintuco formation. The conglomerates of the Mulichinco formation indicate the beginning of a transgressive period, which remained during the deposition of the lower part of the Agrio formation. A small overdeeping of the basin produced the deposition of sandstones of the Avilé member in a probable near-shore bar environment. It follows a regressive period with the deposition of the upper part of the Agrio formation.

The Neocomian sea finally receded, giving way to the deposition of (1) evaporites (limestones, gypsum, anhydrite, celestite, halite) in a subaqueous environment with an evident marine influence; (2) mudstones alternating with marls and gypsum, in a supratidal environment with lagoons, sabkhas, and mud flats; and (3) an alternation of sandstones and alluvial red mudstones that conform the Huitrín formation. These beds are overlain by red sandstones that correspond to the continental rise of the area.

The Mendoza sector of the basin (Fig. 4–2) is characterized by the deposition of the Cuyo group over the pre-Jurassic platform. The Cuyo group is composed of conglomerates and sandstones of the El Freno formation, sandstones, siltstones, and limestones of the Puesto Araya formation, psamites and pelites together with marls and limestones of the El Codo formation, and gypsiferous bodies of the Tábanos formation.

The second subcycle is represented by the Puchenque formation (Dessanti, 1973), and the Lotena formation. The first one comprises from bottom to top a basal conglomerate, sandstones with lumachellic beds, limestones, siltstones, and a polymictic conglomerate. Here the Lotena formation presents similar characteristics to that in the Neuquén sector. This unit is conformably overlain by limestones with corals and ammonites, and marls of the La Manga formation considered by Stipanicic (1969) as being Lower to Middle Oxfordian and equivalent to the Calizas del Calabozo formation in the area of Bardas Blancas (Dessanti, 1973). Overlying the Calizas del Calabozo formation and providing an example of the marginal sector of the basin is the thick evaporite unit of the Auquilco formation.

Following this unit, in the Lower as well as in the lower part of the Middle Kimeridgian (Stipanicic, 1969), the Tordillo formation was deposited with a clastic sequence represented by arkosic sandstones, conglomerates, and shales, with some pyroclastic material. The Tordillo formation reaches a considerable thickness to the west of the large brachyanticline structure of the Sierra Azul.

A new marine transgression (Andean cycle) is represented by a sequence, several meters thick, of limestones alternating with pelites. These sediments contain an ammonite fauna that indicates an age span ranging from the Titonian to the Hauterivian. The sequence corresponds to the Mendoza formation (Dessanti, 1973), equivalent to the Mendoza group (Neuquén section), and it is subdivided into three members: Vaca Muerta, Chachao, and Cienaguitas (Leanza et al., 1977). Representing the regression of the Neocomian sea, it follows the impure limestones, gypsum, salt rock, and potash of the Huitrín formation.

Thick continental red beds were deposited in the Upper Cretaceous (Diamante and Salas formations). The Malargüe formation, composed of fossiliferous limestones and marls, sandstones, and tuffs represents a marine-to-continental transitional environment.

The Mesozoic sediments are affected by an andesitic vulcanism (agglomerates, breccias, and lavas), basic lavas, laccolithic intrusions, dykes, and other irregular bodies of mesosilicic composition that belong to the Molle formation (Oligocene), Plutonitas Andinas formation (Oligocene), and Palauco formation (Miocene–Pliocene).

THE ORE FIELDS

Figures 4–1 and 4–3 show that the celestite and barite deposits from a nearshore environment are found only in certain stratigraphic units: Tábanos, Auquilco, and Huitrín formations. The Tábanos formation hosts the Loncopué barite ore field and the Cuchillo Curá celestite ore field. Related to the Auquilco formation are the Mallin Quemado, Isidora, and Compadrito barite deposits and the Luthema, Monteagudo, and La Salida celestite deposits. The Huitrín formation is known to host mostly celestite deposits. The Arroyo Nuevo deposit, which is being studied at present, is emplaced in shales either of the Los Molles formation or of the lower part of the Tábanos formation.

These deposits have been sporadically mined for the last 40 years, generally in a rudimentary form, producing more than 200,000 tons.

ORE FIELDS RELATED TO THE CUYANO SUBCYCLE

Arroyo Nuevo Mine. This deposit, the largest in the study area, is located 21 km south of the town of Andacollo (Fig. 4–1) and has been mined for the last 20 years. The deposit consists of a barite layer intercalated in shales, which underlie the evaporites of the Tábanos formation and probably corresponds to the Los Molles formation. The barite bed is faulted and intruded by ande-

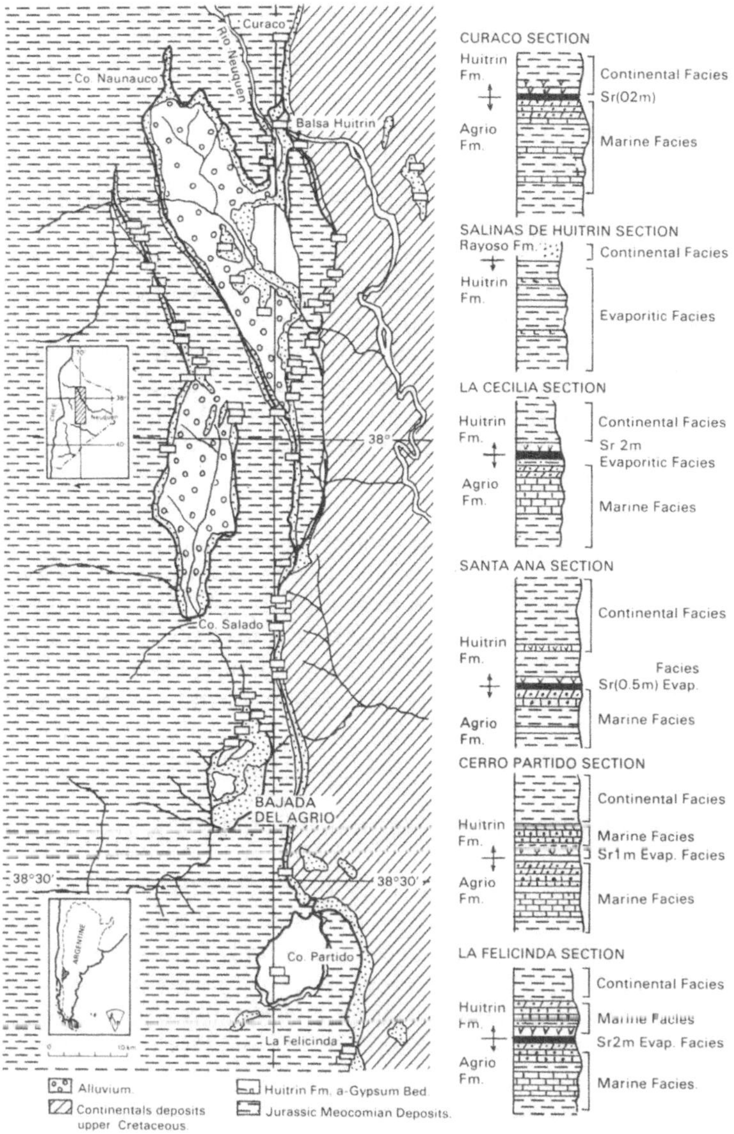

Figure 4–3. Location of celestite occurrences and ore fields (□) related to the Huitrín formation in Neuquén Province.

sitic dykes. The barite is grayish-white, coarse-grained, although some fine-grained occurs, and presents a tabular habit. The fine-grained barite is commonly intercalated with abundant chloritic and clayey materials together with some lenses of limestone. The thickness of the barite layer ranges from 1.5 m to 2.5 m. This deposit is mined underground by the chamber-and-pillar method and has been developed for more than 2000 m.

Samples have a chemical composition ranging from 92.0 percent to 99.01 percent $BaSO_4$ and 0.01 percent to 0.07 percent $SrSO_4$.

Loncopué Ore Field. This barite district is located approximately 7 km east of the town of Loncopué (Fig. 4–1). The stratabound deposits are related to the Tábanos formation, whereas the associated veins are placed in the Lotena formation.

The Tábanos formation is made up of limestones, locally interbedded with gypsum and barite beds. Overlying these are the conglomerates and sandstones of the Lotena formation. These rocks are intruded by Tertiary plutonic rocks of the Los Molles formation.

The barite layers are 0.6 m to 1 m thick and extend for more than 300 m, as was discovered during the underground mining of La Rosita mine. Some shales and chert are interbedded with barite in the Dos Guanacos mine, whereas some clay is present in the barite layers of La Rosita mine. Related to these stratabound deposits are unconformable bodies of barite, hosted by the overlying Lotena formation. These veins are 0.2 m to 1 m thick and about 30 m deep and are exposed for 100–300 m. Several deposits in the district are mined on a small scale.

Cuchillo Curá Ore Field. In this locality, the evaporite facies of the Tábanos formation is represented by limestones grading southward into gypsum. The celestite beds, with an areal extension of 7 km and thicknesses up to 2 m, are dipping to the north. The ore is characterized by textures of "diagenetic crystallization rhythmites" (DCR) (Fontboté and Amstutz, 1983), and it is common to observe a late recrystallization due to remobilization, with the formation of big celestite crystals.

Ore Fields Related to the Auquilco Formation

Mallin Quemado Ore Field. This ore field is located about 50 km north of Zapala. Subhorizontal barite layers are found in this ore field, conformably overlying the Auquilco gypsum and underlying the continental sandstones of the Tordillo formation. Tectonics together with the great plasticity of the underlying gypsum bed produced a rather complicated structure in the deposit, with abundant breccia zones containing clasts and blocks of barite, sandstone, and gypsum. The original barite layer was probably several hundred meters long, about 100 m wide, and 1–3 m thick.

Some unconformable barite bodies related to the barite layers are found in the area, hosted in the overlying sandstones of the Tordillo formation. These veins are up to 1 m thick, several tens of meters long, and up to 50 m deep. Although the barite layers are rather pure, the chemical analyses yielded values ranging from 84 percent to 90 percent $BaSO_4$. The veins are intergrown with quartz, galena, and, in some cases, with a few chalcopyrite. This deposit has been mined on a small scale for the last 40 years. Within the last five years, mining has intensified by means of underground works.

Luthema Monteagudo Mines. The Luthema celestite mine is located on the southeast flank of the Sierra Azul brachyanticline, about 120 km south of the town of Malargüe (Fig. 4–1). It has been mined since 1950 in a trench 90 m long, 15 m wide, and about 12 m deep. The celestite layers are conformable with the Jurassic sedimentary rocks, which consist of gray marly limestones with abundant fossils (La Manga formation) and the overlying gypsum bed of the Auquilco formation. Some Tertiary andesites of the Los Molles formation intrude the sediments.

The stratabound celestite layer crops out along 200 m in the direction of strike and has a subhorizontal dip. The chemical analyses yielded values ranging from 70–80 percent $SrSO_4$ and 5–10 percent $BaSO_4$. The yellow-brown color is due to the presence of abundant iron hydroxides. Several textures can be found such as breccias, rosette-shaped aggregates up to 30 cm in diameter, and celesto-barite crystals. Layers of fine-grained celestite alternating with fine laminated marls are also found (Fernandez et al., 1987).

The Monteagudo mine is located 500 m to the west of the Luthema mine. The deposit consists of a celestite layer interbedded in limestones that outcrops for only 50 m. It has similar characteristics to Luthema, with big crystals and aggregates covered by limonites. Monteagudo was mined in the early 1950s.

El Compadrito Mine. This mine is located about 60 km south of the town of Malargüe (Fig. 4–1). It consists of a barite layer, with little celestite, 3.5 m to 7 m thick and outcrops for about 200 m. This deposit has been mined sporadically for the last 30 years and is the main barite deposit of Mendoza. The mine is located on the north side of the Sierra Azul brachyanticline. The barite layer overlies the limestones of the Caliza del Calabozo formation (Fig. 4–4) and has a "zebra" texture formed by fine-grained barite with silty-clayey ma-

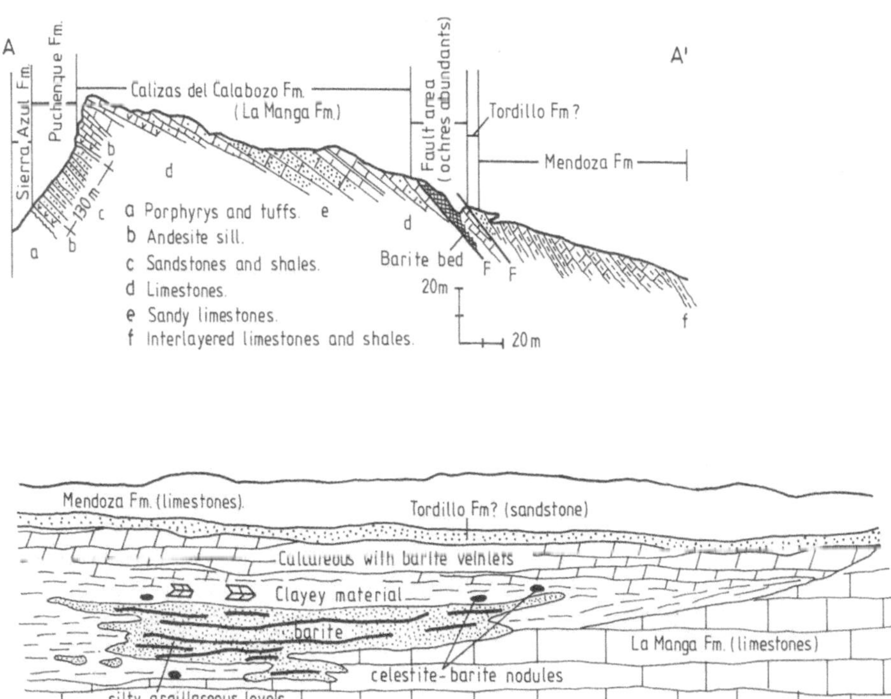

Figure 4–4. Top: Geological section, El Compadrito mine area. Bottom: Stratigraphic location of barite layer of El Compadrito mine.

terial and interbedded layers of limestones with abundant iron hydroxides (Figs. 4–5, 4–6, 4–7). In the pelitic material, mixed-layered clays were identified as allevardite and corrensite, which indicates a continental–littoral environment (Schalamuk et al., 1987). A sandstone bed equivalent to the Tordillo formation overlies the barite layers and is followed by a thick sequence of marls, limestones, and pelites of the Mendoza group, which is well represented in the mine area with an abundant ammonite fauna. In the vicinity of the deposit, andesite–dacite intrusions of the Los Molles formation occur, conformable and unconformable with the limestones.

La Salida Mine. This deposit is located about 40 km south of the El Compadrito mine, near the Coihue-Co creek, and it was only rudimentarily mined. The barite layer, emplaced in the same sequence as the El Compadrito deposit but here in a folded structure, outcrops for only 40 m and is 0.70–1.10 m thick. The barite is fine-grained and associated with silty–clayey material; in part, there are dissolution cavities and stalactite structures 15–25 cm long.

La Isidora Mine. This barite deposit is located southwest of Mendoza, approximately 60 km west of the town of Malargüe. The Puchenque formation is represented in this area by limestones with abundant shale interlayers. Overlying these are the thick evaporite levels of the Auquilco formation, made up of gypsum and anhydrite. The sedimentary rocks are intruded by Tertiary plutonites.

The La Isidora mine has been only rudimentarily mined. It is emplaced in the upper limestone beds, close to the Auquilco formation. The deposit con-

Figure 4–5. General view of the El Compadrito mine, showing the stratigraphic sequence.

Figure 4-6. El Compadrito mine. Zebra texture showing barite interlayered with silty-argillaceous materials.

Figure 4-7. Zebra texture, in situ, in the mine.

sists of a lenticular barite bed conformable to the Jurassic calcareous sediments. The bed can be followed for 140 m in the direction of strike. It has a maximum thickness of 5 m, and it pinches out into gypsum facies southward and into a limestone facies northward. Internally, the mineralized bed presents different textures: In the lower part, the barite shows a marked banding with the development of DCR textures in which the sulfate tends to develop a tabular-acicular habit. In contrast, in the upper part, barite is present in sparry, tabular crystals, up to 5 cm long, randomly intergrown. The rosette-shaped aggregates are frequent, mainly in the uppermost part of the barite bed.

ORE FIELDS RELATED TO THE HUITRÍN FORMATION

Along the 70° longitude, between Zapala and Curacó, Neuquén Province, several celestite deposits occur only at certain stratigraphic positions. Several districts are known, some kilometers long, in which the celestite beds wedge out to outcrop again at a distance of 10–20 km. The most representative districts are Bajada del Agrio, Cerro Salado, Cerro Partido, Naunauco, Sierra de Chorriaca, Continental, and Balsa Huitrín (see Fig. 4-1; Brodtkorb et al., 1975, 1982, 1985).

Several sections, cut through the upper member of the Agrio and Huitrín formations, permit us to characterize the evolution of the evaporite facies as a typical regressive sequence of the Neocomian sea (Fig. 4–8). The upper part of the underlying lithofacies is cemented by sparite and, for their textural and petrographic characteristics, are considered as intertidal sandstones, corresponding to a beach environment. Overlying the sandstones are carbonate bands that show a conspicuous algal lamination, grading laterally into more irregularly stratified limestones. In many areas, a cyclic alternation of carbonates and celestite, locally with DCR textures, is observed. In the Santa Bárbara mine, which belongs to the Bajada del Agrio ore field, the celestite bed grades laterally into intraclastic limestones with a brecciated texture containing gypsum nodules. Due to the dissolution of the nodules, this limestone shows a typical cavern texture with voids filled with secondary micrite. The facies changes occur in less than 1 km. The textural and petrographic characteristics of the celestite facies would indicate an intrasupra- to supratidal environment of formation in which the sediments were exposed to a subaereal environment for some periods of time. Following the celestite layer is a gypsum bed ranging from 2 m to 10 m thick. The gypsum is impure and presents frequent facial variations. The gypsum bed, which presents a nodular texture, indicates deposition in a subaerial environment. Due to the irregular surface, some portions of ancient tidal channels or coastal lagoons remained, where the water temporarily remained, allowing the subaerial precipitation. The celestite beds have thicknesses ranging from a few centimeters to 2 m, and the set of evaporite facies can reach a thickness of 12 m.

Following the evaporites are silty dolostones intercalated with mudstones and red claystones with remobilized gypsum levels as well as crystalline and massive halite beds. These beds are overlain by continental red beds (Fig. 4–8).

The amount of existing deposits together with the characteristics common to some of them allow them to be grouped into three types, based on mineralization geometry and textures:

1. Syndiagenetic, massive celestite beds several meters thick and some hundreds of meters long, which formed in lagoons by evaporation processes;
2. Banded, layered deposits (a) formed in supratidal environments with alternations of celestite beds with limestones, algal mats, and/or

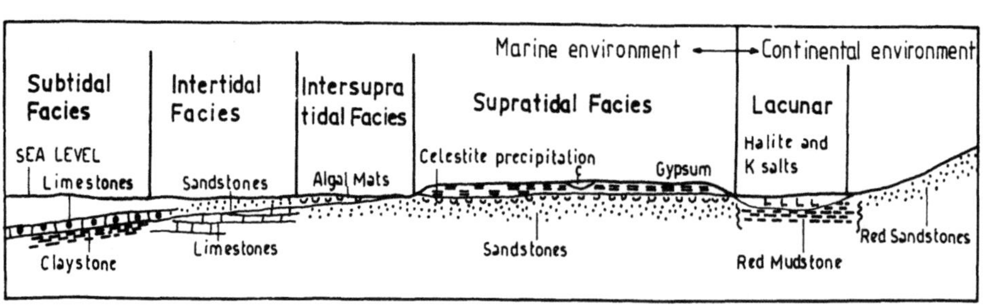

Figure 4–8. Evaporitic cycle of the Huitrín formation.

nodular gypsum; (b) celestite beds with DCR (see Figs. 4-10 and 4-13;

3. Epigenetic barite veins related to types 1 and 2.

The first type of deposits include the Cerro Salado and Naunauco ore fields. These deposits are characterized by layers conformable to the stratification, and their present location depends on the local tectonics. The celestite has a granular texture with sparry habit, and its color changes in the different deposits from yellowish-white to grayish-brown. The second type of deposit includes the Bajada del Agrio, Continental (Figs. 4-9, 4-10, and 4-11), and Balsa Huitrín ore fields. These are early diagenetic conformable layers. The upper part of the layers presents dissolution voids (Fig. 4-12) filled with neoformed crystals, geodes, and stalactites (Fig. 4-13). Mudcracks with evidence of subaqueous sliding have been recognized in the Santa Ana mine (Bajada del Agrio ore field), (Fig. 4-14). In the Balsa Huitrín ore field the celestite-bearing layer is 6-7 m thick and extends for 1000 m in the direction of strike. The third type, epigenetic veins, is best represented by the San José mine. The veins fill a system of secondary tension faults, subparallel to a main fault. The filling is made up of barite with colloform textures that present a downward flow aspect. It is believed that this deposit originated from the overlying barium-bearing celestite bed, which is absent on this flank of the anticline due to erosion. The different solubilities of $BaSO_4$ and $SrSO_4$ produced the precipitation of barite and the migration of $SrSO_4$.

Figure 4-9. Continental ore field (Cecilia mine). General view showing the mine workings along a celestite bed.

Figure 4-10. Continental ore field. Diagenetic crystallization rhythmite texture of celestite.

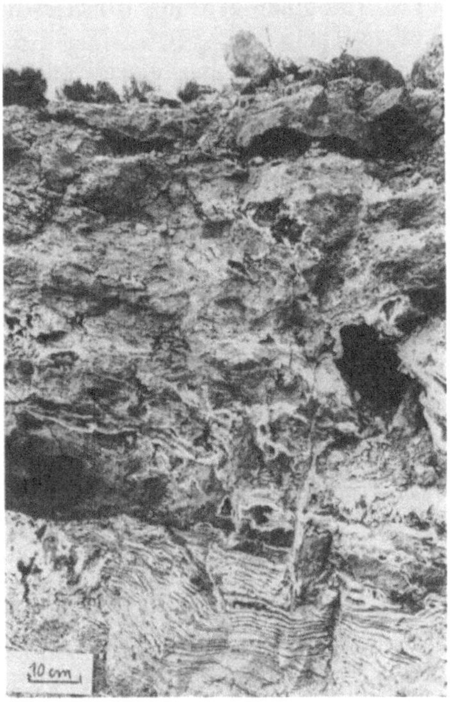

Figure 4-11. Continental ore field. Celestite bed with diagenetic crystallization rhythmites in the lower section and remobilized texture in the upper section.

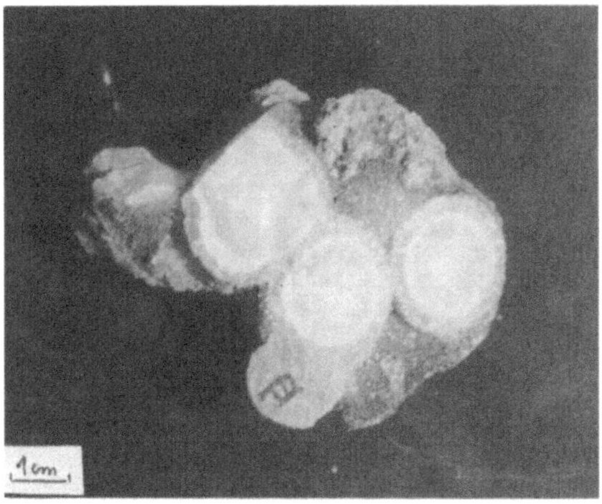

Figure 4-12. Continental ore field. A celestite stalactite that grew up in a large dissolution cavity.

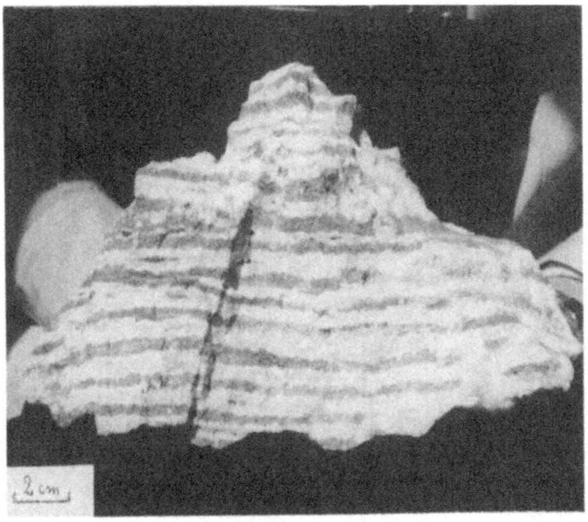

Figure 4-13. Bajada del Agrio ore field (Santa Ana mine). Diagenetic crystallization texture showing the first generation (gray), the second generation with bipolar celestite crystals (white), and the no-filled spaces of the third generation.

Due to their reserves and quality, the Naunauco, Continental, and Balsa Huitrín ore fields are considered the most important from an economic point of view. The deposits associated with the Huitrín formation contain geologic reserves of more than 1 million tons.

HYDROTHERMAL DEPOSITS

The Omega mine is related to Tertiary magmatism and consists of a barite vein hosted by a diorite-adamellite intrusive stock (Cerro Serrucho). The known length of the vein is 70 m and it ranges from 0.7 m to 1.2 m thick. The ore is made up of either massive white barite or tabular crystals 3-5 cm long, randomly intergrown and with abundant limonite impregnations. Minor veins, approximately parallel to the main ore, are found in the surrounding areas.

Figure 4-14. Bajada del Agrio ore field (Santa Ana mine). Disturbed layers of limestone (gray) in a celestite matrix (white).

Table 4-1. The $^{87}Sr/^{86}Sr$ Ratios of Different Minerals and Rocks of the Neuquén-Mendoza Basin

Locality	*Mineral or Rock*	$^{87}Sr/^{86}Sr$	$\delta \pm$	*Sr (ppm)*	*Related to*
Huantraico	andesite	0.7035	0.002	565	Tertiary
Luthema	andesite	0.70472	0.00003		
Collipilli	andesite	0.70454	0.00004		
Loncopué	andesite	0.70455	0.00003		
Omega	barite (vein)	0.70516	0.00002		
Bajada del Agrio Ore Field					Huitrín
Sta. Barbara	celestite layer	0.7072	0.0004		Formation
	barite (rem.)	0.7070	0.0005	7000	
	gypsum	0.7077	0.0002	1040	
Santa Ana	gypsum	0.7077	0.0002	1100	
Cerro Salado Ore Field					Lower
El Olvido	celestite layer	0.7072	0.0005		Cretaceous
La Corrida	celestite layer	0.7072	0.0005		
Cerro Salado	limestone	0.7078	0.0003	350	

Mallin Quemado Ore Field					
Achalay	barite (breccia)	0.70785	0.00005	17000	Auquilco
	barite (vein)	0.70788	0.00004	13580	Formation
	gypsum	0.70710	0.00003	1580	
	anhydrite	0.70707	0.00003	1760	
Rio Agrio	barite (vein)	0.70788	0.00003	16500	
Luthema	celestite (layer)	0.70722	0.00005		
	gypsum	0.70720	0.00005		Upper Jurassic
	limestone	0.70721	0.00005		
El Compadrito	barite (layer)	0.70838	0.00004		
	barite (layer)	0.70831	0.00001		
	gypsum	0.70721	0.00004		
Isidora	barite (layer)	0.70662	0.00003		
	barite (layer)	0.70665	0.00002		
	gypsum	0.70662	0.00004		
	gypsum	0.70665	0.00003		
	limestone (bottom)	0.70755	0.00002		
	limestone (top)	0.70759	0.00002		
Loncopué Ore Field					
					Tábanos Formation
Rosita	barite (layer)	0.70746	0.00004		
	limestone (bottom)	0.70680	0.00003		Middle Jurassic
Florcita	barite (vein)	0.70709	0.00003		

Note: The isotopic values were analyzed by Prof. Mario Barbieri in the Institute of Geochemistry at the University of Rome, Rome, Italy.

Only limited mining of the deposit has been done by means of an open pit. Isotope data on the Omega mine, one of the few hydrothermal barite deposits of the study area, are included for comparative purposes.

Isotope Data

In the last few years the knowledge of the $^{87}Sr/^{86}Sr$ isotope ratios of sediments and the oceans has grown rapidly. Based on the Sr content of fossils and carbonatic rocks, Peterman et al. (1970), Veizer and Compston (1974), and Burke et al. (1982) have shown that significant changes in the $^{87}Sr/^{86}Sr$ ratios of sea water occurred during the Phanerozoic.

The isotope values of celestite, barite, gypsum, and anhydrite and of underlying and overlying limestones from different deposits are shown in Table 4–1. The values of hydrothermal barite veins together with Tertiary andesites from the study area are included for comparison. It can be observed from Figure 4–15 that these values correspond with those presented by Peterman et al. (1970), Veizer and Compston (1974), and Burke et al. (1982) for these different ages.

Conclusions

In order to understand the genesis of these deposits, the behavior of the Sr and Ba cations should be explained separately.

Celestite is the most important strontium mineral in the exogenous cycle and precipitates mainly in the evaporite-basin environment. The Sr content of present sea water is approximately 8 gr/t, from these about 62 percent is available for the formation of celestite and 38 percent is included in other evaporite minerals (Müller, 1962). The Ba average content of celestite is 0.48 percent. Precipitation of celestite from sea water occurs in the boundary between the calcium carbonate and the calcium sulfate precipitation when the sea water reduction reaches approximately 1/4 of the original volume (between 1/3 and 1/5). On the other hand, gypsum begins to precipitate when sea water reaches 1/3.35 times its original volume, and for this reason celestite is often found in close association with gypsum.

If the Sr content of sea water is 8 g/t, celestite is formed. However, if sea water is diluted with continental water, the Sr content decreases to only 4 g/t and Sr is in part added to calcium sulfate.

Recent studies point out the importance of different processes taking place in the sub-, intra-, and supratidal facies, in coastal and continental sabkhas, and during diagenesis. The work of Kinsman (1969) is considered a precursor on the subject. In his opinion, sabkha diagenesis involves the interstitial emplacement of evaporite minerals within the host sediments, but brine pans may accumulate more soluble evaporites because of the greater evaporation losses. According to Kinsman, celestite is a common early diagenetic mineral of the coastal sabkhas and abundant in areas of intense dolomitization. The Sr released during the replacement of aragonite (7000–8000 ppm Sr) by dolomite (600–700 ppm Sr) is added to interstitial water and forms celestite.

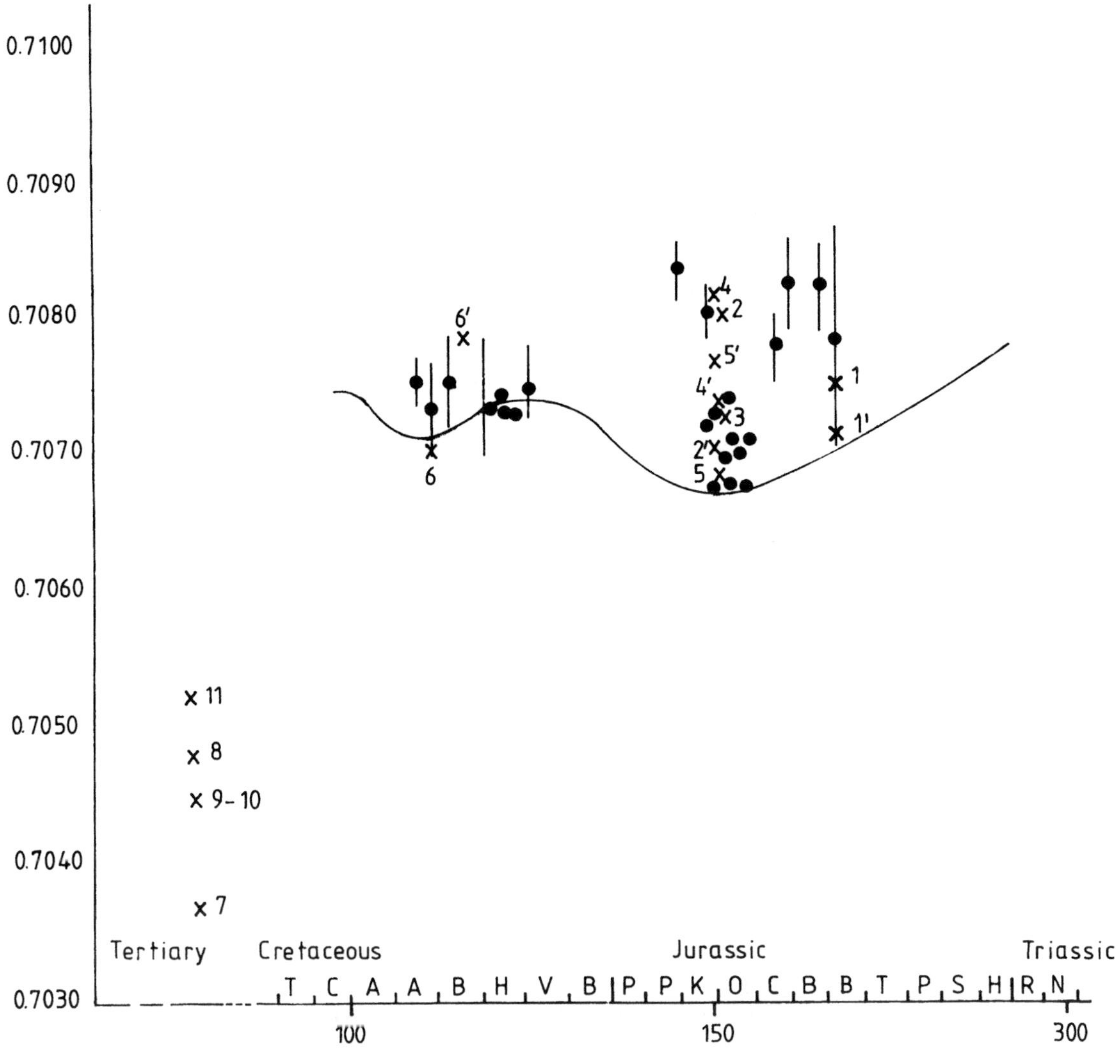

Figure 4–15. $^{87}Sr/^{86}Sr$ ratios from celestities, barites, and andesites after values in Table 4–1.

• data from Peterman et al., 1970; ⫯ data from Veizer and Compston, 1974; wavy line indicates mean values from Burke et al., 1982. Xs represent the author's values from the sources as follow: 1. Loncopué ore field; 2. Mallin Quemado ore field; 3. Luthema mine; 4. Compadrito mine; 5. Isidora mine; 6. Huitrín formation ore fields; 7. Huantraico andesite; 8. Luthema andesite; 9. Colipilli andesite; 10. Loncopué andesite; 11. Omega mine

Olaussen (1981) presented three models to explain the formation of celestite in a subtidal to supratidal facies in the Wenlock of Norway: (1) early diagenetic dolomitization of aragonite mud releases Sr, which may react with the brine to form celestite; (2) Sr enrichment of the interstitial waters produced by transformation of aragonite into calcite (200–400 ppm Sr); (3) late diagenetic dolomitization in which the Sr released may precipitate as late-stage authigenic celestite in geodes and veins.

As seen in Chapter 3, the close association of carbonates, celestite, and gypsum in evaporite environments, similar to those described here, is common to this type of deposit.

Celestite deposits related to the three evaporite cycles are known to occur in the study area. The Cuchillo Cura ore field (Neuquén Province) corresponds to the Tábanos formation (Middle Jurassic), whereas the Luthema and Monteagudo mines (Mendoza Province) correspond to the Auquilco formation (Upper Jurassic). The most important celestite ore fields, Bajada del Agrio, Cerro Salado, Continental, Balsa Huitrín, and Naunauco, correspond to the Huitrín formation (Lower Cretaceous). The Naunauco and Cerro Salado deposits constitute examples of massive deposits several meters thick and hundreds of meters long that were probably formed in a lagoon or salt pan. Deposits with DCR are fairly common. They are formed by a first generation of celestite and calcium carbonate crystals randomly oriented and by a second generation of bipolar crystals located at both sides of the first generation. A third generation of minerals filling the interstices is not present. The Cuchillo Curá ore field, which extends intermittently for 7 km, and the north sector of the Bajada del Agrio ore field (Santa Ana mine), provide the most conspicuous examples of these. The Continental ore field presents beds 1 m to 2 m thick in which the lower part is formed by DCR, whereas the upper part is epigenetically obliterated by dissolution and recrystallization in cavities. The presence of stalactites showing a precipitation according to the solubility order, that is, a barite core surrounded by celestite layers of different colors, is frequent mainly in big cavities.

Depending on their position in the tidal surface, these deposits show gradational facies.

The Ba content of sea water is low (approximately 0.06 ppm; Puchelt, 1967, 1978). Several modes of concentration and formation of barite layers may occur, the most important being barite deposition in coastal areas and barite deposition in deep ocean regions.

In the first case, the source of barite can be the weathering of basement rocks, which then migrates toward the oceans adsorbed in clays (Puchelt, 1967; Hanor and Chan, 1977). According to Fuchs (1984), the weathering depends on the nature of the basement, the morphology of the continent, and the climate. The ions are selectively trapped in particular levels of the weathering profile and set free when the former equilibrium is disrupted.

Recent knowledge about the transformations in a tidal flat, the behavior of sea water and ground water recharges, and the hydrodynamic potential of evaporite pumping (Hsü, 1984) allowed the proposition of new models for the formation of barite deposits. Taking into account the Sr and Ba content of sea water, Fuchs (1980) made the following postulations about the Sr:Ba ratios: the relative amount of available Ba in sea water is less than 1 percent of Sr. The initial content should be higher than 22 ppb of Ba in order to precipitate barite and celestite separately. If not, Ba precipitates as a member of the celestite–barite series. This ratio controls the barite connected directly to evaporite differentiation processes, which will have characteristically high Sr contents (10–12 percent).

In the second case—deposits formed at depth—the origin is considered to be volcanic exhalations or hot springs. These deposits can either be monomineralic or contain a certain amount of sulfides and are frequently hosted by shales. Late epigenetic barite deposits also occur in karstic environments.

With respect to the barite stratabound deposits of Neuquén and Mendoza, those belonging to the Loncopué ore field and Compadrito mine can be considered as formed in a coastal–lagoonal environment. No evaporites are associated with the Arroyo Nuevo mine, shales being the main lithology. This would indicate a deeper depositional environment for the barite of this deposit. Mallin Quemado is considered to have formed in conditions similar to Loncopué and Compadrito. However, the intense tectonic deformation that affected this deposit modified most of its original characteristics. For this reason, several genetic interpretations have been suggested for its formation.

A syndiagenetic deposition for these barite deposits is evidenced by the geologic setting, especially in reference to the location in the sedimentary sequence (Fig. 4–16), together with the orebody geometry, the mineral textures, and, in certain cases, the rhythmic deposition in alternation with silty-clayey material.

It can be concluded that the Sr isotope values of the celestite and barite stratabound deposits correspond to those of the Jurassic and Cretaceous sea.

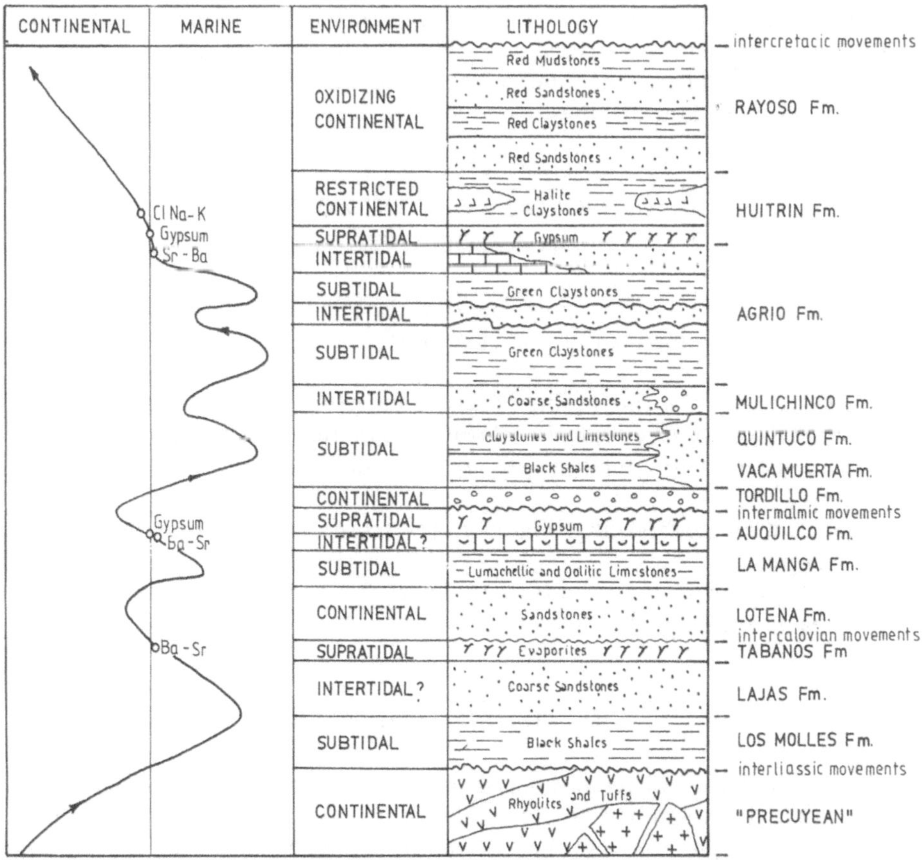

Figure 4–16. Sedimentary sequence of the area and location of barite and celestite occurrences.

It can also be observed that the range of isotope values for celestite and barite, as well as those for overlying and underlying sediments (gypsum and limestones) is very narrow.

LA HELVECIA ORE FIELD

The La Helvecia ore field (Brodtkorb and Brodtkorb, 1980) contains stratabound ores of lead, zinc, and barite emplaced in Paleozoic sedimentary rocks (Fig. 4–17). It is located some 30 km from Guandacol in the province of La Rioja, on the western side of the Cerro Urcushun that has a slope of 30–40°W (Fig. 4–18). This is coincident with the dip of the mineralized bed, sporadically exposing lead, zinc, and barium minerals over a surface of 1 km^2. This area was exploited for the secondary zinc minerals, smithsonite and hemimorfite, and for barite from 1930–1960, without exploiting the primary sulfides. The main host rock is the San Juan limestone of Middle Ordovician age, which is paraconformably overlain by the sandstone of the Volcan formation of Carboniferous age. In between, there are local breccia conglomerates. Andesite dikes of Triassic age intruded the area.

In the La Helvecia mine there is an exploration gallery 385 m long, a vertical shaft, and a crosscut toward the west (Fig. 4–17). In the San Juan limestone there are other evidences of mineralizations to the north and east over a distance of about 15 km, but these have not yet been explored in detail.

LITHOLOGY OF THE ZONE INVESTIGATED

The different lithological types found in the mine workings and on the surface have been grouped as follows (Fig. 4–19):

Type 1. Dark gray, fine-grained massive limestone: oligoterrigenous microsubesparite (the typical rock of the San Juan limestone);
Type 2. Brecciated conglomerate arenites and lithic breccia conglomerates;
Type 3. Calcitic breccia conglomerates (clasts of microsubsparite of Type 1);
Type 4. Yellowish-gray sandstones: quartzose and mica-quartzose arenites and wackes (Volcan formation).

TYPES OF MINERALIZATION

In the study area, different types of mineralizations are associated with different geologic events and in close relation to geomorphologic and climatic processes. Figure 4–19 shows the different types. They have been divided into primary (1, 2, and 3), remobilized (4), and secondary (5).

1. The first manifestation of mineralization in a chronological sense (Type I) is present in the micrites of the San Juan limestone. It is

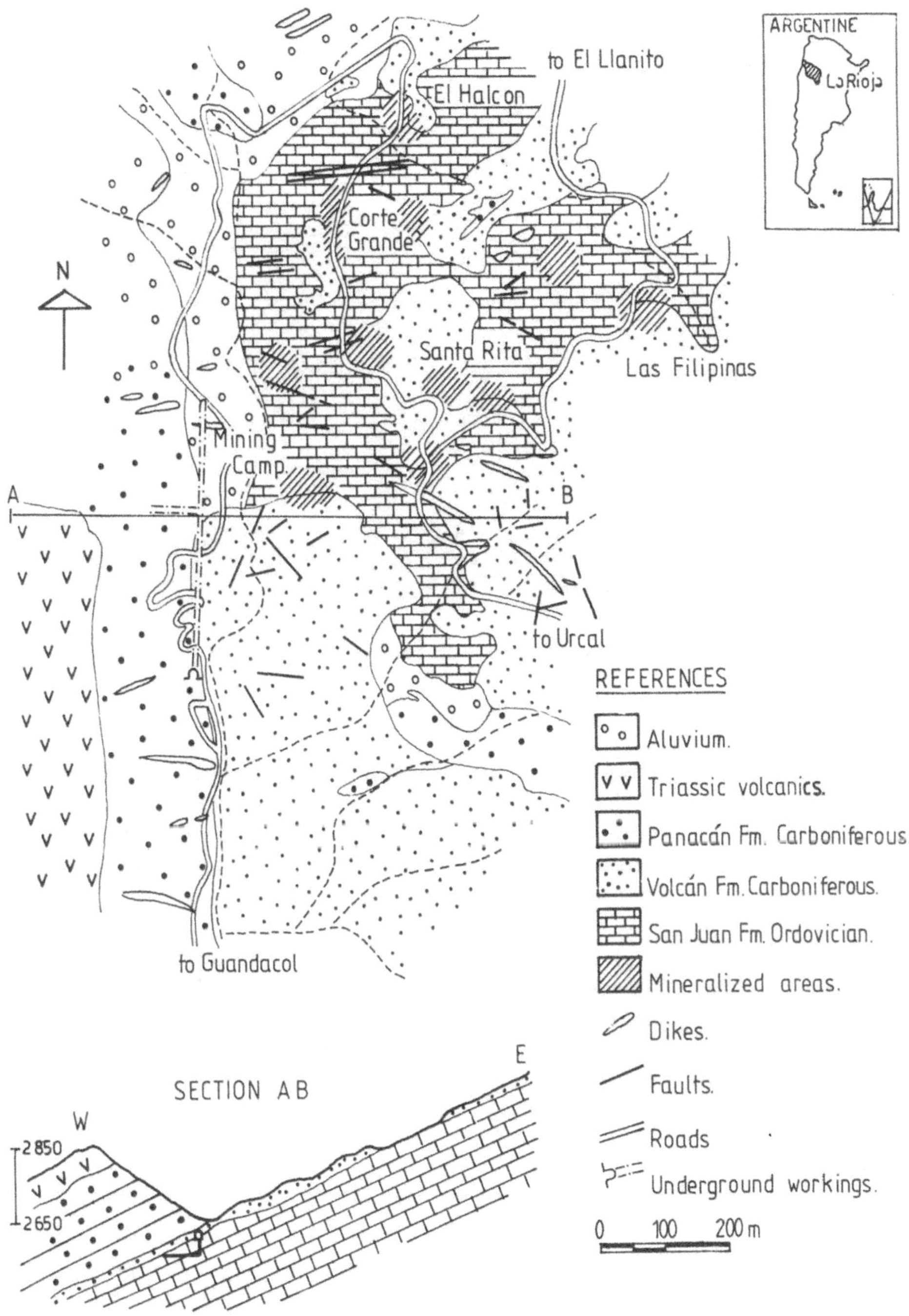

Figure 4–17. Geology of the La Helvecia mine, La Rioja Province, Argentina.

Figure 4–18. General view of the slope of Cerro Urcushun and La Helvecia ore field.

composed of generally fine disseminated grains that often are found aligned along the bedding. They are of galena, sphalerite, and minor chalcopyrite, tetrahedrite and barite. When the association of grains are larger, they often occur as nodules or rosaries. This mineralization is observed mainly in the crosscut at level −30 that goes through 50 m of micrite.

2. The second type of mineralization (Type II) may be clearly observed in the gallery, between 386 m and 341 m, limited by two faults. The primary textures are load casts and coloidal precipitations of ZnS. The first are located in conglomeradic arenites. The coloidal precipitates of ZnS are interpreted as Schalenblende in a state of gel over a free surface of sedimentation.

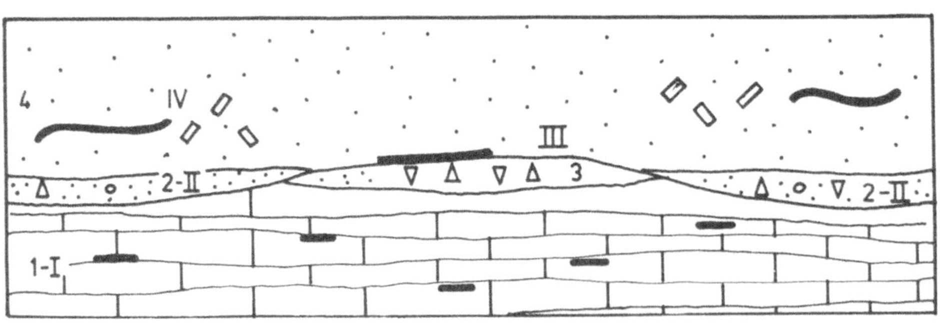

Figure 4–19. Sketch of main rock units and types of mineralizations.

3. The third type of mineralization (Type III) is common. It appears in the unconformity between the San Juan and Volcan formations, to be more precise, between the calcitic breccia conglomerates and the sandstone. The mineralization consists fundamentally of coarse-grained spatic barite, surrounded by some nodules of galena and scarce sphalerite. The barite ore bodies are tabular and concordant with the sediments of the floor and top; they are up to 30 m long and 2 m thick. These orebodies do not always end suddenly along the unconformity, but rather narrow out as the mineral grains are finer and more sporadic, increasing later in size to form another body of larger dimensions. The galena is altered on the surface to cerussite, which preserves it from further oxidation. Microscopic laths and fine stringers of barite were also found in the overlying sandstones (Type IV).
4. On the southern slope of the Cerro Urcuschun, which falls abruptly toward the Guandacol river, whose limestone beds have been disturbed by intrusion, there are veins and ore shoots of barite unconformable with the San Juan limestone, not far from the overlying sandstone. It is understood that these discordant veins were produced by remobilization.
5. Finally, the secondary mineralization, which, though not of genetic interest, has been up to the present the main mineralization exploited.

SEDIMENTATION ENVIRONMENTS AND GENETIC CONSIDERATIONS

It is interesting to mention the paper by Assereto et al. (1976) that relates many Pb, Zn, and barite deposits in Italy with emergent surfaces, considering the stages of emergence as important for the concentration of minerals.

The evaluation of the identifiable structures and textures within the La Helvecia ore field draw the following conclusions: the Type I mineralization (in micrites) is stratiform, whereas the mineralization of Type II is stratabound, limited to a "defined lithological facies." In this mineralization stage, there are geopetal textures that give evidence of the deformation of the sulfurated muds, and also a certain rhythm of the sulfides within the wackes and arenites. There are also botryoidal textures of sphalerite, and although they are not exclusive indicators of a sedimentary origin, their arrangement within the beds and their syneretical texture indicate such an origin. The mineralization of Type III is located in direct relation to the unconformity between the limestones of the San Juan formation and the sandstones of the upper member of the Volcan formation. Diagenetic effects are suggested by the notable development of the barite and also the preferential trend of the galena nodules to locate themselves in the periphery of the barite crystals. In between the Ordovician and the Carboniferous sedimentation, the relief was irregular, with highlands and lowlands. Some zones of negative relief were filled by the lithic and calcilitic conglomerates, arenites, and wackes proceeding from the emergent areas. During the filling of the lowlands, the positive relief continued being subject to karstic type conditions forming calcitic breccias.

Some authors state that the precipitation of sulfides takes place in euxinic, hypersaline environments, whereas others suggest that in the meeting of fresh and salt water there is an important interphase with deposition of cations.

Two galena samples of Type II and Type III were analyzed by the U.S. Geological Survey (Brodtkorb and Brodtkorb, 1982) for lead isotopes.

Sample	$Pb^{206/204}$	$Pb^{207/204}$	$Pb^{208/204}$
Type II	18.084	15.585	37.975
Type III	17.967	15.551	37.731

The calculated ages with the Stacey–Kramers method produced the following results:

Sample	$Age^{207/206}Pb$	$Age^{208/204}Pb$
Type II	380 ±80 m.y.	360 ±80 m.y.
Type III	400 ±80 m.y.	390 ±80 m.y.

This means that the main mineralizations occurred between the Ordovician and Carboniferous, when the area was emergent and the brecciated conglomerates were deposited.

REFERENCES

Amstutz, G. C., M. E. Cisternas, L. L. Diaz, Ll. Fontboté, J. Frutos, Ch. Mayer, S. Th. Schmidt, and A. Wauschkuhn. 1985. Relaciones entre algunos yacimien tos de Ag, Zn, Fe y Ba y las secuencias Marinas Tras Arco del Jurásico y Cretácico Inferior en el Norte de Chile. *4th Cong. Geol. Chileno,* vol. 4, pp. 435–451.

Angelelli, V. 1941. El yacimiento de baritina del cerro Mallín Quemado, Dpto. Picunches, Neuquén. *DNGM Inf. inéd.* **25**:1–30.

Assereto, R., L. Brigo, C. Brusca, P. Omenetto, and P. Zuffardi. 1976. Italian ore mineral deposits related to emersion surfaces, a summary. *Mineralium Deposita* **11**(2):170–179.

Barbieri, M., M. K. de Brodtkorb, S. Ametrano, and V. Ramos. 1981. Datos isotópicos de Sr^{87}/Sr^{86} relacionado a los yacimientos de celestina y baritina de la F. Huitrín, Prov. de Neuquén. *84th Cong. Geol. Arg.,* vol. 2, pp. 787–796.

Barbieri, M., M. K. de Brodtkorb, and A. Brodtkorb. 1984. Nuevos datos isotópicos de Sr^{87}/Sr^{86} de yacimientos de baritina de la prov. del Neuquén. *9th Cong. Geol. Arg.,* vol. 5, pp. 59–65.

Bengochea, A. L., P. J. Maiza, and G. R. Mas. 1984. Estudio geológico del yacimiento de celestina Mina Luthema, Depto Malargue, prov. de Mendoza, Argentina. *9th Cong. Geol. Arg.,* vol. 7, pp. 302–312.

Brodtkorb, M. K. de, V. Ramos, and S. Ametrano. 1975. Los yacimientos estratoligados de celestina-baritina de la F. Huitrín y su origen evaporítico. Prov. de Neuquén, Argentina. *2nd Cong. Iberoamericano de Geol. Ec. Bs. As.,* vol. 2, pp. 143–168.

Brodtkorb, M. K. de, and A. Brodtkorb. 1980. La Helvecia: a stratabound Pb-Zn-

barite deposit, Argentina. In *5th IAGOD Symposium Proceedings,* ed. J. D. Ridge. Stuttgart: E. Schweizerbart'sche Verlagsbuchhandlung.
Brodtkorb, M. K. de, and A. Brodtkorb. 1982. Datos isotópicos de plomo de la mineralización de la Mina La Helvecia, prov. de La Rioja. *Asoc. Geol. Arg. Rev.,* **37**(3):358–361.
Brodtkorb, M. K. de, V. Ramos, M. Barbieri, and S. Ametrano. 1982. The evaporitic celestite–barite deposits of Neuquén, Argentina. *Mineralium Deposita* **17**:423–436.
Brodtkorb, M. K. de, I. A. Schalamuk, M. Barbieri, S. Ametrano, R. Fernandez, R. Etcheverry, and E. Aragon. 1985. Los yacimientos de baritina y celestina del Mesozoico de Mendoza y Neuquén, Argentina. *4th Cong. Geol. Chileno,* vol. 3, pp. 156–183.
Burke, W. H., R. E. Denison, E. A. Hetherington, R. B. Koepnick, H. F. Nelson, and J. B. Otto. 1982. Variation of sea water Sr^{87}/Sr^{86} throughout Phanerozoic time. *Geology* **10**:516–519.
Dessanti, R. N. 1973. Descripción geológica de la Hoja 29b, Bardas Blancas, Prov. de Mendoza. *Serv. Nac. Min. Geol. Bol.* **139**:1–70.
Digregorio, J. H. 1978. Estratigrafía de las acumulaciones Mesozoicas. In *Relatorio Geología y Recursos Nat. del Neuquén,* ed. E. O. Rolleri, 7th Cong. Geol. Arg., pp. 37–65.
Digregorio, J. H., and A. Uliana. 1980. Cuenca Neuquina. In *Geología Regional Argentina,* ed. J. C. Turner, pp. 985–1032. Córdoba, Argentina: Academia de Ciencias.
Fernandez, R., R. Etcheverry, I. B. Schalamuk, and E. Aragon. 1987. Mina Luthema y Monteagudo, Geología y génesis. In press.
Fontboté, L., and G. C. Amstutz. 1980. New observations on diagenetic crystallization rhythmites in the carbonate facies of the Triassic of the Alpujárrides (Betic Cordillera, Southern Spain). Comparison with other diagenetic rhythmites. 1st Symposium on Diagenesis, Barcelona 1980. *Inst. Inv. Geol. Barcelona Rev.* **34**:293–310.
Fontboté, L., and G. C. Amstutz. 1983. Facies and sequence analysis of diagenetic crystallization rhythmites in stratabound Pb-Zn-(Ba,F) deposits in the Triassic of Central and Southern Europe. In *Mineral Deposits of the Alps and of the Alpin Epoch in Europe,* ed. H. J. Schneider, pp. 347–358. Berlin: Springer-Verlag.
Fuchs, Y. 1980. Sur l'importance des eaux connées pour la formation de certains gisements en milien évaporitiques. *Cent. Rech. Expl. Prod. Elf Aquitaine Bul.***6**(1):433–443.
Fuchs, Y. 1984. Migration of fluids during diagenesis: An ore-forming process in carbonate rocks. In *Syngenesis and Epigenesis in the Formation of Mineral Deposits,* eds. A. Wauschkuhn, C. Kluth, and R. A. Zimmerman, pp. 287–293. Berlin: Springer-Verlag.
Groeber, P. 1946. Observaciones geológicas a lo largo del meridiano 70. *I Hoja Chos Malal. Soc. Geol. Arg. Rev.* **1**:177–208.
Gulisano, C. A., A. R. Gutierrez Pleimling, and R. E. Digregorio. 1984. Análisis estratigráfico del intervalo Tithoniano-Valanginiano (Formaciones Vaca Muerta, Quintuco y Mulichinco) en el SO de la provincia del Neuquén. *9th Cong. Geol. Arg.,* 1, pp. 221–235.
Hanor, J. S., and L. H. Chan. 1977. Non-conservative behavior of barium during mixing of Mississippi River and Gulf of Mexico waters. *Earth and Planetary Sci. Letters* **37**:242–250.
Hsü, K. J. 1984. A nonsteady state model for dolomite, evaporite and ore genesis. In *Syngenesis and Epigenesis in the Formation of Mineral Deposits,* eds. A. Wauschkuhn, C. Kluth, and R. A. Zimmerman, pp. 275–286. Berlin: Springer-Verlag.
Kinsman, D. J. J. 1969. Modes of formation, sedimentary associations and diagnostic features of shallow-water and supratidal evaporites. *Am. Assoc. Petrol. Geol. Bull.* **53**:830–840.
Lambert, L. R. 1956. Descripción geológica de la Hoja 35b. Zapala (Neuquén). *DNGM Bol. Buenos Aires* **83**:1–95.

Leanza, H., H. Marchese, and J. C. Riggi. 1977. Estratigrafía del Grupo Mendoza con especial referencia a la F. Vaca Muerta entre los paralelos 35° y 40° lat. S., Cuenca Neuquína-Mendocina. *Asoc. Geol. Arg. Rev.* **32**(3):190–208.

Llambías, E., and L. Malvicini. 1978. Geología, petrología y metalogénesis del área de Colipilli, prov. del Neuquén, Rep. Argentina. *Asoc. Geol. Arg. Rev.* **33**(4):257–276.

Lyons, W., L. Muñiz, and Associates. 1978. Planta o plantas regionales de tratamiento y beneficio de minerales de Ba, Neuquén. Consejo Federal de Inversiones, Buenos Aires.

Müller, G. 1962. Zur Geochemie des Strontiums in ozeanen. Evaporiten unter besonderer Berucksichtigung der sedimentaren Coelestin-lagerstatte von Hemmelte-West. *Geol. Beiheft* **35**:1–90.

Olaussen, S. 1981. Formation of celestite in the Wenlock, Oslo Region, Norway. Evidence for evaporitic depositional environments. *Jour. Sed. Petrology* **51**:37–46.

Oliveri, J., B. Zakalik, and M. Juarez. 1964. Estudio geológico económico preliminar de yacimientos de baritina de la prov. del Neuquén. Inst. de Inv. Mineras, Univ. Nac. Cuyo. *DNGM. Inf. inéd. 1046,* 151p.

Peterman, U. E., C. E. Hedge, and H. A. Tourtelot. 1970. Isotopic composition of strontium in sea water throughout Phanerozoic time. *Geochim. Cosmochim. Acta* **34**:105–120.

Puchelt, H. 1967. Zur Geochemie des Bariums im exogenen Zyklus. *Sitzungsber. Heidelb. Akad. Wiss. Math. Nat. Kl.* **4**:85–205.

Puchelt, H. 1978. Barium. In *Handbook of Geochemistry,* ed. K. H. Wedepohl, Berlin: Springer-Verlag, pp. A-1-56-0-2.

Schalamuk, I., R. Fernandez, R. Etcheverry, and E. Aragon. 1987. Mina El Compadrito, geología y génesis. In press.

Stipanicic, P. N. 1969. El avance en los conocimientos del Jurásico argentino a partir del esquema de Groeber. *Asoc. Geol. Arg. Rev.* **24**(4):367–388.

Veizer, J., and W. Compston. 1974. Sr^{87}/Sr^{86} composition of sea water during the Phanerozoic. *Geochim. Cosmochim. Acta* **38**:1461–1484.

5

Brazilian Stratabound Barite Ore Fields

J. Cassedanne

Brazilian barite estimated reserves are running to 50 million metric tons. Deposits are mainly located in Minas Gerais and Bahia. With the exception of the Camamu mine hosted by Mesozoic rocks, all other economic deposits are in igneous or metamorphic rocks belonging to the pre-Paleozoic or in carbonatite intrusions. They are in veins, layers, and stockworks of variable sizes (Fig. 5-1).

Barite is essentially used in Brazil as drilling mud (35,000 metric tons in 1979). Ore from the bigger deposit of Araxá (Minas Gerais), with 47 million tons of indicated reserves, is found in a carbonatic complex contemporaneous with Paraná volcanism and cannot be used in oil wells because of its high radioactivity.

Stratabound deposits hosted by post-Precambrian sediments are examined first. Among these, the Camamu mine is the only deposit of which detailed reports are available. A short abstract of the other occurrences, broken down by age follows. Description of stratabound barite deposits hosted by pre-Paleozoic rocks starts with that of Água Clara prospect, the best known. The other occurrences are summarized by geographic or geologic areas.

AM = Amazonas, BA = Bahia, CE = Ceará, GO = Goiás, MA = Maranhão, MG = Minas Gerais, MS = Mato Grosso do Sul, MT = Mato Grosso do Norte, PA = Pará, PB = Paraíba, PR = Paraná, PI = Piauí, RN = Rio Grande do Norte, RS = Rio Grande do Sul, SP = São Paulo.

Figure 5-1. Brazilian stratabound barite deposits location map.

STRATABOUND BARITE DEPOSITS HOSTED BY POST-PRECAMBRIAN FORMATIONS

CAMAMU DEPOSIT

The Camamu deposit lies in two small islands, about 3 km from the coastline, near 14°S in the state of Bahia. A narrow channel separates the islands, which extend 5 km from northeast to southwest. Their width varies from 0.3 km to 2.5 km (Fig. 5-2).

The Camamu basin belongs to the taphrogenic units that border the Brazilian shield. It is contiguous to the Recôncavo and Almada basins in the south and the Sergipe–Alagoas basin to the north. Located between the Valença and Itacaré faults, the Camamu basin shows only a narrow outcrop near the coastline and in the above-mentioned islands.

The main features of regional geology were outlined by Viana et al. (1970), Ghigone (1979), Bittencourt et al. (1979), Viana (1980), Ojeda y Ojeda (1981), and Asmus (1982*a*, 1982*b*). These authors showed that sedimentary sequences from the marginal Brazilian basins are related to the rifting evolution that took place during the partition of the African and South American plates.

The first reference to the barite deposits was made by Campos (1922), as large crystals in the Ilha Grande. Leonardos (1934) and Erichsen and Hernano (1940) indicated the possible use of the mineral, while Leonardos and Oliveira (1940) published a regional setting compilation. A short feasibility report was written by Trajano (1945). The barite crystals were examined by Tavora Filho (1946). Bodenlos (1948) published the first economic geologic report on the deposits. Bruni compiled paper on barite in 1973. Abreu (1965), Suszczynski (1975), Tesch et al. (1976), and Teixeira Netto (1977) published hypotheses regarding the deposit formation. Campos and Dardenne (1984) suggested a genetic model for the barite zone.

REGIONAL STRATIGRAPHY

Stages overlying the basement of Rio da Serra, Aratu, Buracica, and Jequiá do not outcrop in this region. They occur below 800 in the BAS-20 oil well, near the limit of the geologic map. Above, the sediments deposited from the Alagoas to the Santonian stage, which represent a continuous sequence, were informally named Camamu group. Two formations were identified by Tesch et al. (1976) and Teixeira Netto (1977): Taipu-Mirim, which is Aptian in age, and Algodões, which is Albian in age. Both are covered by Recent sediments.

The Taipu-Mirim formation is a particular facies of the Alagoas stage, representing a transition zone between the continental Marizal formation and the thick evaporitic beds deposited on the continental shelf and correlated with the Paripuera evaporites. During the lower Aptian, the "Paripuera gulf" allowed the ocean waters to come up to the small evaporitic basins (Ojeda y Ojeda, 1981). Later, during the upper Aptian, this gulf disappeared and a wider basin took its place with the Ipura evaporite beds. Chemical sediments

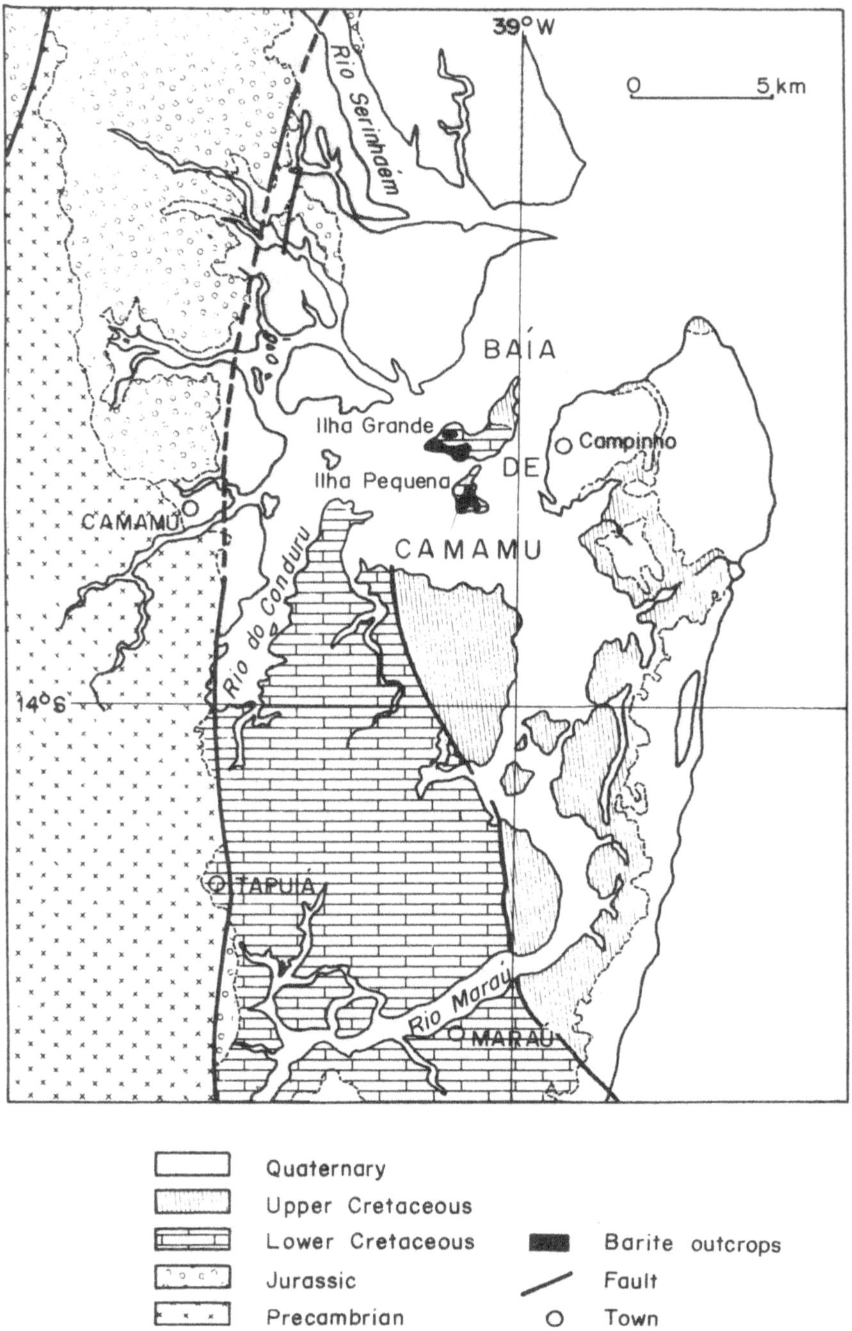

Figure 5-2. Geologic map of the Camamu area (after Teixeira Netto, 1977).

associated with carbonaceous and pyritic shales were deposited in this borderland, with sporadic clastic intercalations. West of the Camamu basin, a local shoal rendered the continental shelf uneven.

Campos and Dardenne (1984) subdivided the above-mentioned formations into three sequences (Fig. 5-3):

Lower Sequence. Terrigeneous, including two facies passing from one to the other; conglomerates and shales. The coarse-grained conglomerates, with granulite, sandstone, and quartz pebbles in a sandy matrix, contain thin intercalated beds of fine-grained micaceous sandstones and greenish argilites. They mainly outcrop in the Ilha Grande as north 40–60° west trending narrow rises. Toward the basin, the conglomerates, sometimes with pyrite and thin carbonaceous beds, laterally turn into dark greenish or brown shales. Seams and nodules of silicified interbedded coquinas also outcrop in the Ilha Grande.

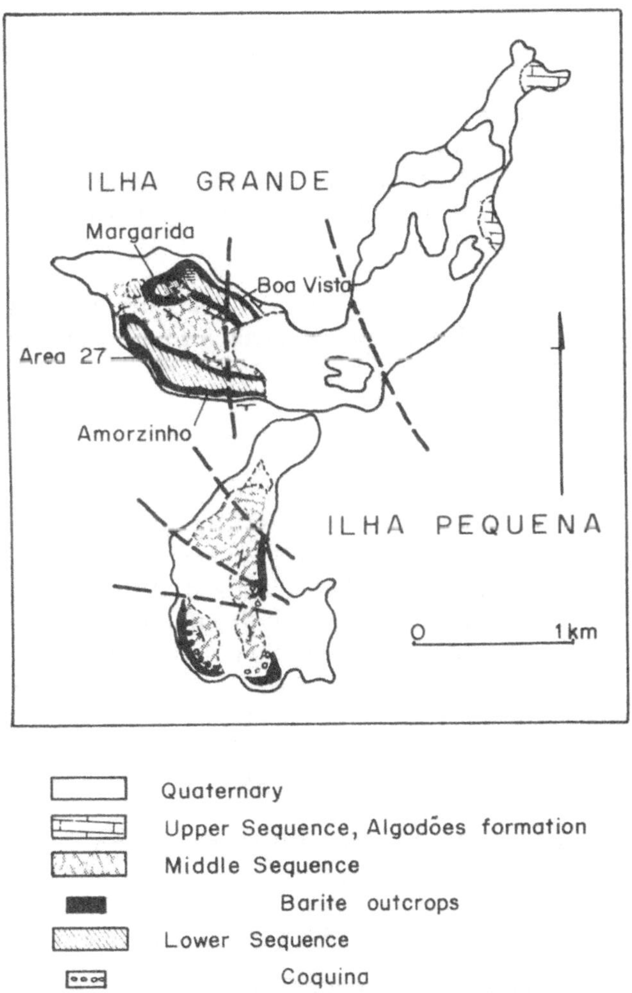

Figure 5-3. Geologic map of the Camamu barite deposit (after Campos and Dardenne, 1984).

Determination of various ostracodes, *Paracypridea obovata obovata* Swain, *P. Langdoni* Krömmelbein, *Reconcava Swaini* Krömmelbein, and *R. Centronodos* Krömmelbein, allows the dating of the lower sequence. It is of the Aratu stage in age and, therefore, older than the Taipu-Mirim formation. An unconformity between this latter and the lower sequence attests to an important stratigraphic break.

Middle Sequence. The base ("transgressive evaporitic series") in the shallower parts of the basin is of fine-grained dolomite, gypsum, and anhydrite intercalated beds in shales that contain argilaceous carbonate seams. In the deeper parts, the anhydrite bed, locally, disappears and a more than 200m thick halite formation takes its place (oil well BAS-20).

The top ("residual evaporitic series") begins with a fine-grained dolomite deposition succeeded by a fibrous gypsum and whitish-gray anhydrite bed, about 30 m in thickness. This sequence passes to a barite layer in the border of the basin. Where the base of the middle sequence was removed by erosion in the outer parts of the basin, the barite layer overlies the conglomerates of the lower sequence. This angular unconformity, perceptible in the mine workings, is underlined by a thin siliceous level stained by iron oxides. Dips of the barite layers vary from 0° to 60°, whereas the underlying conglomerates are horizontal or slightly dipping.

A unit made of green shales, slates, and a few conglomerates conformably overlies the barite layer. It mainly outcrops in the Ilha Pequena and to the south of the Ilha Grande.

Upper Sequence. Fossiliferous saccharoidal dolomites of the Algodões formation overstep the underlying sequences. Outcrops are restricted to the north of the Ilha Grande. Geologic mapping indicates that contacts with the Taipu-Mirim formation are made by faulting.

THE BARITE LAYER

The barite occurs as beds (three in places) and lenses in the middle sequence, where it forms a main layer, the only one worked. Another 0.3 m thick layer sporadically outcrops above the main one, like thin lenses and veinlets made of large crystals resulting from late reprecipitations (Figs. 5–4 and 5–5).

Thickness of the main layer is variable: from zero to several centimeters along old elongated underlying conglomerate ridges to 6–8 m in the depressions between the ridges. The mean is 4 m. A thin kaolinic level often separates the conglomerates from the barite layer, which overlies a crust of iron and manganese oxides.

A vertical zonation may be observed from base to top in the main barite bed (Tesch et al., 1976):

- Spherulitic barite, in thin layers, with good outcrops in the Ilha Pequena,
- Banded (or tabular) heavy vugged, dipping barite (Fig. 5–6). Vugs are sometimes coated by bitumen associated with pyrite, marcasite, and galena;

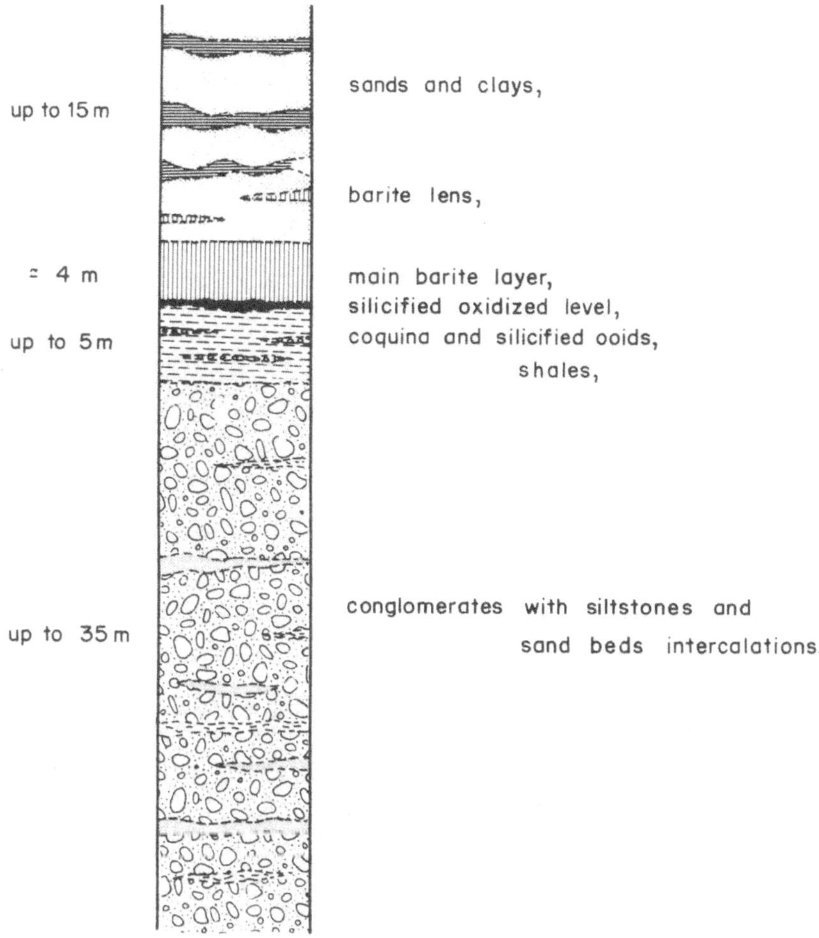

Figure 5–4. Lithostratigraphic section of Ilha Grande barite deposit (after Campos and Dardenne, 1984).

- Botryoidal barite, in coarse concretions, reminding one of oncolites or stromatolitelike structures. Vugs may be important in volume.

Commonly, the ore is whitish when free of bitumen, or reddish or blackish when stained by iron or manganese hydroxides. Clays, cryptocrystalline silica, quartz, and iron oxides (Ilha Pequena) are always present. Campos and Dardenne (1984) did not find the gypsum and celestite reported by Tesch et al. (1976) and Teixeira Netto (1977).

Barite occurs in large milky saccharoidal, microcrystalline, or amorphous crystals. Clusters of small acicular or platy crystals are common. Texture of the ore is variable: hard, massive, soft, or sandy with a network of quartz veinlets.

A 50,000 metric ton ore placer lies to the south of Ilha Grande as hard barite

Figure 5-5. Main barite layer of the Camamu deposit (''27'' area workings).

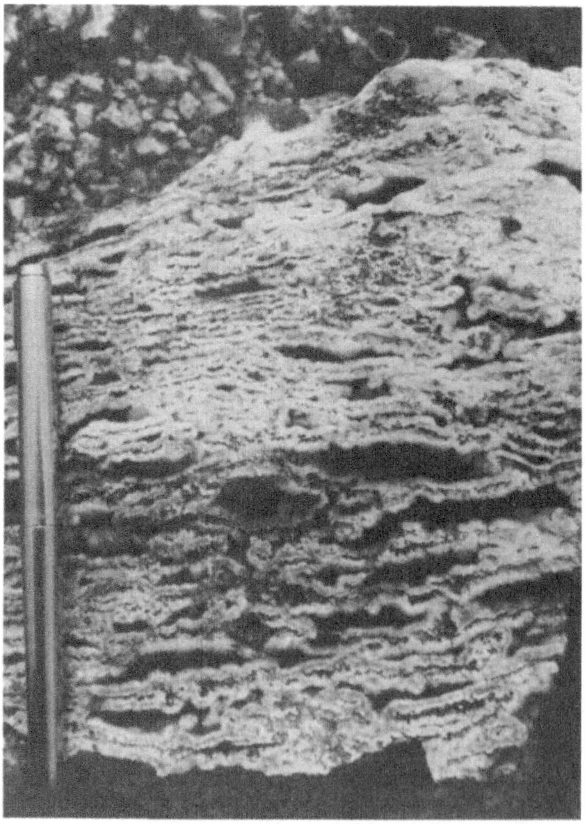

Figure 5-6. Banded vugged barite in the Camamu deposit (Amorzinho area).

boulders and gravels mixed with sandstone. It was built by the demolition of the cliffs that hang over the beach.

Mining of the barite bodies is by open pit with a 70 percent recovery of the ore. Reserves are running to 0.7 million metric tons, with a yearly production of about 50,000 metric tons. The ore is washed and crushed to eliminate the clay impurities. It is sold with a density higher than 4.2.

GEOCHEMISTRY AND ISOTOPE DATA

Absence of calcium sulfate relicts was confirmed by chemical, petrographic, and diffractometric analyses by Campos and Dardenne (1984), who also examined the Ba/Sr ratios and oxygen and sulfur isotopes.

The Ba/Sr ratio shows an enrichment of strontium in the early phases (spherulitic barite), which probably reflects values closely related to that of the calcium sulfate precipitated from sea water. On the contrary, other barium ratios show lower values, testifying to a gradual impoverishment resulting from a series of dissolutions and reprecipitations during the ore formation.

The strontium content (about 2000 ppm) is distinct from that of barite, considered as evaporitic in origin (Fuchs, 1980; Brodtkorb et al., 1982). However, it is closely related to that observed in the deposits where barite shows an evident epigenetic link with evaporitic sequences, in which calcium of sulfate is replaced by barium (Fuchs, 1978, 1980). The registered values of $\delta^{18}O = 13$ ppm and $\delta^{34}S = 15$ ppm in the lower barite level probably correspond to anhydrite precipitation from sea Aptian sulfate. This mineral was subsequently replaced by barite.

Modification of the above-mentioned values in the upper barite levels puts forward bacterial reduction phenomena during their formation, with a correlative enrichment in ^{18}O and ^{34}S whose greatest values reach $\delta^{18}O = 15.6$ ppm and $\delta^{34}S = 26$ ppm.

GENESIS OF THE BARITE DEPOSIT

According to Bodenlos (1948), the barite deposit arises from replacement of arkosic sandstones by barium-rich hypogene waters of low temperature. Replacement of dolomitic limestone by barite related to hydrothermal manifestations associated with basaltic magmatism is maintained by Abreu (1965). Suszczynski (1975) classifies the deposit as sedimentary, stratabound, supergenic, zoned, and differentiated. Tesch et al. (1976) establish a narrow relationship between the barite mineralization and a peculiar sedimentary sequence of the Taipu-Mirim formation. The barium concentration in the sea water of this sequence should be the result of weathering of the Tapuia hornblende porphyritic trachytes that contain 3000 ppm Ba. Teixeira Netto (1977) suggests that rising Ba-rich solutions replaced the gypsum. In order to sustain his hypothesis, Teixeira Netto lists the following evidence:

- The barite crystallization is identical to that of the gypsum deposits. The texture of Camamu barite resembles that of the synchronous evaporites from Sergipe basin (gypsum and anhydrite from the Mariricu formation);

- The preservation of enteromorphic structures is caused by anhydrite swelling in the presence of water;
- The orthorhombic barite needle clusters contrast with the granular texture displayed by the basement barite veins;
- The peculiar barite location on the border of an evaporitic basin.

Finally, Campos and Dardenne (1984) propose that the barium origin is the result of selective weathering of K-feldspars, where K is isostructurally substituted by Ba and weathering is made easy by an arid climate. The barium origin should probably be either in the basement granulite feldspars or in the slates and conglomerates from clastic sequences. A magmatic origin is discarded by Campos and Dardenne because the two main phases of alkaline vulcanism (135–130 and 80–60 m.y.) are far from the time of mineralization of the Camamu deposit. In order to justify their sedimentogenetic hypothesis, Campos and Dardenne (1984) point out that form and thickness of the main barite layer are strictly conditioned by the conglomerate's paleotopography. They also report slumpings on the ridge conglomerate slopes and synsedimentary packing faults, where throw is greater toward the depressions.

All the above-mentioned authors agree that during Aptian time calcium sulfate precipitated from sea water into nearby small basins, as evidenced by the associated clastic sediments. Pyrite and anhydrite occurrences, near the barite deposit, were reported from oil wells located laterally eastward. Direct barite precipitation, which only occurs in very peculiar evaporitic conditions, did not occur in the Camamu area, for barite of this type always overlies anhydrite beds.

The proposed Campos and Dardenne genetic model is analogous to that of Fuchs (1978, 1980, 1981), who states that the barite mineralization occurs in a "mixture zone" between a continental environment, where K-feldspars are leached, and a confined evaporitic environment, where gypsum and anhydrite are precipitated. Indispensable conditions for replacement of Ca of sulfate by Ba from conated waters are realized. This process is slow and develops over a long time period in various stages. So, perhaps, the Camamu deposit formation may be:

Rift Phase–Aratu Stage. Sedimentation takes place during a tectonic phase (Ilhas group). Quiet tectonic period with local sedimentation break. Deposition carries on in other Recôncavo areas. Silicification of outcropping rocks (coquinas) of the lower sequence.

Evaporitic Phase–Alagoas Stage (Aptian). African and South American plates continue to part during the retreat of the sea, marked by evaporitic conditions. Deposition of the Taipu-Mirim formation of which border sediments are rich in clastics and organic matter.

Post-Alagoas Stage (Albian). The plates continue to part during the spread of the sea over land areas: deposition of the Alagoas formation. Beginning of the replacement of Ca by Ba and remobilization of Fe and Si. Active faulting and partial emersion of deposit.

Post-Alagoas Phase to Recent. Continuation of the preceding phase with important faulting and uncovering. Mineralization continues far from these areas. Late silicification and general weathering.

OTHER STRATABOUND BARITE OCCURRENCES HOSTED BY POST-PRECAMBRIAN FORMATIONS

CRETACEOUS

Apodi Group. The sedimentary sequence of the Potiguar basin is known as the Apodi group and comprises the Jandaira formation at the top (limestones and dolomites laterally passing to gypsum) and Açu formation at the base (fine and coarse clastic sediments). White to gray platy barite crystals occur in thin beds between these formations. Barite was deposited in a shore environment from shallow bay waters, protected from the open sea by a coral barrier (Beurlen, 1967). Main outcrops are located in Felipe Guerra, Apodi, and Jaguarana (Rio Grande do Norte State). To the north of Upanema, between 12 km and 13 km of the Upanema-Mossoró road, barite occurs as blady crystals up to 5 cm in length in detrital boulders precipitated from the base of the limestone cliffs. $SrSO_4$ barite content is 0.80 percent (Castro et al., 1977).

Santana Formation. Eluvial barite boulders are found in Simões township (Piauí State), where an evaporitic sequence with lead and zinc occurrences is reported (Farina, 1978). Barite contains 59.5 percent BaO and 0.01 percent Sr.

Gramame Formation. Platy crystal barite clusters, up to 4 cm in diameter, associated with pyrite and calcite in a limestone bed of the Gramame formation, are reported by Rao (1968), near João Pessoa (Paraíba State).

JURASSIC

In the Tucano basin, three facies belong to the Marizal formation: (1) a fluvial one, composed of medium- to coarse-grained sandstones, sometimes conglomeratic, whitish, orange or reddish; (2) a deltaic one, composed of conglomerates with centimetric polymictic pebbles occurring in beds, lenses, and channels cutting up the underlying sandstones; and (3) a lacustrine, of rhythmic reddish clays (Souza, 1984).

Fazenda Barra. Recently discovered, this prospect located in the Recôncavo basin, lies on the northern flank of the Aporá ridge, near the border of the South Tucano basin. It is located southeast of Ouriçangas and is hosted by Marizal formation sediments (Beckel, 1985).

Regional formations from top to bottom include:

- The Barreiras formation, overlying, in nonconformity, the lower sediments, makes up the main heights,

- The Marizal formation, 70 m to 80 m in thickness, hosts the barite deposit. The latter as iron oxide cemented sandstones, constitute small hills like the Morros Pelado and Urubu. The Marizal formation is composed of polymictic sandstones with rare clay intercalations several centimeters thick. Multidecimetric to metric conglomerates lenses occur near the base of the sequence. The subhorizontal beds, like those of the Recôncavo basin, are dislocated by two fault systems.

Sandstones that vary from whitish to yellowish are medium to coarse grained. They are mainly made of rounded to subrounded quartz grains, with some weathered feldspars and rock fragments, and a high porosity reaching up to 40 percent. When not healed by barite or iron oxide, the rock is crumbly. Channels and cross bedding are common.

Barite, with some clay and iron oxides, occurs as a spathic filling between the rock grains and includes some lightly corroded clasts.

Barite mineralization is very irregular with sudden lateral variations in the outcrop scale. Variation along the vertical depth is not noticeable. A few galena occur locally in the basal conglomerate. Lead and zinc are reported in variable amounts, reaching up to 1.7 percent Pb and 4.9 percent Zn in part of the underlying Ilhas group.

Barite mineralization took place during early diagenesis, before rock lithification, for barite crystals and clusters separate the sandstone grains. The deposit should be Alagoan in age, almost contemporaneous with the Marizal formation.

Up to 22 m in depth, ore reserves are valued at 370,000 metric tons with a 37.3 percent SO_4Ba grade and are located in six areas never more than 2 km apart. Average ore zone thickness is 6 m.

PERMIAN

Motuca Formation. Seams and barite plates up to 1 cm thick are reported from an argilaceous siltstone in the middle part of the Motuca formation, Permian in age. The mineralized sequence is about 6 m thick (Lima and Leite, 1978).

Nova Olinda Formation. Two barite occurrences in the Igarapé Xituba area (state of Pará) are reported by Rezende and João (1980): one made of lost massive boulders, and the other outcropping in the Igarapé do Garimpo valley in association with a siliceous brecciated rock. They should be the result of "primary concentrations of stratabound barite in a sedimentary sequence." The latter comprises from top to bottom:

- Recent alluvium and colluvium,
- A silicified zone, sometimes spotted by sulfides, Cretaceous in age,
- Penatecua diabase occurring as sills and dykes, Jurassic to Cretaceous in age,
- Nova Olinda formation, gray and buff shales, sandstones, limestones, and siltstones with halite and gypsum lenses. Lower to middle Permian in age,

- Itaituba formation, sandstones, siltstones, and limestones of Upper Carboniferous age.

LOWER PALEOZOIC

In the Rio Juca area, northwest of Ceará State, Suszczynski (1975) reports a stratabound barite layer, associated with pyrite, in a silty and detrital sequence of the Cococi region, lower Paleozoic in age. According to Farina (1974), this occurrence is a discordant barite vein.

STRATABOUND BARITE DEPOSITS HOSTED BY PRE-PALEOZOIC FORMATIONS

ÁGUA CLARA DEPOSIT

Southwest of São Paulo and northwest of Paraná, folded and metamorphosed sequences running northeast–southwest to east–west, outline a wide synclinorium, open toward the northeast, where barite and fluorite (associated with some sulfides) are present on both flanks as two narrow belts. The eastern belt stretches over 170 km from Rio Branco do Sul to Registro, passing Bocaiuva do Sul. The western one, more than 70 km in length, runs from Rio Branco do Sul to Itapirapuã, passing Cerro Azul. Sedimentation advanced from south to north, whereas the Ba enrichment is from northeast to southwest. Economic barite prospects are restricted to Paraná.

REGIONAL GEOLOGY

Metamorphosed rocks of the Setuva and Açungui groups overlie an old basement. The Açungui group was described by Derby (1878) as "a noncrystalline metamorphic series belonging to the homonymous area." It was later defined by Oliveira (1927). Basal sequences where barite is hosted were first named the Setuva formation by Bigarella and Salamuni (1956), and further separated from the Açungui group by Marini (1970) on the basis of a nonconformity between the two sequences. Ebert (1971) raised the Setuva formation to Setuva group. See Andrade e Silva (1984) for a discussion of the regional stratigraphy.

Fritzsons et al. (1982) distinguish from top to bottom (Fig. 5-7).

Upper Proterozoic	Açungui group	Capiru formation Votuvera formation
Middle Proterozoic	Setuva group	Água Clara formation Perau formation, where barite occurrences are located
Middle to Lower Proterozoic	Pre-Setuva complex	
Middle Proterozoic to Archean	Crystalline complex	

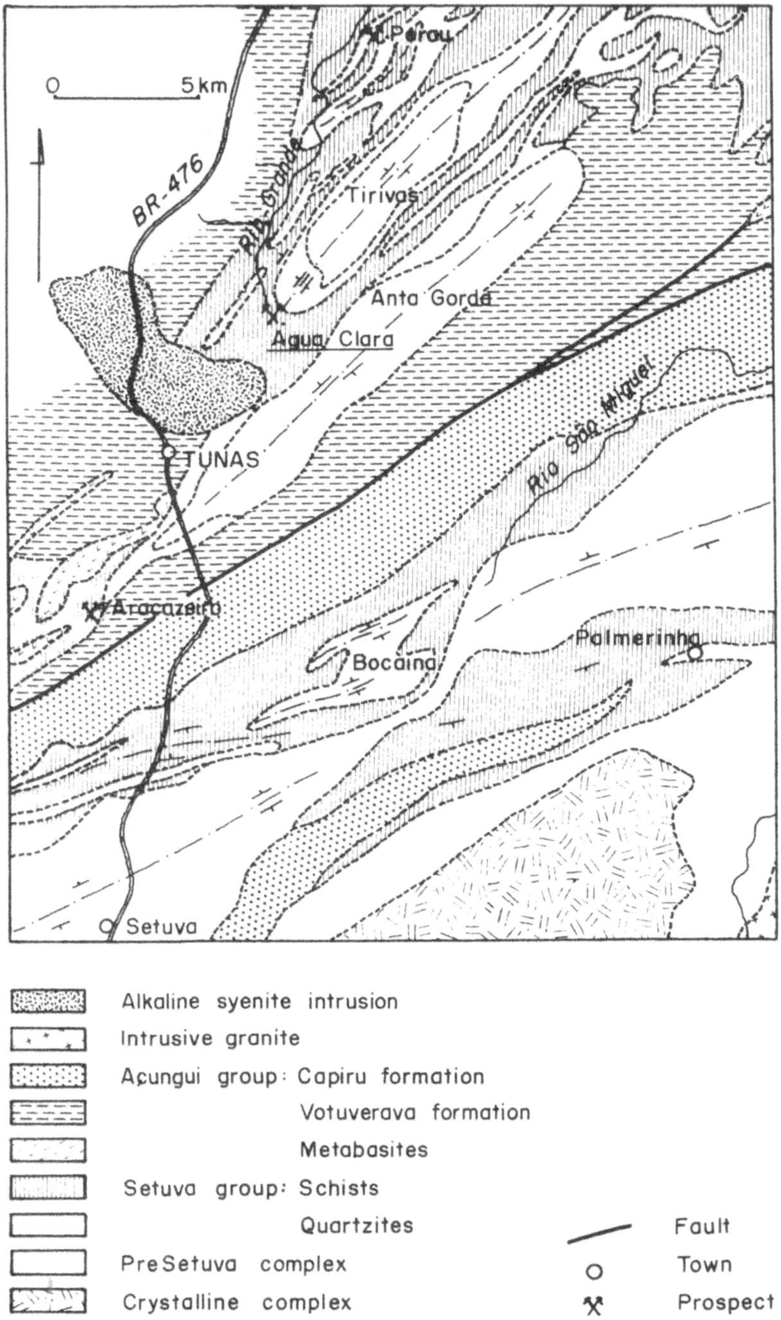

Figure 5-7. Geologic map of the Água Clara area (after Fritzsons et al., 1982).

The basement is leucocratic nebulitic migmatites without banding and granite gneiss. There is no preferential trend.

The pre-Setuva complex comprises migmatites, gneiss, ultrabasic rocks, and high-grade metamorphosed quartzite and runs northeast–southwest. Outcrops of this complex are in anticline cores.

The Setuva group includes carbonate and iron metamorphosed sediments, quartzites, and micaschists with interbedded amphibolites. The presence of small stratabound sulfide occurrences above quartzites, and barite and magnetite layers associated with carbonate rocks, mainly in the Água Clara and Perau prospects, suggest for the lower sequence, after the pre-Setuva faulting, a deposition during the early development of a basin. A pelitic sedimentation with contemporaneous magmatism originating the amphibolite sills followed. Setuva rock was faulted during Transamazonic orogeny.

The Açungui group is of low metamorphic extinites with occasional stromatolitic dolomites, marbles, quartzites, and silt and clay metamorphosed sediments. These rocks are intruded by granitic porphyric plutons, syn- or post-tectonic, with contact metamorphism aureole. Intrusions belong to the Brazilian orogeny (Pontes, 1982).

Longitudinal faults associated with strong cataclasis are Cambro-Ordovician in age.

Dykes and intrusions of basaltic, alkaline, and ultrabasic rocks result from the Mesozoic magmatism along northwest–southeast fractures. Near Água-Clara, trachytic sills and dykes are associated with the Tunas alkaline intrusion.

The Pariquera-Açu, Guabirotuba, and Alexandra formations are scattered sedimentary mantle (conglomerates, sands, and clays), Plio-Pleistocene in age. Holocene and Recent alluvium fills valleys and depressions.

Axis of the rather asymmetric main folds run northeast–southwest with a southwest plunge. Competent formations are weakly folded as opposed to the incompetent ones that are intensely followed. A second fold system north 40° east running and plunging 8° southwest occurs in superposition. In the Setuva and Açungui groups, the sliding tectonics turn parallel foliation and stratification. Normal and reverse faults are of weak throw. The metamorphic grade increases from northwest to southeast. The Setuva formation belongs to the amphibolite to greenschist facies and the Açungui group to the greenschist facies.

THE ÁGUA CLARA PROSPECT

Almost all papers covering barite deposits in the state of Paraná refer to the Água Clara prospect, which lies about 1 km to the south of the homonymous hamlet, in Rio Branco do Sul Township. This prospect belongs to a narrow mineralized bed coming from the south (Campo Chato) and continuing toward the northeast, to the Perau prospect (Pb, Zn, Cu; Silva et al., 1982). It is located on the southwestern flank of the Ribeirão da Barra anticline (Andrade e Silva and Schöll, 1982).

Near the prospect, the pre-Setuva complex is made up of mylonite, biotite, and quartz biotite gneiss at the base and micaschists at the top.

The Setuva group comprises a chemical sedimentary sequence overlaid by a vulcanic sedimentary (clastic and chemical) sequence. From base to top, the following units may be observed in the neighborhood of the prospect (Andrade e Silva, 1984):

- Quartzites with chert intercalations.
- 34 m. Metamorphosed dolomites with clay and chert intercalations up to 1 m thick. Lead, zinc, and copper minerals are scattered locally.
- 6 m. Metamorphosed dolomites with lenses and centimetric seams of white pure barite. The outcrop stretches for more than 1 km toward the north.
- 6 m. Metamorphosed dolomites with centimetric micaschists, chert intercalations, and quartz veinlets
- 15 m. Quartz micaschists, partly calcareous, with lenses and centimetric to decimetric barite layers associated with magnetite and pyrite, itabirite seams, and cherts.
- Main barite layers.
- 7 m. Quartz micaschists in part calcareous.
- More than 20 m. Quartz micaschists, chlorite muscovite schists, and phyllites with two orthoamphibolite intercalations showing texture relicts and igneous minerals.

A short adit was opened around 1960 (galeria Vila) and trenches and two adits (Clara Cruz and Água) were opened in 1974 to search for lead minerals. Two new adits (Taboa and Paiol) were opened in 1976 from the barite outcrops, producing 180 metric tons of barite ore. Mining stopped in 1977 (Fig. 5-8).

In the Paiol adit, the main barite layer, parallel to the wall rock bedding, 0.6 m to 1 m thick, was followed for 40 m, with a 10° north strike and a 20° northwest dip. It passes to a magnetic layer laterally. The barite is light to dark gray, fitted, granoblastic with a 0.01 cm to 1 cm grain size. Darker beds contain magnetite and rare hematite spots or pyrite seams. The strontium content of the barite is high.

Bigarella (1948) and Marini and Leprevost (1967) believe that the Água Clara deposit is hydrothermal. However, Marini and Leprevost accept a partially sedimentary origin. Deposits from the Setuva formation are sedimentary (Barbour and Oliveira, 1979), but with partial remobilization for Andrade e Silva and Schöll (1982). Andrade e Silva (1984) points out the syngenetic character of the stratabound Água Clara deposit, located in the transition zone between two sequences, and concludes that the presence of barite in marble and the lack of typical vulcanic minerals suggest a sedimentary origin. On the contrary, Fagundes et al. (1984) support the hypothesis that barite space and time association with fluorite, banded-iron formation, and sulfides suggest various precipitation facies from exhalative fluids from the sea floor. The barite facies should be located between the lower sulfide facies and the upper banded-iron facies, produced by the opening related to amphibolites and chlorite muscovite schists resulting from metamorphosed vulcanic tuffs.

In the Campo Chato prospect, 12 km to the south–southwest of the Água Clara deposit, barite occurs as veins and three layers in a 9 m thick quartzite

Figure 5–8. Paiol adit portal at the Água Clara prospect. Main barite layer outcrops on the right (after Beckel, 1985).

unit, stratigraphically below the Água Clara sequence. Barite, sedimentary in origin, was remobilized during metamorphism. Roseira (barite in quartzite) and Araçazeiro, both in Rio Branco do Sul Township, are small Ba–Pb–BIF prospects in the neighborhood, as is Tabuleiro (Cerro Azul Township).

Other Stratabound Barite Deposits Hosted by Pre-Paleozoic Formations

"Quadrilatero Ferrífero"

Stratabound metamorphosed barite (and fluorite) occurrences are known from the carbonate sequences of the Minas supergroup (Middle Proterozoic in age), where they lie in two narrow belts. The main one, where barite is uppermost, outcrops in the south near Ouro Preto from Antônio Pereira to Ribeirão do Carmo near Mariana. To the north of the "Quadrilatero Ferrífero," the Sabará and Caeté (Taquarassu) occurrences are similar but without economic importance. The main prospects near Ouro Preto are of economic importance, according to Lacourt (1938):

Igreja Velha. Igreja Velha lies 1 km to the southeast of Antônio Pereira. There, a 1 m to 1.8 m thick fine-grained whitish marble presents barite lenses and seams that are several centimeters thick. Locally, they constitute more than 40 percent of the rock. At the base, a 0.3 m thick, sandy white barite layer occurs. Thin barite lenses are intercalated 2 m above in another 1 m thick marble. Hematitic schists overlie the carbonate sequence. In 1938 1800 metric tons were produced. Reserves are small, only 12,000 metric tons.

Timbopeba. In the neighborhood, Timbopeba is a prospect upon barite boulders, whose wall rock was fully weathered.

Chacrinha, Chácara do Cintra, and Bom Jesus. These are all places on a barite layer that extends over 2200 m. Chacrinha, 200 m from the Padre Faria church, is a prospect with a white pure barite layer near the top of a dolomitic marble bed. Chácara do Cintra lies near the Ouro Preto–Itacolomy road, on the right side of Ribeirão do Carmo. The outcrop, 1 m to 2 m thick, is formed by white saccharoidal barite seams, several centimeters thick, alternating with marble. Bom Jesus, near Mariana, is similar to the Chácara do Cintra prospect but contains pyrite (Fig. 5–9).

In the same region, du Toit (1927) reports a barite lens interbedded in dolomitic marble near the Itacolomy station.

West–southwest of Araxá, an 0.8 m thick lens, in addition to simple and complex discordant barite veins, occurs in the Croatá mine (Fazenda dos Agudos). Wall rock is quartzitic, belonging to the Araxá Group (Precambrian B in age).

BAMBUÍ BASIN

The Bambuí group, more than 570 m.y. old, mainly outcrops in the Rio São Francisco valley (states of Minas Gerais and Bahia) and near Brasilia (state of Goiás). It includes thin conglomerates and thick quartzites and siltstones

Figure 5–9. Thin barite seams in marble from the Bom Jesus prospect, near Mariana (after Lacourt, 1938).

and important carbonate formations, with stromatolites, where Pb–Zn–fluorite and barite deposits are hosted. Fluorite always occurs in the upper part of the carbonate sequence and barite in the lower ones. Locally, Pb–Ag mineralizations are intercalated between the barite and fluorite levels. Exceptional sedimentary conditions, in connection with buried hills, are marked by facies changes near the ore deposits. They are associated with a regressive series in the upper parts of sedimentary cycles (Beurlen, 1971).

Barite occurrences are located along 370 km from Januaria in the south (Minas Gerais) to Paratinga in the north, near the Riacho Brejo Velho, left tributary of the Rio São Francisco. Axis of this belt, which reaches up to 150 km in width, near Correntina, Santa Maria da Vitória, and Montalbânia, is parallel to the São Francisco River, but displaced to the west. Carbonate beds are slowly dipping north.

As sedimentation prograded from south to north, the richer barite occurrences lie near Januaria. They are a stratabound, white metric barite layer in a series of dolomites, limestones, and siltstones. Disseminated and discordant barite veins and veinlets as irregular lenses are also known (Suszczynski, 1975).

Near Irecê, in Bahia State, a discontinuous light gray to pinkish dolomitic bed, containing oolites, includes a few galena and barite. This latter is sedimentary in origin, according to Pedreira and Torquato (1983), but mainly occurs as discordant veins. The dolomite belongs to the B1 unit of Misi (1979).

Near Melancias, in the same state, a horizontal 1 m thick banded barite layer comprises barite and calcite ribbons alternating with dark dolomite, ankerite, galena, sphalerite, pyrite, chalcopyrite, and covellite. Wall rocks are laminated dark gray limestones (Misi, 1978).

CHAPADA DIAMANTINA

Several barite occurrences are reported from the clastic low-grade metamorphosed sediments of the Chapada Diamantina group in Bahia State. Barium resulting from weathering of an old continent would be provided by sedimentation and deposited as the matrix of sandstones. Later metamorphism moved the barite into specific levels (Suszczynski, 1975).

The Chapada Diamantina group is part of the Espinhaço supergroup (Precambrian B, 1.7 to 1.1 b.y. old) that lies in angular unconformity on older Precambrian gneisses and is covered, also in angular unconformity, by the Bambuí group formations. The Chapada Diamantina group, several thousand meters thick, consists of extrusive volcanic rocks in the lower members, quartzites, schists, diamond-bearing conglomerates, and phyllites in the middle units, and limestone in the upper part. The group is divided into six units (Pedreira, 1976; Bruni and Schobbenhaus, 1976) and is folded into wide anticlines trending north to north-northwest, which are intersected by transverse and longitudinal faults.

At the base, the Rio dos Remédios complex, a long westward dipping volcanic sequence hosts several stratabound barite deposits. It is a metamorphic felsic flow, a few hundred meters thick, with local quartzites, conglomerates, and schists. The main deposits are:

Baraúna. This prospect, near the village of the same name, is 26 km distant from Seabra, head of the township to which it belongs. A north–south barite vein, dipping 45–60° east, is known to be 500 m in length and 80 m in height. It is interbedded in quartzite of the Chapada Diamantina group.

Mining is by an open pit, 20 m in height, at 1170 m. The barite layer, 5 m wide in outcrop, pinches out at the base of the open pit where its width varies from 1 m to 2 m. At the 1120 m level (Fig. 5-10), a 230 m long adit was built in the barite layer with shafts, which showed that quartz progressively replaces barite in depth.

Parallel veins and barite seams occur in the wall rocks that are cross-checked by perpendicular or irregular milky quartz veinlets. Light gray barite is saccharoidal with disseminated and nodular quartz and blue patches of oxidized copper minerals. Reserves are running to 100,000 metric tons. The deposit is classified as hydrothermal by Bruni (1973).

Saco Redondo (3 km to the north) and Velame, in the neighborhood, are similar to the Baraúna prospect.

Santa Luzia. This deposit lies 8 km south of Ibitiara (Fig. 5-11). A north-south, westward dipping or vertical 0.5–3 m thick barite layer is interbedded in metamorphosed extrusive volcanic rocks and hematite schists. Mining extends over a 300 m long strip, but small barite outcrops are known along 2 km in the same trend (Fig. 5-12). On both sides, small divergent veins protrude

Figure 5-10. Portal at 1120 m level of the Baraúna prospect. The level stretches along the barite layer.

Figure 5-11. Looking southward in the Santa Luzia deposit.

Figure 5-12. Mining the Santa Luzia barite layer.

from the barite layer into the wall rocks in which lenses and seams of barite occur. The barite is whitish, saccharoidal, or in large blades with some translucent or transparent crystals but without druses. Locally, abundant specks or undulated hematite sheets stain the ore. Quartz is disseminated or in clusters with centimetric milky or transparent crystals. Spots of malachite and azurite are rare. Drilling shows that barite laterally passes to quartz in depth. Reserves are running to 1 million metric tons.

The deposit should be hydrothermal (Schobbenhaus, 1969). Rather, it appears to be volcanic sedimentary in origin, later metamorphosed with partial remobilization.

There is a small stratabound Aguada prospect in the neighborhood and, 40 km to the south, the Joazeiro deposit, similar to that of Santa Luzia, occurs. There, the stratabound barite bed, 3 m thick, stretches over 600 m, with a 45° east dip. Reserves are 1.7 million metric tons.

Itapura. This deposit, in Miguel Calmon Township, lies on the eastern flank of the Serra de Jacobina, near its southern end. Five small open pits along a 6 km extension mine several discordant veins and a main stratabound barite layer, 2.2 m thick and dipping 75° east. Wall rock is an argilaceous quartzite belonging to the Chapada Diamantina group (Bruni, 1973). The barite is whitish, sometimes reddish or greenish, and associated with manganese oxides in fissures. Total reserves are 1.4 million metric tons.

"NORDESTE"

Cunha e Silva (1983) reports a belt of stratabound barite occurrences, associated with fluorite, in the upper Precambrian quartzites that overlie the scheelite-bearing skarns of the Piranha Province (Florânia, Parelhas, São Joâo do Sabugi, in Rio Grande do Norte State, and Caopoã, Poão, Riachuelo, and Trapiá near Santa Luzia in Paraíba State).

REFERENCES

Abreu, S. F. 1965. *Recursos minerais do Brasil,* vol. 1. São Paulo: E. Blücher, pp. 210–214.

Andrade e Silva, C. G. de. 1984. Discussão sobre o controle litoestratigráfico e a gênese da jazida de barita de Água Clara, Estado do Paraná. *Anais 33rd Congress bras. Geol., Rio de Janeiro,* vol. 7, pp. 3704-3709.

Andrade e Silva, C. G. de, and M. V. Schöll. 1982. Metalogênese da barita de Água Clara no âmbito da evolução da Formação Setuva. *Anais 32nd Congress bras. Geol., Salvador,* vol. 3, pp. 1017–1023.

Asmus, H. E. 1982*a*. Geotectonic significance of Mesozoic–Cenozoic magmatic rocks in the Brazilian continental margin and adjoining emerged area. *Actas Quinto Congress Latino-Americano Geol., Buenos Aires,* pp. 761–779.

Asmus, H. E. 1982*b*. Significativo geotectônico das feições estruturais das bacias marginais brasileiras e áreas adjacentes. *Anais 32nd Congress bras. Geol., Salvador,* vol. 4, pp. 1547–1557.

Barbour, A. P., and M. A. F. de Oliveira. 1979. Pb, Zn, Cu e Ba no distrito do Perau. Módelo para sulfetos no Vale da Ribeira. *IG-USP São Paulo Bol.* **10**:97–119.

Beckel, J. 1985. Depósitos estratiformes de barita no Brasil. Relat. Curso de Mestrado UFRJ, 13p. (unpublished)

Beurlen, H. 1971. Mineralizações de chumbo, zinco e fluorita no Grupo Bambuí. *5th Simpósio Geol. do Nordeste,* Bol. esp. 1, Recife, SBG Nordeste, pp. 27–28.

Beurlen, K. 1967. *Geologia da região de Mossoró.* Col. Mossoroense, Sér. C., vol. 18. Rio de Janeiro: Pongetti, 173p.

Bigarella, J. J. 1948. Estudos preliminares na Série Açungui. II. Rochas calcáreas. *Arq. Biol. Tech. Curitiba* **3**:201–354.

Bigarella, J. J., and R. Salamuni. 1956. Estudos preliminares na Série Açungui. V. Estruturas organogêneas nos dolomitos da Formação Capiru (Estado do Paraná). *Dusenia, Curitiba* **7**:317–323.

Bittencourt, A. C. S. P., G. S. Vilas Boas, J. M. Flexor, and L. Martin. 1979. Geologia dos depósitos quaternários no litoral do Estado da Bahia. SME.CPM. Geologia e recursos minerais da Bahia. *Textos básicos,* Salvador **1**:1–21.

Bodenlos, A. J. 1948. Barite deposits of Camamu Bay, State of Bahia, Brazil, *U.S. Geol. Surv. Bull. 960-A,* 33p.

Brodtkorb, M. K. de, V. Ramos, M. Barbieri, and S. Ametrano. 1982. The evaporitic celestite–barite deposits of Neuquen, Argentina. *Mineralium Deposita* **17**:423–436.

Bruni, C. E. 1973. Perfil analítico da barita. *DNPM, Rio de Janeiro, Bol.,* no. 3, 44p.

Bruni, M. A. L., and C. F. Schobbenhaus. 1976. *Carte geológica do Brasil ao milionésimo.* Folha Brasilia SD-23. MME-DNPM, Brasilia, 162p.

Campos, E. G., and M. A. Dardenne. 1984. Geologia e geoquímica do depósito de barita de Camamu, Bahia. *Anais 33rd Congress bras. Geol., Rio de Janeiro,* vol. 3, pp. 1144–1162.

Campos, L. F. G. 1922. Salitre e baritina, Brasil. *DNPM-SGM, Rio de Janeiro, Bol.,* no. 4, 30p.

Castro, C. de, C. Faria, and P. J. Duarte. 1977. Ocorrências de barita no Grupo Apodi, RN. Atas VII Simp. Geol. Nordeste, Campina Grande (PB), *SBG Recife,* no. 6, pp. 489–499.

Cunha E Silva, J. da. 1983. Zonação polimetalífera da região da Borborema. Estados do Rio Grande do Norte e Paraíba. *Min. and Met.* **47**(445):24–36.

Derby, O. A. 1878. A geologia da região diamantífera da Província do Paraná no. Brasil. *Arch. Mus. Nad.,* Rio de Janeiro **111**:89–98.

Du Toit, A. L. 1927. A geological comparison of South America with South Africa. *Carnegie Institution of Washington Publication No. 381,* 157p.

Ebert, H. 1971. Observações sobre a litologia e subdivisão do "Grupo Setuva" no Estado do Paraná: com sugestões à tectônica geral do "geossinclínio Açungui." *Anais 25th Congress bras. Geol., São Paulo,* vol. 1, pp. 131–146.

Erichsen, A. I., and A. Hernano. 1940. Geologia do litoral bahiano ao Sul de Salvador. *Min. and Met.* **5**(25):18–20.

Fagundes, P. R., M. A. Dardenne, and A. T. C. Veiga. 1984. Evolução da mineralização de fluorita no depósito de Sete Barras, Paraná. *Anais 33rd Congress bras. Geol., Rio de Janeiro,* vol. 8, pp. 3755–3769.

Farina, M. 1974. Panorâmica da tipologia dos depósitos minerais do Nordeste do Brasil. *Anais 28th Congress bras. Geol., Pôrto Alegre,* vol. 6, pp. 80–93.

Farina, M. 1978. Considerações sobre o modo de ocorrência da barita no Nordeste e aspectos metalogêneticos correlatos. *Anais 30th Congress bras. Geol., Recife,* vol. 4, pp. 1478–1485.

Fritsons, O., Jr., G. F. Piekarz, and D. Falcade. 1982. Geologia e potencial econômico do Grupo Setuva (PR). *Anais 32nd Congress bras. Geol., Salvador,* vol. 3, pp. 987–1001.

Fuchs, Y. 1978. Concentrations métalliques et milieux confinés: sur un exemple de relation entre une minéralisation barytique et un milieu à évaporites. *Sci. Terre* **22**(2):127–146.

Fuchs, Y. 1980. Sur l'importance des eaux connées pour la formation de certains gisements en milieux évaporitiques. *Cent. Rech. Expl. Prod. Elf Acquitaine Bull.* **4**(1):433–443.

Fuchs, Y. 1981. Les minéralisations des milieux carbonatés de la plateforme continentale. *Soc. Geol. Ital. Mem.* **22**:35–51.

Ghigone, J. I. 1979. Geologia dos sedimentos fanerozóicos da Bahia. SME.CPM. Geologia e recursos minerais da Bahia. *Textos básicos,* Salvador **1**:23–74.

Lacourt, F. 1938. Baritina e pirita no Município de Ouro Preto, Minas Gerais. *Min. and Met.* **11**(11):298–301.

Leonardos, O. H. 1934. Bário. *DNPM-SFPM,* Rio de Janeiro, **2**:3–20.

Leonardos, O. H., and A. I. de Oliveira. 1940. *Geologia do Brasil.* Com. bras. do Centenário de Portugal, Rio de Janeiro, 472p.

Lima, E. de A. M., and J. F. Leite. 1978. Projeto estudo global dos recursos minerais da bacia sedimentar do Parnaíba. Integração geológica-metalogênetica: relatório final da etape III. *CPRM/DNPM,* Recife, vol. 1.

Marini, O. J. 1970. Geologia da folha Rio Branco do Sul (PR). Tese de Doutoramento FFCL-Rio Claro, São Paulo, 190p. (unpublished)

Marini, O. J., and A. Leprevost. 1967. Principais reecursos minerais do Grupo Açungui no Estado do Paraná. *Paran. Geoc., Curitiba, Bol. no. 23–25,* pp. 157–181.

Misi, A. 1978. Ciclos de sedimentação e mineralizações de chumbo-zinco nas sequências Bambuí (Supergrupo São Francisco), Estado da Bahia. *Anais 30th Congress bras. Geol., Recife,* vol. 6, pp. 2548–2561.

Misi, A. 1979, O Grupo Bambuí no Estaado da Bahia. SME.CPM. Geologia e recursos minerais da Bahia. *Textos básicos,* Salvador **1**:119–154.

Ojeda y Ojeda, H. A. 1981. Estrutura, estratigrafia e evolução das bacias marginais brasileiras. *Rev. bras. Geoc.* **11**(4):257–273.

Oliveira, E. P. de. 1927. *Geologia e recursos minerais do Estado do Paraná.* Mon. VI, Min. Agr. e Com. SGMB, Rio de Janeiro, 172p.

Pedreira, A. J. de C. L. 1976. *Carta geológica do Brasil ao milionésimo.* Folha Salvador SB-24, MME-DNPM, Brasilia, 126p.

Pedreira, L. M., and J. R. Torquato. 1983. Geologia da área compreendida entre a Gruta dos Brejões e a Serra da Babilônia, Município de Morro do Chapeú, Bahia. *Rev. bras. Geoc.* **13**(2):119–126.

Pontes, J. B. 1982. Geologia e potencialidade da Formação Água Clara. *Anais 32nd Congress bras. Geol., Salvador,* vol. 3, pp. 1002–1016.

Rao, A. B. 1968. Barite roses (desert roses) in limestone formations near João Pessoa, Paraíba State. *Jour. Min.,* Recife **6**:37–39.

Rezende, N. G. A. M., and X. S. J. João. 1980. Ocorrência de Cu, Pb, Zn, Ba e F na sinéclise Amazônica—o exemplo do Iguarapé Xitubá (PA). *Anais 31st Congress bras. Geol., Camboriú,* vol. 3, pp. 1713–1724.

Schobbenhaus, C. F. 1969. Mapa geológico preliminar da região sententrional da Serra do Espinhaço—Bahia Central (Brasil). *Anais 23rd Congress bras. Geol., Salvador,* pp. 75–86.

Silva, C. R. da, A. Takahashi, and F. Batolla, Jr. 1982. Geologia e mineralização da região Perau-Água Clara, Vale da Ribeira, PR. *Anais 32nd Congress bras. Geol., Salvador,* vol. 3, pp. 1024–1036.

Souza, W. S. T. de. 1984. Estudo das unidades deposicionais na pesquisa de mineralizações stratabound: un exemplo de aplicação na Formação Marizal, Bahia. *Anais 33rd Congress bras. Geol., Rio de Janeiro,* vol. 3, pp. 1106–1114.

Suszczynski, E. F. 1975. A área mineralizada das Ilhas Grande e Pequena, no litoral da Bahia, em Camamu. In *Recursos minerais reais e potenciais do Brasil e sua metalogenia,* pp. 424–430. Rio de Janeiro: Interciências.

Tavora Filho, E. 1946. Estudo morfológico de baritina de Camamu. *Est. bras. Geol.,* Rio de Janeiro **1**(2):47–59.

Teixeira Netto, A. S. 1977. Barita de Camamu. Considerações sobre a gênese. *Tecn. Petrobrás Bol.,* Rio de Janeiro **20**(2):77–92.

Tesch, N. A., F. O. Moraes, P. E. L. Silva. 1976. *Projeto Marauito, prospecção de rochas oleígenas e barita.* Relatório final, CPRM-SUREG, Salvador, 190p.

Trajano, R. B. 1945. *Barita da Ilha Grande de Camamu, Bahia, Brasil.* LPM, Rio de Janeiro, Publication No. 16, pp. 11–43.

Viana, C. F. 1980. Cronoestratigrafia da margem continental brasileira. *Anais 31st Congress bras. Geol., Camboriú,* vol. 2, pp. 832–843.

Viana, C. F., E. G. Gamajr, I. A. Simões, J. A. Moura, J. R. Fonseca, and R. J. Alves. 1970. *Revisão estratigráfica da Bacia do Recôncavo/Tucano.* Rel. Direx 1381. Petrobrás, Salvador RPBA-SERSTRA.

6

Stratabound Barite Ore Fields of North America (Excluding Arkansas)

Sandra H. B. Clark and Forrest G. Poole

The purpose of this chapter is to summarize available information about the stratabound barite ore fields of North America, with the exception of those in Arkansas (which are described in Chapter 7). Our discussion is organized by geographic area rather than by genetic type because, although many different genetic types of deposits occur within each area, patterns of distribution of deposit types are regional. The term *stratabound* is applied to barite deposits confined to specific stratigraphic units. For the purposes of this discussion, the stratabound deposits are divided into the following genetic types.

Type 1: Syngenetic–diagenetic stratiform barite that was deposited parallel to stratification in volcanic settings;
Type 2: Syngenetic–diagenetic stratiform barite that was deposited parallel to bedding in sedimentary rocks;
Type 3: Epigenetic stratabound barite that was deposited generally discordant to bedding in layers, cavity fillings, and veins;
Type 4: Residual barite concentrations in colluvium and soil formed by weathering and erosion of previously deposited barite.

Type 1 barite deposits (Table 6–1; Fig. 6–1) consist of beds adjacent to massive sulfide bodies in volcanogenic massive sulfide ore deposits and barite disseminations in the sulfide ore. These deposits are primarily in terranes that formed in island-arc and other sea-floor volcanic environments and were later transported and accreted to the margins of the North American continent.

Type 2 barite deposits extend from western Alaska through western Canada and the western conterminous United States and into northwestern Mexico.

Table 6–1. Stratabound Barite Ore Fields, Deposits, and Occurrences in North America by Location and Deposit Type

	Deposit Type			
	Syngenetic-Diagenetic, Stratiform			
Location	*In Volcanic Environments*	*In Sedimentary Environments*	*Epigenetic Stratabound*	*Residual Concentrations in Soil*
Western North America				
Southeastern Alaska	Castle Island(?)		Lime Point(?)	
	Admiralty Island(?)			
	Glacier Creek(?)			
Northern Alaska		Red Dog		
		Nimiuktuk		
		Drenchwater Creek		
		Atigun Canyon		
		Sheenjak River		
Seward Peninsula			Sinuk River(?)	
Alaska Range	Chitsia Mountain(?)			
	Tonzona River(?)			
Yukon and Northwest Territories		Macmillan Pass District		
		Akie District		
		Anvil District		

British Columbia	Lynx Britannia		Sulphur Creek	
Western Conterminous United States		Northeastern Washington Central Idaho Nevada Southern California		
Mexico		Central Sonora		
Eastern North America				
Newfoundland	Buchans			
Nova Scotia			Walton	
Virginia	Lynchburg(?)			
South Carolina–Georgia	Kings Creek Lincolnton–McCormick			
Central North America and the Appalachian Mountains		Valley and Ridge occurrences Arkansas	Upper Mississippi Valley district Illinois–Kentucky Central Missouri Southeastern Missouri Central Kentucky Central Tennessee Sweetwater district Cartersville district Southern Virginia	Central Missouri Southeastern Missouri Southern Virginia Eastern Tennessee Georgia–Alabama

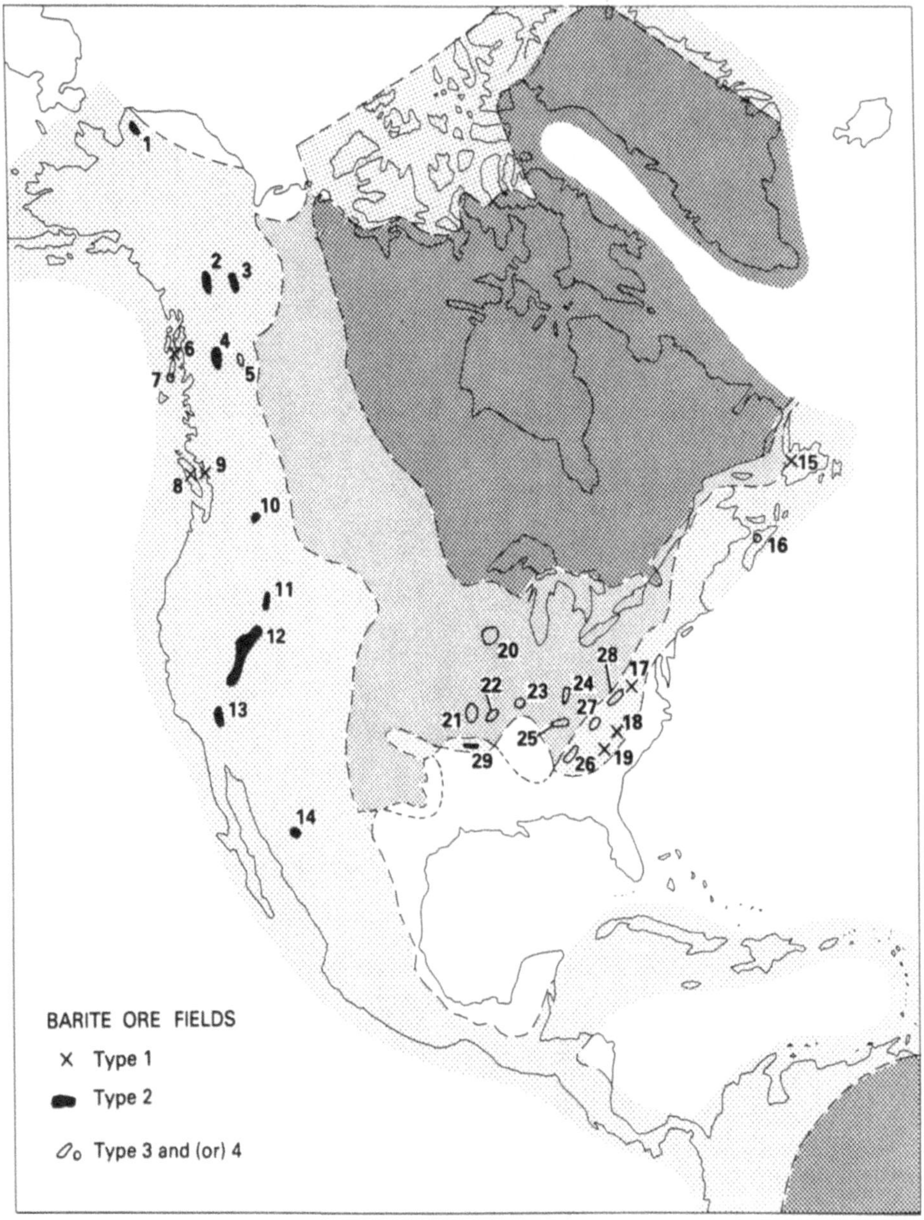

Figure 6–1. Location of the major stratabound barite ore fields of North America in relation to major Phanerozoic tectonic elements.

Dark gray indicates shield areas; medium gray indicates stable cratonic platform areas; light gray indicates Paleozoic and younger orogenic areas and includes outer cratonic platform, continental shelf and slope, and ocean basin settings. Areas covered by coastal plain deposits are white. Barite deposits and districts shown are (1) Red Dog–Drenchwater Creek; (2) Anvil; (3) Macmillan Pass; (4) Akie (Gataga); (5) Sulphur Creek; (6) Castle Island; (7) Lime Point; (8) Lynx; (9) Britannia; (10) northeastern Washington; (11) central Idaho; (12) Nevada; (13) southern California; (14) central Sonora; (15) Buchans; (16) Walton; (17) Lynchburg; (18) Kings Creek; (19) Lincolnton–McCormick; (20) upper Mississippi Valley; (21) central Missouri; (22) southeastern Missouri; (23) Illinois–Kentucky; (24) central Kentucky; (25) central Tennessee; (26) Georgia–Alabama; (27) eastern Tennessee; (28) southern Virginia; and (29) Arkansas.

Numerous occurrences and small deposits of this barite also are present elsewhere on the North American continent. With the exception of the Arkansas deposits, most of the commercially important type 2 deposits are in the Paleozoic siliceous deep-water marine rocks of the highly deformed allochthonous terranes of the western Cordillera.

The major resources of type 3 barite in North America are carbonate-hosted epigenetic deposits in rocks that were deposited in shallow marine water on the stable cratonic platforms or on continental shelves. These deposits commonly are associated with or are similar to Mississippi Valley-type lead–zinc deposits. The carbonate-hosted barite generally occurs as veins or stockworks, but the deposits, like the Mississippi Valley-type deposits, are considered to be stratabound because their mineralization is restricted to specific stratigraphic units or intervals. Also included in type 3 deposits are barite-cemented sandstones and siltstones.

Type 4 residual barite deposits formed by concentration of barite in unconsolidated material overlying preexisting barite deposits in bedrock. Little or no information is available on the bedrock source for most residual deposits, but most are concentrated in soils overlying bedrock composed of limestone and dolostone. Many of the type 4 deposits probably are developed from type 3 carbonate-rock-hosted deposits; this supposition is based on analogy with bedrock deposits in the same geologic settings.

WESTERN NORTH AMERICA

ALASKA

Barite deposits are known in the Brooks Range of northern Alaska, the Seward Peninsula, the Alaska Range, east-central Alaska, and southeastern Alaska (Fig. 6–2) and occur in Proterozoic to Upper Triassic host rocks (Bundtzen and Henning, 1978). The largest are the Mississippian to Permian(?) barite–sulfide deposits of the western Brooks Range.

Barite has been produced only from the Castle Island (type 1?) deposit (850,000 tons mined between 1963 and 1980) and the Lime Point (type 3?) deposit (several large test shipments mined in 1915) in southeastern Alaska. Isolated elliptical masses of barite that range up to 25 m thick in the Castle Island deposit are underlain by and are conformable with Upper Triassic siliceous metasedimentary rocks and pillow lavas (Herreid et al., 1978; Bundtzen and Henning, 1978). The barite is associated with copper, lead, and zinc minerals. At Lime Point, small amounts of barite were produced from massive barite–dolomite lenses in pre-Ordovician marble. Barite also has been reported from other areas in southeastern Alaska. It occurs with lead, zinc, and precious metals in Paleozoic volcanic and sedimentary sequences on Admiralty Island (type 1?). Significant resources (type 1?) are present in barite–silver–base metal prospects near Glacier Creek where barite lenses and barite-rich beds that range up to 10 m thick and 600 m long are hosted by Paleozoic metavolcanic and metasedimentary rocks (MacKevett et al., 1974; Bundtzen and Henning, 1978).

The largest barite–sulfide deposits known in Alaska are the recently discovered type 2 deposits along Red Dog Creek in the western Brooks Range (Lange et al., 1985; Moore et al., 1986). Estimated reserves are 77 million metric tons

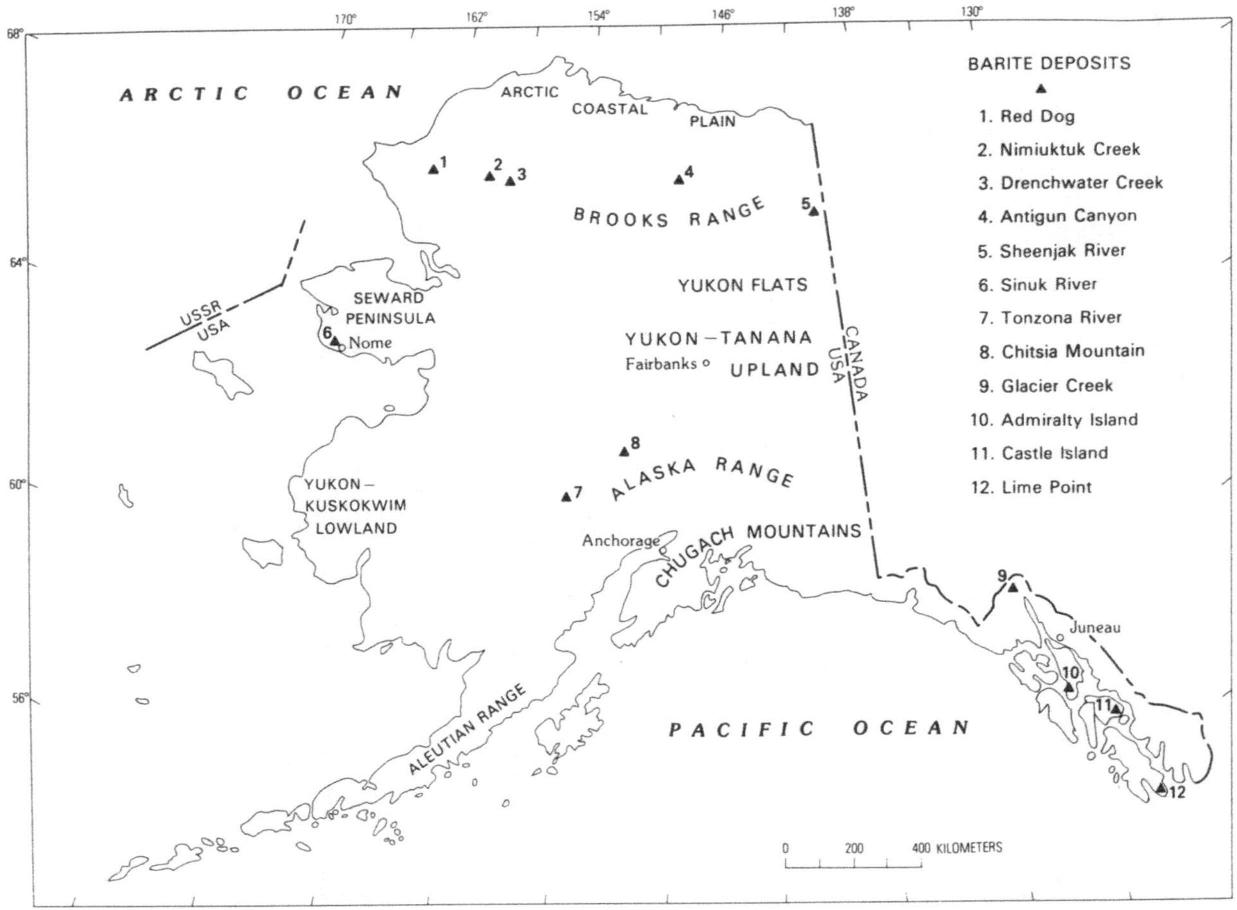

Figure 6–2. Location of barite deposits in Alaska.

of 17 percent zinc, 5 percent lead, and 82 grams per metric ton (g/t) of silver (Moore et al., 1986) in the Main deposit; additional significant reserves occur in the adjacent barite-rich Hilltop deposit. The estimated size of the barite resource has not been published. The Red Dog deposits are accumulations of silica, barite, and sulfides in Lower Mississippian to Permian(?) siliceous black shale and chert. The deposits are in a series of imbricate thrust sheets. The Main deposit is a nearly flat elongated stack of 1600 m long and 150 m to 975 m wide lenses that have an internal structure that mimics the thrust sheets (Moore et al., 1986). The Hilltop deposit is smaller and has a similar internal structure. Barite is concentrated toward the top and periphery of the deposits and is commonly coarse grained and massive to poorly bedded.

The Red Dog deposit formed from submarine exhalations in an early stage of a long-lived, second-order, starved basin during a time of tectonic instability and locally elevated heat flow (Moore et al., 1986). Mineralization consisted of precipitation on the sea floor and in unconsolidated sediment below the sediment-water interface (Schmidt and Zierenberg, 1987).

Other occurrences of type 2 barite as lenses, layers, and nodules in Mississippian to Permian black shale and chert sequences of the Brooks Range have been reported from Nimiuktuk (Mayfield et al., 1979), Drenchwater Creek (Nokleberg and Winkler, 1978; Lange et al., 1985), Wulik (Bundtzen and Henning, 1978), and Atigun Canyon (Paris, 1977). At the Nimiuktuk prospect, stratiform barite without associated sulfides is in the same geologic setting as the Red Dog deposits (Einaudi and Hitzman, 1986; Mayfield et al., 1979). Barite (type 2?) also occurs northwest of the Sheenjek River in northeastern Alaska where beds and lenses up to 6 m thick and 30 m long are interbedded with chert, shale, and mafic sills (Brosgé and Reiser, 1968).

On the Seward Peninsula of western Alaska, barite occurs with fluorite, lead and zinc sulfides, and associated gold and silver at the Sinuk River prospects (type 3?) (Herreid, 1966; Brobst et al., 1971). The introduced minerals generally occur as streaks and pods along foliation of the host rocks in thrust sheets of Proterozoic marble and schist (Nome Group).

Possible type 1 barite deposits have been identified in the Alaska Range near Chitsia Mountain and near the Tonzona River (Bundtzen and Henning, 1978). Near Chitsia Mountain, type 1(?) barite–galena–pyrite lenses up to 3 m thick and 100 m long are hosted by Mississippian metarhyolite tuff and porphyry. Stratiform barite (type 1?) that ranges up to several meters thick and is of an undetermined strike length near the Tonzona River is hosted by Devonian(?) phyllite.

CANADA

One of the largest type 2 zinc–lead–barite belts in the world has been discovered recently in the Paleozoic Selwyn basin and its southern extension, the Kechika trough in the Canadian Cordillera of the Yukon, Northwest Territories, and northeastern British Columbia (Fig. 6-3; Carne, 1976, 1979; MacIntyre, 1982; Carne and Cathro, 1982; Goodfellow, 1985; Dawson, 1985). The amount of barite is estimated to be more than 100 million metric tons (Caine, 1981). Estimated zinc-lead reserves are about 686 million metric tons averaging 4.6–7.8 percent zinc, 2.0–7.1 percent lead, and as much as 85 g/t silver.

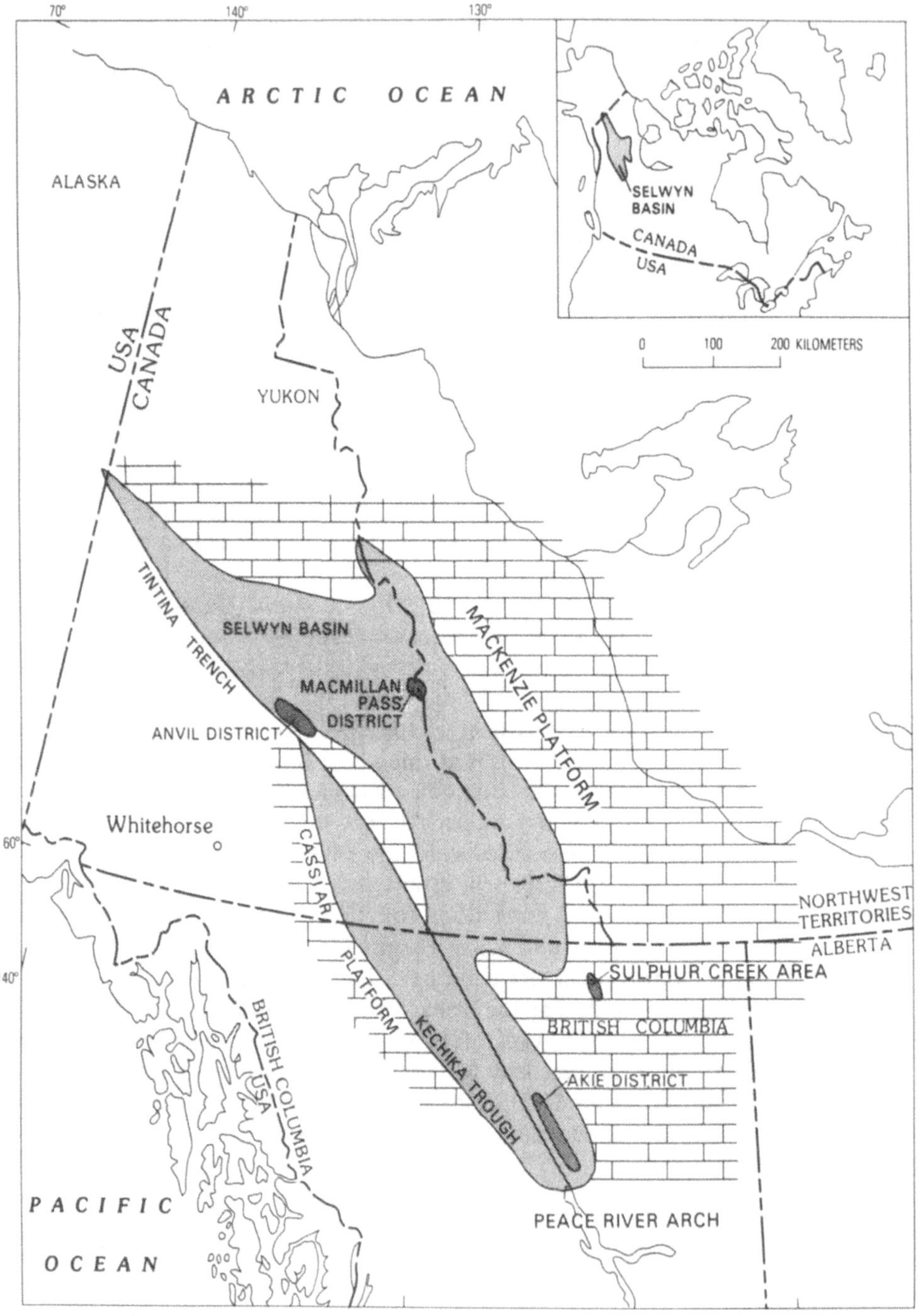

Figure 6–3. Location of major stratiform barite–sulfide and sulfide occurrences (dark gray) in the Selwyn basin (light gray) of western Canada and carbonate-rock-hosted barite deposits near Sulphur Creek (dark gray). Carbonate platform areas are patterned (after Carne and Cathro, 1982).

The most economically significant discoveries are the Tom and Jason deposits in the Macmillan Pass district and the Cirque and Driftpile Creek deposits in the Akie (Gataga) district. Other important barite deposits of the Selwyn basin include Cathy, Bar, Oro, Pete, GHMS, Gargantua North, Jeff, and Tea in the Macmillan Pass district (Dawson and Orchard, 1982) and Bear, Mount Alcock, Fluke, Pie, Elf, Kwadacha, and Sika in the Akie district (MacIntyre, 1982, 1983).

The Selwyn basin was a deep-marine basin that formed by subsidence accompanying rifting of the western continental margin of North America. The basinal rocks, which are primarily chert, shale, coarse-grained clastic sedimentary rocks, and minor volcanic rocks, are bordered by carbonate-shelf rocks (the MacKenzie and MacDonald platforms) on the north and east and by transform faults and the Cassiar platform on the west. The Selwyn basin and the bordering shelf carbonates are part of the Cordilleran miogeocline, in which a thick prism of sedimentary rocks of Proterozoic to Middle Jurassic age were deposited along a relatively stable continental margin that was affected by intermittent crustal extension (Abbott et al., 1986). The zinc–lead–barite deposits vary greatly in age, ranging from Cambrian (Anvil deposits) to Mississippian (Tea and nearby deposits). The deposits are in an imbricate series of allochthons that detached from crystalline basement as a result of a Middle Jurassic collision of an island arc with the miogeocline (Tempelman-Kluit, 1979).

In the Macmillan Pass district, type 2 barite deposits occur mainly in Devonian shales and siltstones comprising part of a thick and extensive blanket of cherty clastic rocks (Carne and Cathro, 1982; Dawson and Orchard, 1982). The bedded barite is in fine-grained clastic rocks (shale and siltstone) that often are bounded above and below by coarser-grained clastic sedimentary rocks and are accompanied by chert and limestone (Dawson and Orchard, 1982). The ore deposits are mineralogically zoned laterally and vertically and have barite enrichment in the upper and distal parts. Mineralization was associated with rifting, during which basin-margin growth faults provided conduits for upward movement of hydrothermal fluids, which were expelled onto the sea floor (Carne and Cathro, 1982; Bailes et al., 1986; McClay and Bidwell, 1986).

Type 2 barite–sulfide deposits in the Akie (Gataga) district occur in rhythmically bedded Middle and Upper Devonian baritic and pyritic carbonaceous black shales, siliceous argillites, and cherts (MacIntyre, 1982). Barite and sulfide minerals are intermixed in contrast to the separation of barite and sulfide minerals in the Tom and Jason deposits (Pigage, 1986). The deposits are in silicified, carbonaceous, thick-bedded shales and ribbon-banded porcellanites. Deposition was in a third order-basin in the Kechika trough. Mineralization was the result of exhalative activity from vent areas associated with regional growth faults. Barite and sulfide minerals precipitated from dense brine pools on the sea floor (Pigage, 1986).

In the Anvil district, near the westernmost exposures of the Selwyn basin, sulfide–barite type 2 deposits occur in Lower Cambrian graphitic phyllite that marks the transition between calcareous and noncalcareous pelitic rocks (Carne and Cathro, 1982). A zone of massive sulfides that contain barite forms the top of the deposits and overlies Lower Cambrian massive pyritic sulfides, pyritic quartzite, and ribbon-banded graphitic quartzite. Barite is much less abundant in the Anvil district than in the Macmillan Pass or Akie (Gataga) districts. Mineralization in the Anvil district is believed to have resulted from exhalative hydrothermal activity (Carne and Cathro, 1982).

In addition to the sedimentary exhalative barite deposits of the Selwyn basin, type 3 barite also occurs as replacement deposits in a Middle Devonian carbonate-shelf sequence at several localities near Sulphur Creek in northeastern British Columbia (Morrow et al., 1978). The barite mineralization is thought to have been the result of mixing euxinic sea water with meteoric ground water near the seaward edge of a large coastal aquifer; this caused the dissolution of the shelf carbonate and the precipitation of barite in solution cavities (Morrow et al., 1978)

Type 1 barite occurs in significant amounts in some polymetallic massive sulfide deposits in southern British Columbia (Dawson, 1985); for example, at the Lynx deposit on Vancouver Island, barite is associated with Permian volcanogenic massive sulfide deposits, and at the Britannia Mine on Howe Sound, barite occurs in the Jane orebody hosted by Upper Jurassic and Lower Cretaceous metavolcanic rocks.

WESTERN CONTERMINOUS UNITED STATES

Type 2 barite deposits of Paleozoic age occur in the eugeosynclinal terrane of the mid-Paleozoic Roberts Mountains allochthon and in Mesozoic and Cenozoic accreted eugeosynclinal terranes in the western United States Cordillera (Fig. 6–4; Papke, 1984; Poole, in press). Barite beds as much as several meters thick occur in rocks of Devonian age in northeastern Washington and northern California, rocks of probable Devonian and Permian age in central Idaho, in Ordovician or Devonian rocks in southern California, and mainly in Ordovician and Devonian rocks in Nevada. Some Proterozoic marginal-marine bedded barite occurs locally in northeastern Washington.

The largest reserves of barite in the United States are the type 2 deposits of Nevada, where barite occurrences are known in rocks from Cambrian through Pennsylvanian age. A belt containing hundreds of type 2 barite deposits extends the length of the state of Nevada and has an average width of 125 km. The Nevada barite belt, which is one of the largest in the world, has reserves that are estimated to be 90 million short tons (Papke, 1984). The barite deposits of Nevada are not associated with significant amounts of sulfide minerals.

The Paleozoic barite forms light- to dark-gray beds that are conformable to bedding in enclosing rocks. Individual barite beds and intervals containing barite beds are lensoidal in shape. This shape may reflect the original geometry of sea-floor vent mounds and basins (Poole, in press). Alignment patterns of barite deposit clusters suggest localization of exhalative vents and troughs along sea floor fault systems. The deposits are mainly in marine siliceous eugeosynclinal rocks that are principally mudstone, siltstone, and chert and minor limestone, dolostone, sandstone, conglomerate, tuff, and mafic lava. The barite beds consist mainly of laminated finely crystalline rock, layers of rosettes, turbiditic barite sandstone that has Bouma divisions, and some debris-flow barite conglomerate. Brachiopods and hollow tubes, believed to represent the remains of worm tubes, occur in some barite beds and are believed to

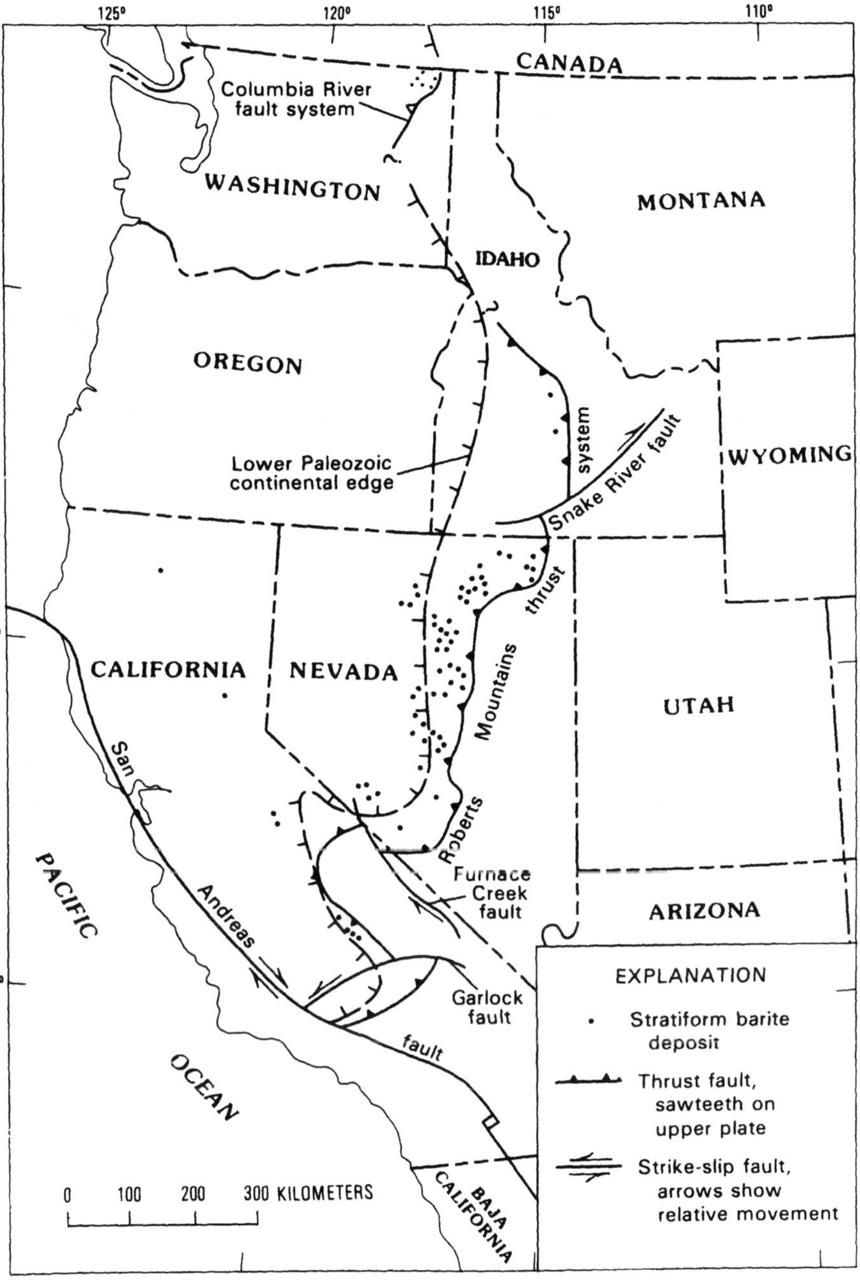

Figure 6-4. Location of stratiform barite deposits in the western United States Cordillera (from Poole, in press).

represent indigenous faunas at or near exhalative vents on the sea floor (Poole, in press). No indigenous benthic fossils have been found in the host rocks, most of which are organic rich and formed under anoxic conditions unfavorable for benthic animal life. The barite deposits formed in relatively deep-water marginal ocean basin and continental rise settings. Extensively fractured oceanic crust provided conduits for ascending barium-bearing hydrothermal acid solutions that discharged onto the sea floor and reacted with sea water sulfate to precipitate barite (Poole, in press).

MEXICO

Type 2 barite deposits of Late Devonian and Early Pennsylvanian age occur in northwestern Mexico (central Sonora) within an allochthonous eugeosynclinal assemblage of Ordovician to Permian rocks (Fig. 6–5; Poole et al., 1983). The barite deposits are hosted by siliceous marine rocks that are principally chert, mudstone, siltstone, and conglomerate and subordinate dolostone, lime-

Figure 6–5. Location of stratiform barite deposits in northwestern Mexico (after F. G. Poole and R. J. Madrid, U.S. Geological Survey, written communication).

stone, and tuff. The barite deposits are located about 90 km east–southeast of Hermosillo, Sonora (Fig. 6–5), in the vicinity of Mazatán (Barita de Sonora mine) and near Cobachi. Barita de Sonora, the only operating barite mine in the region, has been producing the higher-grade (greater than 4 specific gravity) Lower Pennsylvanian barite sandstones and conglomerates. Large reserves of the generally lower-grade Upper Devonian barite occur in the Barita de Sonora mine area and near Cobachi. Upper Devonian barite deposits consist of gray fine-grained barite, rosette barite, and barite sandstone, conglomeratic barite, and barite-cemented sandstone and conglomerate in beds as thick as 1 m. Lower Pennsylvanian barite deposits at Barita de Sonora consist of gray fine-grained barite, rosette barite, barite sandstone and conglomerate in beds as thick as 4 m. Most of the barite beds are turbidites exhibiting Bouma divisions and debris-flows in submarine-fan systems that record southwestern transport directions (Poole and Madrid, 1986).

EASTERN NORTH AMERICA

CANADA

The Buchans deposit in the Dunnage terrane (oceanic lithosphere) of Newfoundland (Fig. 6–6) is a type 1 volcanogenic massive sulfide deposit of the Kuroko type in which barite is common (Dawson, 1985; Thurlow et al., 1975; Swanson et al., 1981). The barite occurs as a major constituent in the sulfide ores and locally forms beds and lenses that cap ore bodies. The barite at Buchans is associated with polymetallic zinc–lead–copper–silver–gold minerals in an Ordovician felsic volcanic sequence. The ores formed by interactions among volcanic hot springs, wall rocks, and sea water on and near the sea floor.

The Walton deposit (type 3) in Nova Scotia (Fig. 6–6) contains massive barite and argentiferous base metals at the base of the Mississippian Windsor Group marine carbonate rocks (Boyle, 1972; Patterson, 1988). The Windsor Group carbonates were deposited in a basin developed on the older (Cambrian–Ordovician) Meguma terrane. In the Walton deposit, barite is a massive, compact, and fine-grained replacement of limestone (Boyle, 1972; Patterson, 1988). Numerous other prospects and occurrences of barite in Nova Scotia (Fig. 6–6), some of which are stratabound, have been described by Felderhof (1978). Most of the stratabound carbonate-hosted occurrences are in the Windsor group.

UNITED STATES

Barite deposits, which have been described as veins, but which may be mostly metamorphosed type 1 deposits, have long been known in rocks of the Piedmont belt near Lynchburg, Virginia (Fig. 6–7). Most barite forms stratiform layers in carbonate, metaclastic, and metavolcanic rocks of the Late Proterozoic Lynchburg Group (R. S. Good and J. F. Conley, Virginia Division of Mineral Resources, oral communication, 1988).

Type 1 barite deposits also are known in the Kings Creek district of the

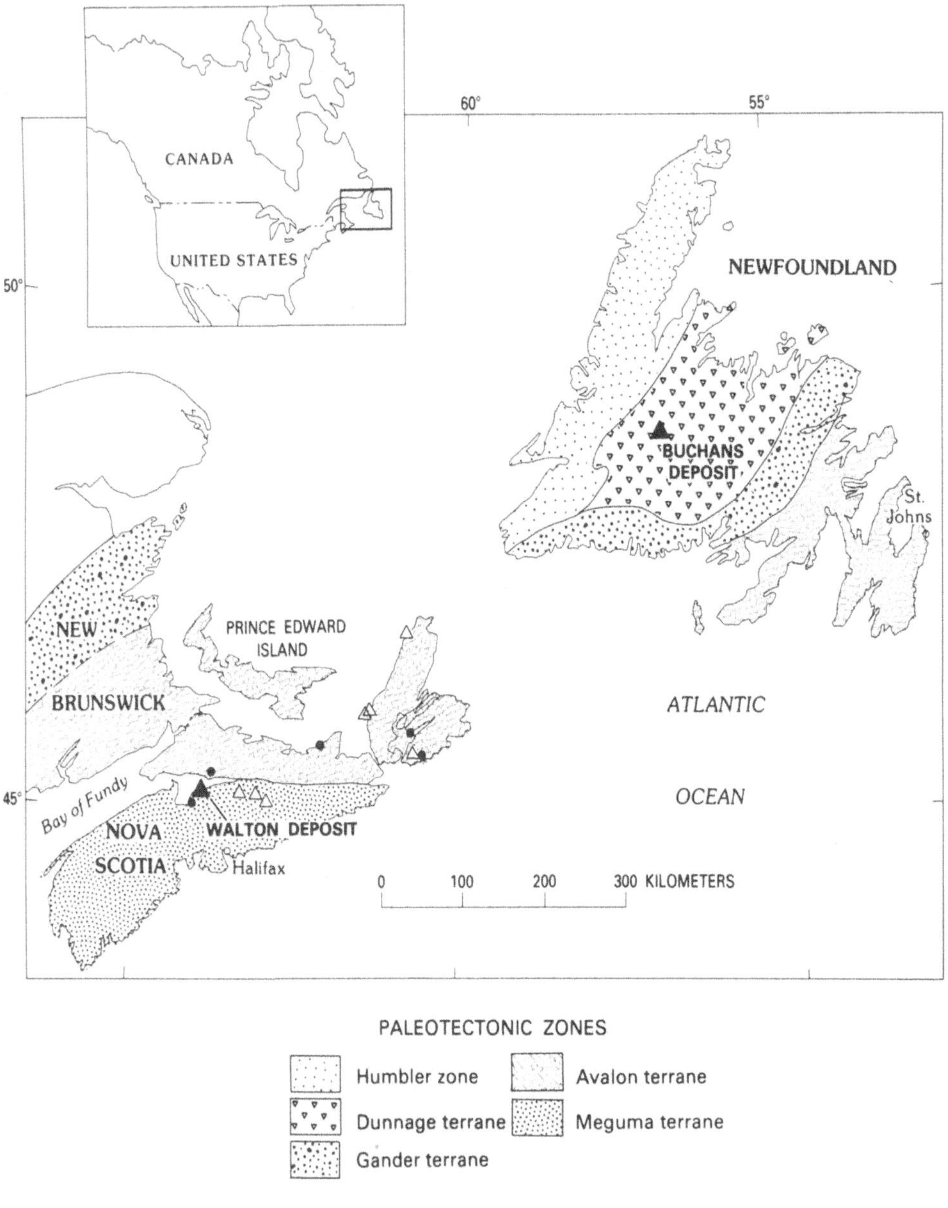

Figure 6-6. Location of the Buchans deposit in Newfoundland and the Walton deposit in Nova Scotia (large triangles) and prospects and occurrences of stratabound clastic-rock-hosted (small solid circles) and carbonate-rock-hosted (small open triangles) barite in Nova Scotia. Locations of prospects and occurrences after Felderhof, 1978; paleotectonic zones after Stephens, 1986.

Kings Mountain belt in South Carolina and in rocks of the Carolina slate belt in the Lincolnton-McCormick area in Georgia and South Carolina (Fig. 6-7; Carpenter, 1982). The Kings Mountain belt is underlain by complexly deformed metasedimentary and metavolcanic rocks. Barite occurs as massive layers and lenses up to 3 m thick that generally are aligned along foliation in quartz sericite schist. The Carolina slate belt consists mainly of a sequence of

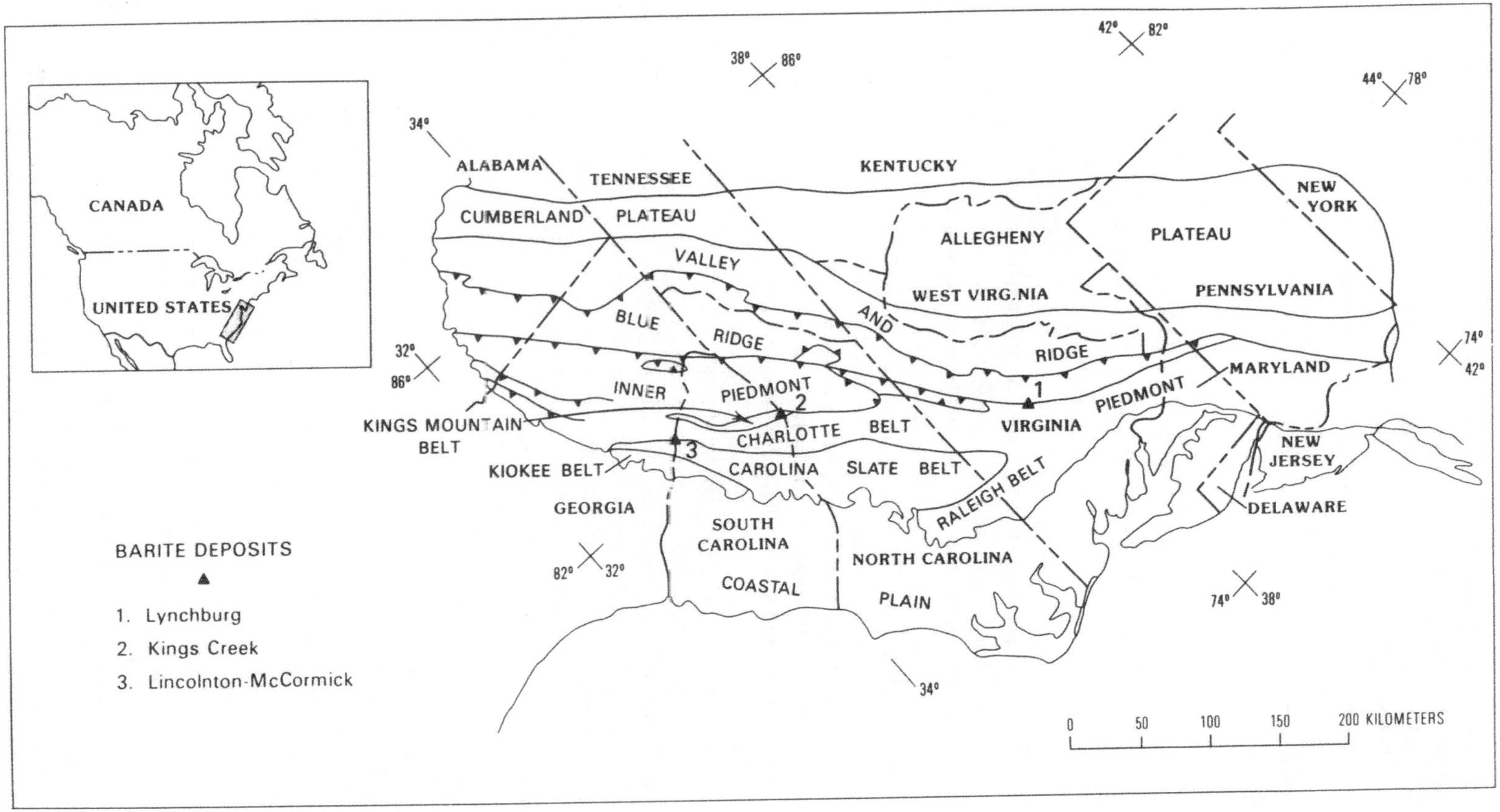

Figure 6-7. Location of stratiform(?) barite deposits in the eastern United States (after Price and Hatcher, 1983).

metamorphosed Cambrian volcanic and sedimentary rocks that were deposited in an island-arc environment. The barite is in the metavolcanic rocks and is associated with zinc–copper–lead–sulfides and gold.

Although the origin of the barite deposits in the Piedmont, Kings Mountain, and Carolina slate belts is uncertain, they could have formed at some distance from their present locations and could have been subsequently accreted to the eastern margin of the North American craton along with their host rocks.

CENTRAL NORTH AMERICA AND THE APPALACHIAN MOUNTAINS

Type 3 barite is associated with many of the carbonate-hosted epigenetic stratabound zinc–lead deposits in the stable cratonic platform areas of North America and in Paleozoic continental shelf carbonate sequences of the folded Appalachians. Barite is generally present only as a minor constituent, but it is the major ore mineral in some deposits. These carbonate-rock-hosted stratabound deposits, which commonly are referred to as Mississippi Valley-type deposits, include the Upper Mississippi Valley district of Wisconsin, Illinois, and Iowa; the central and southeastern Missouri districts; the Illinois–Kentucky district; the central Kentucky district; and the central and eastern Tennessee districts.

In the southeastern Missouri barite district, which contains types 3 and 4 barite, residual accumulations of barite in soils have produced at least 12 million tons of barite since mining began in 1850. Type 3 bedrock mineralization of Upper Cambrian and Lower Ordovician dolostones is characterized by open-space filling of vugs and fractures (Kaiser et al., 1987). Mineralization was controlled by intersection of faults and zones of intense jointing with permeable lithologies (Kaiser et al., 1987).

Most of the barite that has been mined in Tennessee was produced from type 4 deposits in the Sweetwater district. In the underlying bedrock, type 3 barite occurs in veins and breccias in the upper part of the Ordovician Knox Group dolostones. The barite is associated with fluorite, sphalerite, galena, and pyrite. Some of the structures that localized barite are thought to have been produced by solution and by tectonic movement (Maher, 1970).

In the Cartersville district of Georgia, barite is produced from type 4 deposits formed by deep weathering. The bedrock source of the barite is type 3 veins, pods, and fracture fillings in Lower Cambrian dolostone (Kesler, 1950; Reade et al., 1980).

Barite has been produced from many other type 4 deposits that overlie carbonate bedrock in Tennessee, Virginia, Georgia, and Alabama (Clark, 1987*a*). Although little is known about the bedrock source of the barite in most of these deposits, it is probable that many of them were concentrated from type 3 occurrences in the underlying bedrock where stratal zones of permeability along with fracture zones controlled movement and deposition of the ore fluids, as in Mississippi Valley-type deposits.

The ore solutions for Mississippi Valley-type deposits generally are thought to have formed by episodic sediment dewatering and expulsion of geopressured metalliferous and petroliferous warm brines from deeply buried basinal black shales during the compaction stage of sedimentary basin evolution (Cathles and Smith, 1983). This pore-fluid expulsion could result in upward migration of metal- and hydrocarbon-bearing ore fluids through conduits into shallow basin margin sediment-hosted settings where ore minerals and/or petroleum were deposited (Poole and Desborough, 1985). Data from fluid inclusions show that the ore fluids were slow-moving, high-density brines; commonly contained hydrocarbons; and generally had ore-forming temperatures of between 100°C and 150°C, but as high as 200°C (Roedder, 1967, 1976; McLimans, 1977). The ore-forming brines migrated out of deep basins along basal aquifers toward the surface or to sites where growth faults or reef structures provided cross channels toward the surface (Cathles and Smith, 1983). Deposition of ore minerals occurred at sites where interconnected openings provided continuous flow of ore solutions. Some deposits were localized by features associated with unconformities, such as paleokarst terranes; some were localized by facies boundaries between platformal carbonates and basinal shales; and some occur in platform carbonates of biostromal character or back-reef and fore-reef settings (Anderson and Macqueen, 1982).

The mineralized areas are associated spacially with settings that are peripheral to sedimentary basins and on the arches between basins (Clark, 1987*a*). Mineralization predominantly took place in carbonate rocks, mainly dolostone, of Cambrian to Middle Ordovician age. Exceptions are the Illinois-Kentucky and Tri-State districts where mineralization took place in Mississippian limestone.

In the Appalachian region, barite deposits and occurrences are more common and larger in the southern states (Virginia, Tennessee, Georgia, Alabama) than in the northern states. The development of thick residual soils in which barite was concentrated is one factor in the distribution pattern. The distribution also may reflect a regional zonation from south to north of barium, zinc, and lead (Clark, 1987*a,* 1987*b*).

In addition to the type 3 and 4 deposits described above, type 2 deposits of barite nodules have been reported from shales and mudstones at numerous localities in Ordovician, Silurian, and Devonian shales of the Valley and Ridge province in the folded Appalachian Mountains (Clark, 1985, 1986; Nuelle and Shelton, 1986). In some places, the nodules coalesce and form beds that resemble the commercial type 2 deposits of the western part of North America; however, no economic concentrations have been found.

SULFUR AND OXYGEN ISOTOPE STUDIES

Sulfur and oxygen isotope values for barite from stratabound deposits provide information on the possible sources and conditions of precipitation of the sulfate in the barite. Values of sulfur and oxygen isotopes have been determined for barite beds and nodules from type 2 deposits and occurrences (Table 6-2; Fig. 6-8) in northwestern Canada (Cecile et al., 1983; Goodfellow and Jonasson, 1984; Mako and Shanks, 1984; Shanks et al., 1987), central Nevada

Table 6-2. Sulfur and Oxygen Isotope Values of Barite from Some Type 2 and 3 Deposits and Occurrences in North America

Location	$\delta^{34}S$ (per mil) Range (number of samples)	$\delta^{34}S$ (per mil) Average	$\delta^{18}O$ (per mil) Range (number of samples)	$\delta^{18}O$ (per mil) Average	Age
	Bedded barite from type 2 deposits and occurrences				
Northwestern Canada	23.2–29.7 (3)	27.0	16.2–19.5 (3)	17.9	Middle to Late Devonian
Nevada	20.9–28.6 (10)	24.4	—	—	Late Devonian
Sonora	— (1)	27.2	—	—	Late Devonian
	21.2–31.8 (2)	26.5	—	—	Early Pennsylvanian
	Nodular barite from type 2 deposits and occurrences				
Northwestern Canada	41.8–53.0 (4)	46.3	20.3–29.1 (4)	23.3	Late Devonian
Nevada	29.1–56.3 (13)	38.7	—	—	Late Devonian
Sonora	— (1)	35.5	—	—	Late Devonian
	— (1)	37.9	—	—	Early Pennsylvanian
Appalachian basin	28.0–47.4 (9)	38.5	19.3–30.4 (4)	24.6	Middle Devonian
	Epigenetic barite (carbonate hosted)				
Missouri	20.5–32.2 (17)	26.7	17.7–19.6 (11)	18.7	Late Paleozoic
British Columbia	24.2–30.1 (12)	27.2	—	—	Devonian

Sources: Data from Cecile et al. (1983) for northwestern Canada; Rye et al. (1978) for Nevada; Mark Huebner, U.S. Geological Survey, written commun. (1981) for Sonora; Nuelle and Shelton (1986) for the Appalachian basin; Kaiser et al. (1987) for Missouri; and Morrow et al. (1978) for British Columbia.
Note: — indicates no data.

(Rye et al., 1978), central Sonora (Mark Huebner, U.S. Geological Survey, written commun., 1981), and the Appalachian basin (Nuelle and Shelton, 1986). Isotope values vary greatly between those that are close to the isotope-age curves of Claypool et al. (1980; Fig. 6-9) and those modified by enrichment in ^{18}O and ^{34}S. Enrichment in the heavy isotopes occurs when ^{32}S and ^{16}O are removed preferentially from the soluble sulfate pool by the action of sulfate-reducing bacteria. The amount of enrichment can reflect a number of factors—the degree to which a system is closed, the isotopic composition of the sulfate pool, and the rates of sulfate reduction, sedimentation, and removal of isotopically light sulfur. In a study of samples from type 2 deposits in northwestern Canada, Cecile et al. (1983) noted that samples that had isotopic values close to the evaporite curve are from thick beds of barite hosted mainly

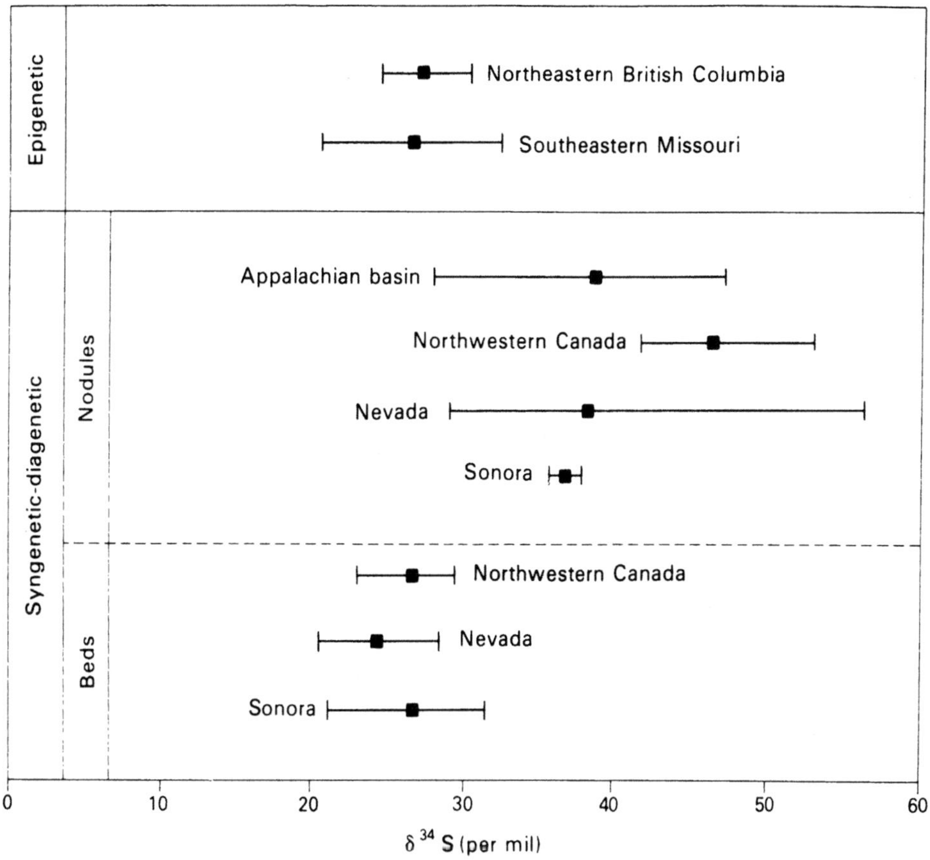

Figure 6–8. The range and average (small squares) sulfur isotope values for some type 2 stratiform and type 3 epigenetic deposits in North America. See Table 6–2 for data.

by organic-rich shales and from thin and nodular beds hosted by organic-free sedimentary strata. Samples enriched in heavy isotopes are from thin or nodular deposits hosted in organic-rich sedimentary strata. The enrichment in heavy isotopes can be explained by a barite precipitation–dissolution process in which barium is cycled between oxygenated and reduced zones in stratified sea water or in wet sediment. The enrichment in heavy isotopes reflects a slow rate of barite precipitation relative to the rate of barite dissolution and reduction (Cecile et al., 1983).

Some barite samples from type 2 zinc–lead–silver sulfide deposits in the Anvil Range of the Yukon Territory, Canada, have sulfur isotope values that are considerably lighter than the sulfur isotope values of contemporaneous sea water (Shanks et al., 1987). The light sulfur values are the result of mixing isotopically light sulfate in the barium-bearing ore fluid with isotopically heavier sulfate in sea water (Shanks et al., 1987).

Sulfur isotope ratios of type 3 barite from the deposits in northeastern British Columbia (Table 6–2; Fig. 6–8) are approximately the same as those of

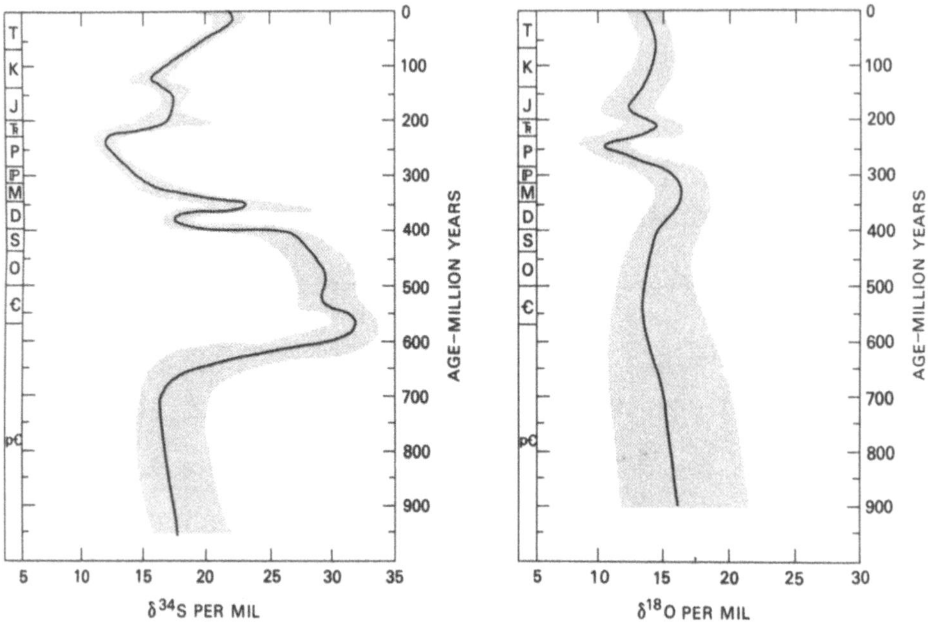

Figure 6–9. Oxygen and sulfur isotope age curves for sulfate. The heavy lines are the best estimate for $\delta^{18}O$ and $\delta^{34}S$ in sulfate minerals in equilibrium with sulfate dissolved in the world ocean of that time; shaded areas are a qualitative estimate of its uncertainty (from Claypool et al., 1980).

Devonian sea water (Fig. 6–9). The sea water mixed with meteoric ground water, which resulted in precipitation of barite (Morrow et al., 1978).

Type 3 barite from the southeast Missouri barite district (Table 6–2; Fig. 6–8) is progressively enriched in ^{34}S with increased distance (up to 150 m) away from a central fault from which fluids vented outward (Kaiser et al., 1987). The pattern of distribution resulted from mixing of isotopically light sulfate of the ore fluid with heavy sulfate from the host dolostone.

SUMMARY AND CONCLUSIONS

The four genetic types or categories of the major stratabound barite ore fields of North America have regional distribution patterns related to geologic setting. Type 1 syngenetic–diagenetic stratiform volcanogenic barite deposits formed in island-arc and other sea-floor volcanic environments. Type 2 syngenetic–diagenetic stratiform barite deposits are in fine-grained sedimentary rocks in deep-water marine settings adjacent to the continental shelf. Type 3 epigenetic stratabound barite deposits generally are hosted by carbonate rocks that were deposited in stable cratonic-platform and continental-shelf environments. Type 4 residual barite concentrations commonly overlie type 3 carbonate-rock-hosted barite deposits or carbonate bedrock; those type 4 barite deposits that lack bedrock exposures generally occur in the same geologic setting as type 3 deposits.

Type 1 deposits in which barite is associated with polymetallic massive sulfides occur in the western Cordillera and eastern accreted terranes of North America. These deposits formed in island-arc and other sea-floor environments and were later deformed, metamorphosed, and accreted to the North American continent. The deposits are related to submarine volcanism and generally form on the flanks of lava domes or near the base of volcanic cones (Dawson, 1985). Volcanogenic deposits that contain barite are in Proterozoic and Ordovician host rocks in eastern North America and in Cambrian to Permian host rocks in western North America; this reflects times of island-arc development or rifting.

The most significant type 2 barite deposits are those of the western Brooks Range in Alaska, the Selwyn basin and the Kechika trough in Canada, and the Roberts Mountains allochthon in Nevada. Most are sedimentary-exhalative deposits that accumulated in deep-water marine environments. The barite is generally in Paleozoic fine-grained clastic marine sedimentary sequences and commonly is associated with cherty or siliceous rocks. Many of the deposits are Devonian, but mineralization also occurred during times of tectonic instability from Proterozoic through Permian time. Some of the barite deposits are associated with zinc–lead–silver sulfide minerals and occur in the upper and peripheral parts of ore zones. Other ore deposits are composed entirely or primarily of barite and have no associated sulfide minerals. These deposits are generally in complexly deformed thrust sheets that were accreted to the North American continent.

A generalized model for the genesis of the sedimentary exhalative deposits is the upward movement of ore-forming fluids along sea floor faults or rift zones and their discharge into sea water in deep-marine basins. The model is analogous to deposits on the modern sea floor that form from the discharge of hydrothermal springs. Barium in ore fluids that discharge in a submarine environment precipitates either on the sea floor or in sediments just below the sea floor. Precipitation results from a decrease in temperature, increase in pH, and oxygenation and dilution of the solution by mixing with sulfate-bearing sea water (Poole, in press). Ore minerals also can precipitate in poorly consolidated sediments below the sea water–sediment interface when some of the fluids do not reach the surface. If barite and sulfide deposits form, then they are generally in the proximity of exhalative vents, in adjacent or nearby second- or third-order basins, and/or at reduction–oxidation interfaces. Factors that determine whether or not barite deposits will be precipitated, and their relation to sulfide deposits, include oxygen level in the sea water, geometry of the basin into which the fluids are vented or have migrated, water depth that controls boiling of the fluids, and rates of sedimentation.

Type 3 barite commonly is associated with and is similar to Mississippi-Valley type lead–zinc deposits. Significant amounts of barite also have been produced from type 4 deposits in unconsolidated material overlying type 3 deposits. The type 3 barite deposits generally are hosted by carbonate rocks that were formed on the Paleozoic stable cratonic platform and continental shelf of the North American continent. An exception is the Walton deposit of eastern Canada, which was deposited in carbonate rocks that formed in a basin developed on terranes accreted to the continent. The ore minerals in type 3 barite deposits primarily filled open spaces in chemically favorable strata in post-Permian time. Barium, base metals, and petroleum were transported along basal aquifers to depositional sites by warm mineralizing fluids driven out of shales in deep sedimentary basins during compaction. Host rocks are primarily, but not entirely, of Cambrian through Middle Ordovician age; but, for most deposits, the age of mineralization is not established firmly.

Sulfur and oxygen isotope values in barite vary greatly and are valuable in interpreting the source of sulfate and conditions in which the barite precipitated. Sulfur isotope values from type 2 deposits tend to be lighter (less than 30 per mil) in beds of fine-grained barite and heavier (greater than 30 per mil) in nodules of barite. The lighter values are close to the value for coeval sea water sulfates. Enrichment in heavy isotopes generally reflects slow rates of barite precipitation relative to the rate of barite dissolution and reduction. In type 3 barite deposits, sulfur and oxygen isotopes generally are close to that of the ore fluid except where modified by interaction with host rocks.

REFERENCES

Abbott, J. G., S. P. Gordey, and D. J. Tempelman-Kluit. 1986. Setting of stratiform, sediment-hosted lead-zinc deposits in Yukon and northeastern British Columbia. In *Mineral Deposits of Northern Cordillera,* ed. J. A. Morin, pp. 1–18. Montreal, Quebec: Canadian Institute of Mining and Metallurgy.

Anderson, G. M., and R. W. Macqueen. 1982. Ore deposits models—6. Mississippi Valley type lead-zinc deposits. *Geoscience Canada* **9**:108–117.

Bailes, R. J., B. W. Smee, D. W. Blackadar, and H. D. Gardner. 1986. Geology of the Jason lead-zinc-silver deposits, Macmillan Pass, eastern Yukon. In *Mineral Deposits of Northern Cordillera,* ed. J. A. Morin, pp. 87–99. Montreal, Quebec: Canadian Institute of Mining and Metallurgy.

Boyle, R. W. 1972. The geology, geochemistry, and origin of the barite, manganese, and lead-zinc-copper-silver deposits of the Walton-Cheverie area, Nova Scotia. *Geol. Survey Canada Bull.* **166**:1–181.

Brobst, D. A., D. M. Pinckney, and C. L. Sainsbury. 1971. Geology and geochemistry of the Sinuk River barite deposit, Seward Peninsula, Alaska. *U.S. Geol. Survey Professional Paper 750-D,* pp. D1–D8.

Brosgé, W. P., and H. N. Reiser. 1968. Geochemical reconnaissance maps of granitic rocks, Coleen and Table Mountain Quadrangles, Alaska: *U.S. Geol. Survey, Open-File Report,* 4 sheets, scale 1:250,000.

Bundtzen, T. K., and M. W. Henning. 1978. Barite in Alaska. *Alaska Div. Geol. Geophys. Surveys Mines and Geol. Bull.* **27**(4):1–4.

Caine, T. W. 1981. The Yukon and Northwest Territories. In *The Industrial Minerals of Canada,* Industrial Minerals Supplement, ed. P. W. Harben, pp. 36–37.

Carne, R. C., 1976, Stratabound barite and lead-zinc-barite deposits in the eastern Selwyn basin, Yukon Territory. *Canadian Department of Indian and Northern Affairs Open-File Report 1976-16,* 41p.

Carne, R. C. 1979. Geological setting and stratiform mineralization, Tom claims, Yukon Territory. *Canadian Department of Indian and Northern Affairs Open-File Report 1979-4,* 30p.

Carne, R. C., and R. J. Cathro. 1982. Sedimentary exhalative (sedex) zinc-lead-silver deposits, northern Canadian Cordillera. *Canadian Inst. Min. Metal. Bull.* **75**(840):66–78.

Carpenter, R. H. 1982. Slate belt of Georgia and Carolinas—Geology, stratigraphy, structure, and exploration potential. In *Exploration for Metallic Resources in the Southeast,* G. O. Allard, and R. H. Carpenter, conveners, pp. 14–18. Athens, Georgia: University of Georgia, Department of Geology and Center for Continuing Education.

Cathles, L. M., and A. T. Smith. 1983. Thermal constraints on the formation of Mississippi Valley-type lead-zinc deposits and their implications for episodic basin dewatering and deposit genesis. *Econ. Geology* **78**:360–372.

Cecile, M. P., M. A. Shakur, and H. R. Krouse. 1983. The isotopic composition of western Canadian barites and possible derivation of oceanic sulfate $\delta^{34}S$ and $\delta^{18}O$ age curves. *Canadian Jour. Earth Sci.* **20**(10):1528–1535.

Clark, S. H. B. 1985. Diagenetic barite nodules in Ordovician and Devonian shales of the Appalachian basin. *Geol. Soc. America Abstracts with Programs* **17**(7):548.

Clark, S. H. B. 1986. The origin of some shale-hosted barite nodules of the Appalachian basin (abstract). *Terra Cognita* **6**(3):514.

Clark, S. H. B. 1987*a*. Zinc, lead, and barium deposits and occurrences in Paleozoic rocks, east-central United States. *U.S. Geological Survey Miscellaneous Investigations Series Map I-1773,* scale 1:2,500,000, 74p.

Clark, S. H. B. 1987*b*. Regional zonation of lead, zinc, and barium in Paleozoic sedimentary rocks of the Appalachian basin. *Geol. Soc. America Abstracts with Programs* **19**(2):80.

Claypool, G. E., W. T. Holser, I. R. Kaplan, H. Sakai, and I. Zak. 1980. The age curves of sulfur and oxygen isotopes in marine sulfate and their mutual interpretation. *Chem. Geology* **28**:199–260.

Dawson, K. R. 1985. Geology of barium, strontium, and fluorine deposits in Canada. *Geol. Survey of Canada Econ. Geology Report 34,* 136p.

Dawson, K. M., and M. J. Orchard. 1982, Regional metallogeny of the northern Cordillera: Biostratigraphy, correlation, and metallogenic significance of bedded barite occurrences in eastern Yukon and western District of Mackenzie. *Geol. Survey of Canada, Current Research, Part C, Paper 82-1C,* pp. 31–38.

Einaudi, M. T., and M. W. Hitzman. 1986. Mineral deposits in northern Alaska—an introduction. *Econ. Geology* **81**(7):1583–1591.

Felderhof, G. W. 1978. Barite, celestite, and fluorite in Nova Scotia. *Nova Scotia Dept. Mines Bull. 4,* 463p.

Goodfellow, W. 1985. Geochemistry unveils "blind" deposits. *GEOS* **14**(1):17–20.

Goodfellow, W. D., and I. R. Jonasson. 1984. Ocean stagnation and ventilation defined by $\delta^{34}S$ secular trends in pyrite and barite, Selwyn Basin, Yukon. *Geology* **12**:583–586.

Herreid, G. 1966. Preliminary geology and geochemistry of the Sinuk area, Seward Peninsula, Alaska. *Alaska Div. Mines and Minerals Geol. Report 24,* 19p.

Herreid, G., T. K. Bundtzen, and D. L. Turner. 1978. Geology and geochemistry of the Craig A-2 quadrangle and vicinity, Prince of Wales Island, southeastern Alaska. *Alaska Div. Geological and Geophysical Surveys Report 48,* 49p.

Kaiser, C. J., W. C. Kelly, R. J. Wagner, and W. C. Shanks III. 1987. Geologic and geochemical controls of mineralization in the Southeast Missouri barite district. *Econ. Geology* **82**:719–739.

Kesler, T. L. 1950. Geology and mineral deposits of the Cartersville district, Georgia. *U.S. Geol. Survey Professional Paper 224,* 95p.

Lange, I. M., W. J. Nokleberg, J. T. Plahuta, H. R. Krouse, and B. R. Doe. 1985. Geologic setting, petrology, and geochemistry of stratiform sphalerite-galena-barite deposits, Red Dog Creek and Drenchwater Creek areas, northwestern Brooks Range, Alaska. *Econ. Geology* **80**:1896–1926.

MacIntyre, D. G. 1982. Geologic setting of recently discovered stratiform barite-sulfide deposits in northeast British Columbia. *Canadian Inst. Min. Metal. Bull.* **75**(840):99–113.

MacIntyre, D. G. 1983. Geology and stratiform barite-sulphide deposits of the Gataga District, northeastern British Columbia. In *Sediment-hosted Stratiform Lead-Zinc*

Deposits, Vol. 9, ed. D. F. Sangster, pp. 85–119. Victoria, British Columbia: Mineralogical Association of Canada Short Course Handbook.

MacKevett, E. M., Jr., E. C. Robertson, and G. R. Winkler. 1974. Geology of the Skagway B-3 and B-4 quadrangles, southeastern Alaska. *U.S. Geol. Survey Professional Paper 832,* 33p.

Maher, S. W. 1970. Barite resources of Tennessee. *Tennessee Division of Geology Report of Investigations 28,* 40p.

Mako, D. A., and W. C. Shanks III. 1984. Stratiform sulfide and barite-fluorite mineralization of the Vulcan prospect, Northwest Territories—Exhalation of basinal brines along a faulted continental margin. *Canadian Jour. Earth Sci.* **21:**78–91.

Mayfield, C. F., S. M. Curtis, I. F. Ellersieck, and I. L. Tailleur. 1979. Reconnaissance geology of the Ginny Creek zinc-lead-silver and Nimiuktuk barite deposits, northwestern Brooks Range, Alaska. *U.S. Geol. Survey, Open-File Report 79-1092,* 20p.

McClay, K. R., and G. E. Bidwell. 1986. Geology of the Tom deposit, Macmillan Pass, Yukon. In *Mineral Deposits of Northern Cordillera,* ed. J. A. Morin, pp. 100–114. Montreal, Quebec: Canadian Institute of Mining and Metallurgy.

McLimans, R. K. 1977. Geological, fluid inclusion, and stable isotope studies of the upper Mississippi Valley zinc-lead district, southwest Wisconsin. Ph.D. dissertation. University Park, Pa.: The Pennsylvania State University.

Moore, D. W., L. E. Young, J. S. Modene, and J. T. Plahuta. 1986. Geologic setting and genesis of the Red Dog zinc-lead-silver deposit, western Brooks Range, Alaska. *Econ. Geology* **81:**1696–1727.

Morrow, D. W., H. R. Krouse, E. D. Ghent, G. C. Taylor, and K. R. Dawson. 1978. A hypothesis concerning the origin of barite in Devonian carbonate rocks of northeastern British Columbia. *Canadian Jour. Earth Sci.* **15**(9):1391–1406.

Nokleberg, W. J., and G. R. Winkler. 1978. Geologic setting of the lead and zinc deposits, Drenchwater Creek area, Howards Pass Quadrangle, western Brooks Range, Alaska. *U.S. Geol. Survey Open-File Report 78-70C,* 16p.

Nuelle, L. M., and K. L. Shelton. 1986. Geologic and geochemical evidence of possible bedded-barite deposits in Devonian rocks of the Valley and Ridge Province, Appalachian Mountains. *Econ. Geology* **81:**1408–1430.

Papke, K. G. 1984. Barite in Nevada. *Nevada Bur. Mines and Geol. Bull.* **98:**1–125.

Paris, C. 1977. Models of barite deposition with reference to the barite occurrence of Atigun Canyon, central Brooks Range, Alaska. Special Topics Paper. Fairbanks, Alaska: University of Alaska, 45p.

Patterson, J. M. 1988. Vertical and horizontal extensions to mineralization—two case histories from Nova Scotia, Canada. In *North American Conference on Tectonic Control of Ore Deposits and the Vertical and Horizontal Extent of Ore Systems,* eds. G. Kisvarsanyi and S. K. Grant. Rolla, Mo.: University of Missouri-Rolla.

Pigage, L. C. 1986. Geology of the Cirque barite-zinc-lead-silver deposits, northeast British Columbia. In *Mineral Deposits of Northern Cordillera,* ed. J. A. Morin, pp. 71–86. Montreal, Quebec: Canadian Institute of Mining and Metallurgy.

Poole, F. G. In press. Stratiform barite in Paleozoic rocks of the western United States. In *7th IAGOD Symposium Proceedings, Luleà, Sweden,* ed. E. Zachrisson. Stuttgart: E. Schweizerbart'sche Verlagsbuchhandlung.

Poole, F. G., and G. A. Desborough. 1985. Metal concentrations in marine black shales. In *U.S. Geological Survey Research in Mineral Resources–1985 Program and Abstracts,* ed. K. Krafft, pp. 43–44. U.S. Geological Survey Circular 949.

Poole, F. G., and R. J. Madrid. 1986. Paleozoic rocks in Sonora (Mexico) and their relation to the southwestern continental margin of North America. *Geol. Soc. America Abstracts with Programs* **18**(6):720–721.

Poole, F. G., B. L. Murchey, and J. H. Stewart. 1983. Bedded barite deposits of middle and late Paleozoic age in central Sonora, Mexico. *Geol. Soc. America Abstracts with Programs* **15**(5):299.

Price, R. A., and R. D. Hatcher, Jr. 1983. Tectonic significance of similarities in the evolution of the Alabama-Pennsylvania Appalachians and the Alberta-British Columbia Canadian Cordillera. In *Contributions to the Tectonics and Geophysics of Mountain Chains,* ed. R. D. Hatcher, pp. 149–160. Geological Society of America Memoir 158.

Reade, E. H., W. R. Power, M. J. Keeners, and D. H. White. 1980. Barite deposits in the Cartersville district, Bartow County, Georgia. In *Excursions in Southeastern Geology,* vol. 2, eds. R. W. Frey and T. L. Neathery, pp. 379–384. American Geological Institute Guidebook for Geological Society of America 1980 Annual Meeting.

Roedder, E. 1967. Environment of deposition of stratiform (Mississippi Valley-type) ore deposits, from studies of fluid inclusions. *Econ. Geology Monograph 3,* pp. 349–362.

Roedder, E. 1976, Fluid-inclusion evidence on the genesis of ores in sedimentary and volcanic rocks. In *Handbook of Stratabound and Stratiform Ore Deposits,* ed. K. N. Wolf, pp. 67–110. Amsterdam: Elsevier.

Rye, R. O., D. R. Shawe, and F. G. Poole. 1978. Stable isotope studies of bedded barite at East Northumberland Canyon in Toquima Range, central Nevada. *U.S. Geol. Survey Jour. Res.* **6**(2):221–229.

Schmidt, J. M., and R. A. Zierenberg. 1987. The Red Dog Pb-Zn-Ag deposit, Alaska—an example of nonexhalative processes in the formation of syngenetic massive sulfides. In *USGS Research on Mineral Resources—1987 Program and Abstracts,* ed. J. S. Sachs, p. 61. U.S. Geological Survey Circular 995.

Shanks, W. C., III, L. G. Woodruff, G. A. Jilson, D. S. Jennings, J. S. Modene, and B. O. Ryan. 1987. Sulfur and lead isotope studies of stratiform Zn-Pb-Ag deposits, Anvil Range, Yukon—basinal brine exhalation and anoxic bottom water mixing. *Econ. Geology,* **82**:600–634.

Stephens, M. B., compiler. 1986. Stratabound sulphide deposits in the Appalachian-Caledonian orogen. *Geological Survey of Canada Map 1649A,* scale 1:5,000,000.

Swanson, E. A., D. F. Strong, and J. G. Thurlow, eds. 1981. The Buchans orebodies—Fifty years of geology and mining, *Geological Association of Canada Special Paper 22,* pt. 1, 350p.

Tempelman-Kluit, D. J. 1979. Transported ophiolite, cataclasite, and granodiorite in Yukon—evidence of arc-continent collision. *Geological Survey of Canada Paper 79-14,* 27p.

Thurlow, J. G., E. A. Swanson, and D. F. Strong. 1975. Geology and lithogeochemistry of the Buchans polymetallic sulfide deposits, Newfoundland. *Econ. Geology* **70**:130–144.

7

BARITE DEPOSITS OF ARKANSAS

R. A. Zimmermann and G. C. Amstutz

Problems of ore genesis are much concerned with age relations of ore and enclosing rocks. For the stratabound barite deposits of Late Mississippian age of the Ouachita Mountains of Arkansas (Fig. 7–1), the problem is one of age relations of textures and structures formed during diagenesis and disruption by later igneous intrusions. Workers proposing models with regard to the genesis of the Arkansas deposits, which are formed in the lowermost part of the Stanley Shale (Late Mississippian), include Parks and Branner (1932), Brobst (1958; 1963), Scull (1958), Zimmermann and Amstutz (1961, 1964*a,* 1964*b*), Hanor and Baria (1977), Zimmermann (1964, 1976), and Okita (1983, in Howard and Hanor, 1987). In this chapter, considerable attention is also given to features not, or only inadequately, covered in the literature for the Arkansas barite deposits, which particularly fix the age of formation for the barite. Such features include (1) morphology of nodules and lenses of barite during diagenesis and resulting from differential densities of barite and enclosing sediment, (2) fossils filled with and replaced by barite, (3) the presence of mudflow breccias and slump structures within both barite- and nonbarite-bearing strata, and (4) crosscutting relations as regards faults and igneous dikes as well as the textures of the intruded rocks resulting from thermal effects.

An overall picture of the stratigraphic situation and the stratigraphic position in which these features occur is given below.

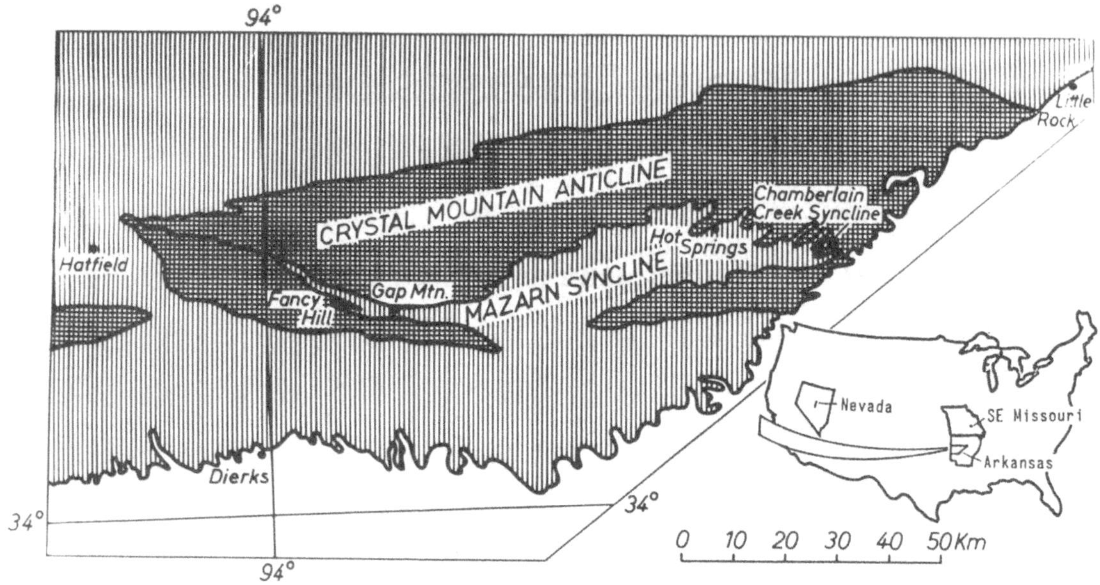

Figure 7–1. The generalized geologic map is of Western Arkansas with barite deposits discussed, including the Chamberlain Creek syncline, Gap Mountain, Fancy Hill, Hatfield, and Dierks deposits. Crosshatched area: Cambrian, Ordovician, Silurian, and Devonian; vertically lined: Carboniferous; blank: Cretaceous and Tertiary. Location map of the United States with barite districts of Nevada, southeast Missouri, and Arkansas.

OUACHITA STRATIGRAPHY, OROGENY, AND RELATIONSHIP TO BARITE

The Paleozoic section from top to bottom is given in Table 7–1, according to Miser and Purdue (1929), Keroher et al. (1966), and other workers.

According to Hass (1950), the Upper Arkansas Novaculite is Devonian–Mississippian in age and the Hatton Tuff and barite is Late Mississippian (Meramecian); according to Frezon and Dixon (1975, p. 181), the Hatton Tuff Lentil in Oklahoma is Early Pennsylvanian in age but, according to Glick (1979), interval C of Mississippian age in Arkansas; the tuffite beds in the Chamberlain Creek syncline (Fig. 7–2) are probably equivalent to the Hatton Tuff (Zimmermann, 1964).

The problem of the genesis, and therefore the problem of the environment of deposition of the stratabound barite deposits in the lowermost part of the Stanley Shale, is made more complex because the bedded deposits occupy a junctional position in the stratigraphy between two (or three, in the case that the Arkansas Novaculite of Devonian–Mississippian age be included as a major subdivision [see above]) major subdivisions of the Paleozoic section. Thus, where major tectonic changes took place: (1) a relatively thin (about 1700 m)

Table 7-1. Paleozoic Sedimentary Rocks in the Barite District of Central and Western Arkansas

Post Arkansas Novaculite (Mississippian to Pennsylvanian) (up to 10,000 m of section) from top to bottom:
Atoka Formation, (sandstone), Middle Pennsylvanian (1860 m)
Jackfork Sandstone, Late Mississippian(?)-Early Pennsylvanian (1860 m)
Stanley Shale, Late Mississippian (1860 m) (Hatton Tuff lies 100 m above the base in the Chamberlain Creek syncline); Hatton Tuff (0-25 m) dated as Late Mississippian or Early Pennsylvanian (see below); barite near base of formation also dated Late Pennsylvanian by Hass (1950)
Pre-Stanley Shale (Cambrian to Mississippian) (about 1700 m) from top to bottom:
Arkansas Novaculite (Caballos Novaculite in Texas) (100-300 m), Middle Devonian-Mississippian
Missouri Mountain Slate, Silurian (15 + 90 m)
Blaylock Sandstone, Silurian (0-465 m)
Polk Creek Shale, Upper Ordovician (0-20 m)
Big Fork Chert, Middle Ordovician (220 m)
Womble Shale, Lower and Middle Ordovician (310 m)
Blakely Sandstone, Lower Ordovician (0-120 m)
Mazarn Shale, Lower Ordovician (310 m)
Crystal Mountain Sandstone, Ordovician(?) (260 m)
Collier Shale, Cambrian (60+ m)

Lower to Middle Paleozoic section made up of partly alternating sandstone, shale, and chert, and partly deposited in a shallow marine environment (Davies and Williamson, 1976) with a rather thick 300 m section of novaculite or metamorphosed chert (Arkansas Novaculite) deposited over a rather great expanse of time in a basin starved of clastics, with depositional environment having ranged from a supratidal and shallow (platform?) in the west in Oklahoma to open marine in the east (Jones and Knauth, 1979) in the upper part of the Lower Paleozoic; and (2) a relatively thick Upper Paleozoic section of turbidite trough deposits of sandstone and shale (up to 10,000 m), beginning first with the Stanley Shale (Late Mississippian) (2000 m thick) followed by Jackfork Sandstone and the Atoka formation.

The orogeny of the Ouachita area began during the Pennsylvanian (Desmoinesian time). Evidence indicates that the source of the thick accumulation of clay and sand turbidites in the barite areas was to the north and east during interval C of the Mississippian (Glick, 1979). A tectonic-sedimentation model involving the part of the old predrift continent (or submarine rise) of Pangaea on the south and east, and, to the north, the Ouachita trough, and midwestern North American continent, is used by workers to explain the compression and folding of the Paleozoic section (investigations here gained considerable momentum during the 1960s and 1970s; Morris, in Wickham et al., 1976, p. 174; Graham et al., 1976; Keller and Cebull, 1973; Lowe, 1975). According to Hanford (1979), a major delta system located along the collision suture zone provided much of the sediment comprising the Stanley group flysch via submarine fans. A model for the Stanley fan system is given by Niem (1976).

The barite beds lie in the lower part of the Stanley Shale, and the formation

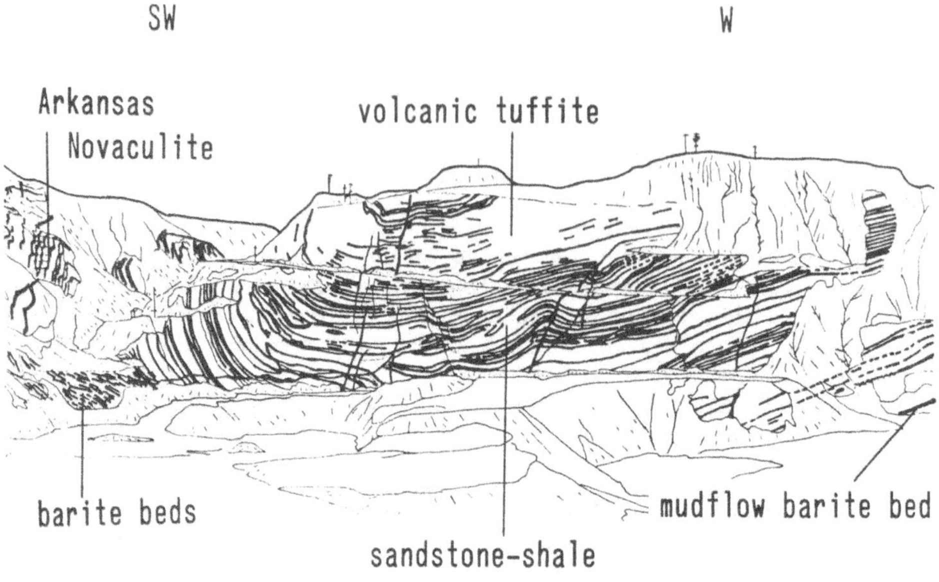

Figure 7-2*a*. Cross section of the Chamberlain Creek syncline. The bedded barite deposits of the Stanley Shale (Missisippian) lie above the Arkansas Novaculite (Devonian–Mississippian) and below the sandstone-shale section of the Stanley Shale. Cross cutting features are alkalic lamprophyric dikes ranging to carbonatite dikes of Late Cretaceous age. About 150 m of stratigraphic section are shown from the base of the barite beds to the uppermost sandstone-shale beds above the volcanic tuffite near the top center of the section.

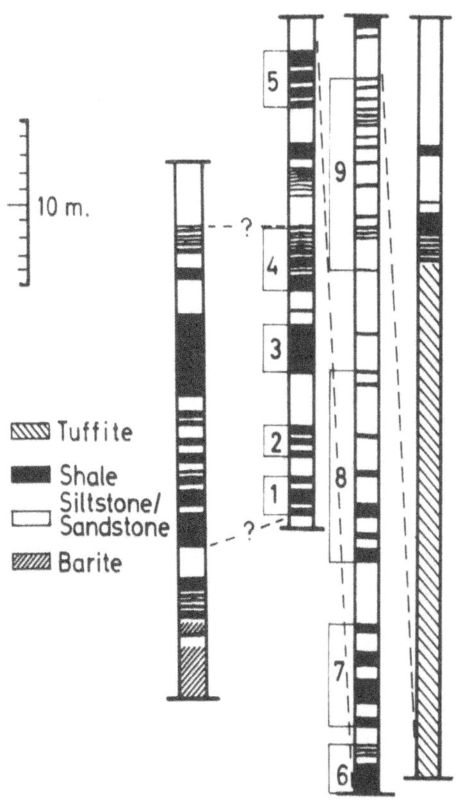

Figure 7-2*b*. Columnar section of the sandstone-shale-tuffite lithology above the barite beds in the Chamberlain Creek syncline. Numbers designate sandstone-shale groups: increase in the amount of sandstone units upward within each group corresponds with thinner intermediate shale units in each group. Shale deposition is, thus, interrupted by influx of sand, each unit of which is in sharp contact with the underlying shale layer but grades upward to the overlying shale bed.

in the Chamberlain Creek syncline (Figs. 7-1 and 7-2) contains not only shale but also slump structures, thin beds of limestone, a novaculite-sandstone-shale-barite breccia bed, rhythmic sandstone-shale turbidites, and volcanic tuffite as well (Zimmermann, 1964). Evidence from the stratigraphy, including observations of algal-like layering and massive barite beds with mudflow breccia from the Fancy Hill deposits, provide criteria for evaluating the origin and genesis of the barite. Beds immediately below the base of the barite section (Chamberlain Creek syncline) include fossil mud volcano chert concretions (see below), depositional facies indicators that do not speak for deep marine conditions in this part of the section.

SYNGENETIC FORMATION AND SEDIMENTARY FEATURES

Syngenetic formation of the barite is indicated by the following features (see Zimmermann and Amstutz, 1961, 1964*a*, 1964*b*); Zimmermann (1964, 1969, 1976, 1982): (1) stratiform-banded nature of the bedded barite, (2) diagenetic compaction features, (3) rhythmicity, (4) slumping and mudflow or turbidite structures, (5) crosscutting features by igneous intrusions, (6) worldwide occurrence of barite in sedimentary rocks of Devonian-Mississippian age, (7) lack of symmetrical distribution in the shales with igneous intrusions, (8) parallelism of the Ouachita deposits with the margin of the Ouachita trough, and (9) low Sr concentrations and Ba/Sr relationships. Perhaps the strongest evidence for a syngenetic origin are disruption textures shortly after sedimentation, since barite of mudflow or turbidite origin, slumps and folds; during diagenesis, textures such as compaction features and rhythms due to initial sedimentation and crystallization differentiation; and, cross cutting and breakage textures during igneous dike intrusions. Disruption textures are so convincing because, of course, something had to be present before a disrupted fragmented texture could form.

The discovery and eventual exploitation and depletion of the chief producing Arkansas barite deposit, the Chamberlain Creek syncline deposit (before the most recent opening up of the Fancy Hill district) (Figs. 7-1 and 7-2), took place over a relatively short time in recent years; Miser and Purdue (1929) may have already known of the Chamberlain Creek syncline deposit near Magnet Cove in Hot Springs County as it was discovered early in the century, but they do not indicate the possibility of barite at the same stratigraphic horizon in their contribution to the geology of the De Queen and Caddo Gap quadrangles (Miser and Purdue, 1929) where many miles of the Arkansas Novaculite-Stanley Shale contact is exposed and where large barite deposits occur at Fancy Hill. Parks and Branner (1932) present a magnificent photograph of the layered barite in the Chamberlain Creek syncline and indicate the possibility of a sedimentary origin. Next, McElwaine (1946*a*, 1946*b*) and Jones (1948) explored the deposits, and Brobst (1958) surveyed the deposits of the United States. Zimmermann and Amstutz (1961, 1964*a*) were the first to give detailed descriptions of the textures and structures for their genetic value; later Zimmermann (1964, 1967, 1969, 1976, 1982), Zimmermann and Amstutz (1983), Hanor and Baria (1977), and Brobst (1973) gave additional evidence pointing to the sedimentary and diagenetic origin of the barite deposits.

Concurrent with the exploration and exploitation of the Arkansas Ouachita Mountain barite deposits, mining in Missouri of Mississippi Valley strata-bound and, to some extent, stratiform karst barite deposits in dolomite was well under way (see Ligasacchi and Amstutz, 1960; Zimmermann and Amstutz, 1983). Barite in karsts there assumed a more coarsely crystalline habit and was associated with coarse-grained galena and sphalerite; it had only to be recovered by gravity separation from chert and produced a white, nearly pure concentrate, useful not only for oil exploration but also for chemical purposes. With more intensive petroleum exploration, improved shipping methods, and advances in flotation technology, the production from the Missouri deposits did not meet the demand, and the regional extent of the large Arkansas and Nevada districts with bedded types of fine-grained gray and commonly less pure barite was realized through exploration. These large bedded deposits resemble the barite-bearing, only slightly metamorphosed massive sulfide–barite deposits of Meggen and Rammelsberg, Germany (Zimmermann and Amstutz, 1964*c*; Amstutz et al., 1971; Zimmermann, 1970). Textures and structures at the mines of Meggen and Rammelsberg compare with those found in the huge barite deposits of Arkansas and Nevada and prove still further the syndiagenetic formation of these deposits. In the Nevada deposits of the Shoshone Range, various rhythmic patterns occur such as repeated upward thickening of the barite layers (Zimmermann, 1969).

NATURE OF THE BEDDED BARITE

RHYTHMIC BANDS

The basic pattern seen in the banded barite is one of interbedded siliceous shale and barite (Fig. 7–3) with other superimposed patterns. The lower contact of the shale is sharp and often humpy; the upper contact grades to barite, which, in turn, grades to calcite followed in sharp contact with the next overlying shale layer. As this sequence is unfailingly repeated, one may speak of a rhythm or rhythmic banding. The contacts may be differentiated in the hand specimen, but are much more easily differentiated with the aid of a hand lens. The banding is probably the basic feature regarded by those researchers who interpret the barite as having formed syndiagenetically. Scull (1958) proposed a genetic interpretation, but his criteria for replacement are either unclear or, in our opinion, do not support replacement. Present work confirms that the geochemical and physical adjustments that took place in the sediments are diagenetic changes; in part, recrystallization and diagenetic segregation and differentiation as described by Pettijohn (1975) for carbonates, Zimmermann (1976) for barite, Amstutz and Park (1971) for the fluorspar deposits of Illinois, and Fontbote and Amstutz (1983) for numerous deposits in Europe and South America. Further in the rhythmic bands of Arkansas features observed under the microscope in thin section such as fine, humpy features along the sharp contact of the shale with the underlying barite, as well as the presence of organic contents in the shale bands, lead one to consider the initial or primary presence of algal layers and barite-bearing sediments later subject to diagenetic crystallization.

Figure 7–3. Rhythmic sequence of shale, barite, and carbonate forming the banded type of barite in the Chamberlain Creek syncline. Each banded pair begins with a siliceous shale layer at the base, grading upward to barite and barite with carbonate. Somewhat similar features observed to the west in the Fancy Hill deposits, where, however, carbonate is absent, point to a shallow water algal original.

On the north slope of the Chamberlain Creek syncline especially, the rhythmic type of bedding is well exposed and persists throughout the lower part of the section, with the baritic layers generally thicker than the thin dense shale layers. The barite layers average about 5 mm or so in thickness. The shale and barite layers have undergone differential weathering on exposure at the surface, and the shale, being less resistant to weathering, forms the grooves, whereas the barite forms alternating ridges. Microscopic observation of the section, illustrated in Figure 7–3, shows the baritic layers to range from 3 mm to 6 mm in thickness, with alternating shale layers approximately 0.5 mm thick. Thin veinlets of calcite occur in places. The upper portions of some clay layers contain radiating masses (0.05 mm in diameter) of barite. These small barite masses are surrounded by parts of the uppermost shale layers. A few pyrite crystals are scattered along the lower parts of the shale layers. The barite layers consist essentially of fine-grained crystalline barite and calcite. Along the lower parts of the layers are scattered idiomorphic barite crystals (0.05 mm long) within a matrix of finer-grained (0.02 mm to 0.03 mm) barite. Near the center of the layer, a rapid accumulation of calcite begins, with the calcite occupying the space between the barite crystals. Barite crystals range up to 0.1 mm in places and the calcite crystals range up to 0.15 mm. These upper calcareous layers are in sharp contact with the succeeding clay layers. The nature of these textures tends to indicate that the barite originated first and the calcite second.

Some of the calcareous barite layers have minor subrhythmic bands within them, with thin local accumulations of calcite in their upper parts below thin bands of shale suggestive of diagenetic crystallization segregation. The particular specimen examined gave a specific gravity reading of 3.8.

Another specimen from the north flank of the Chamberlain Creek syncline at about the same stratigraphic horizon showed the thin clay partings of a similar type of rhythmite to become stylolitic in places. The stylolitic seam consists of granular calcite and a very fine-grained material, the composition of which is uncertain.

Figure 7–4 is a 1:1 photograph of a thin section from beds showing alternating barite and shale layers. Microscopic examination has shown the barite layers to consist of intergrown crystals of granular barite and quartz. The grains average 10μm in diameter. Barite is by far more abundant. The quartz crystals are usually only partly developed, with the larger crystals occurring along the shale layers. In some cases, the quartz crystals are themselves intergrown. Some crystals are truncated by the shale layers. Complete crystals with idiomorphic faces on all sides are rare, although some occur within small partially developed barite ray patterns, forming a plumose texture. The texture and mineralogy is similar to numerous other occurrences of layered and lenticular barite and quartz in the Chamberlain Creek syncline.

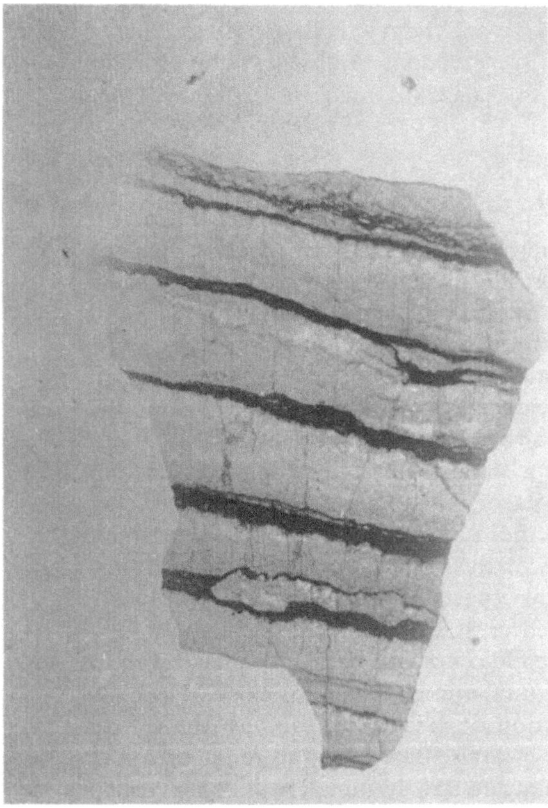

Figure 7–4. Rhythmically layered barite (white) and shale from the Chamberlain Creek syncline. Also shown here is the frequently observed top–bottom feature in beds of interlayered barite and shale, consisting of humpy, or bubbly, upper surfaces of the barite layers and often the gradual transition of shale into barite at the lower part of the barite layers. Compare with Figure 7–3.

Rhythmic layering is known from other barite occurrences; in the Nevada barite belt, a second type of rhythm consisting of a repetition of upwardly thickening barite is superimposed on the primary type of alternating barite and clay.

RHYTHMIC NODULAR AND LENTICULAR BARITE

Along the marginal portion of the Chamberlain Creek deposit layers of dense bedded barite usually grade downward into beds with abundant barite nodules, with the larger nodules, usually in the lower part of the bed, reflecting the high sedimentary accumulation of barite, or finely granular barium sulfate as it is known to occur in present-day oceans (Dehairs et al., 1980), coupled with sedimentary and/or later diagenetic growth of the nodules. Beds with large nodules are most often underlain by a relatively thick bed of siltstone or claystone. The middle of Figure 7–5 illustrates a layer of dense barite underlain by a nodular barite layer. These beds are located on the steeply dipping south flank near the nose of the syncline.

The most prevalent type of rhythmic layering is where an interlayering occurs between beds of nodular or lenticular barite and silt or shale beds. Figure 7–6*a* shows the rhythmic nature of this type of layering. Going from the scale

Figure 7–5. Layers of barite (high relief) and clay. Top of beds is to the left. Clay layer (right of center below hammer) grades upward to barite nodules and dense intergrowth of nodular barite, with top of barite bed forming nearly flat but slightly nodular surface. Weathered pit exposure, nose of Chamberlain Creek syncline.

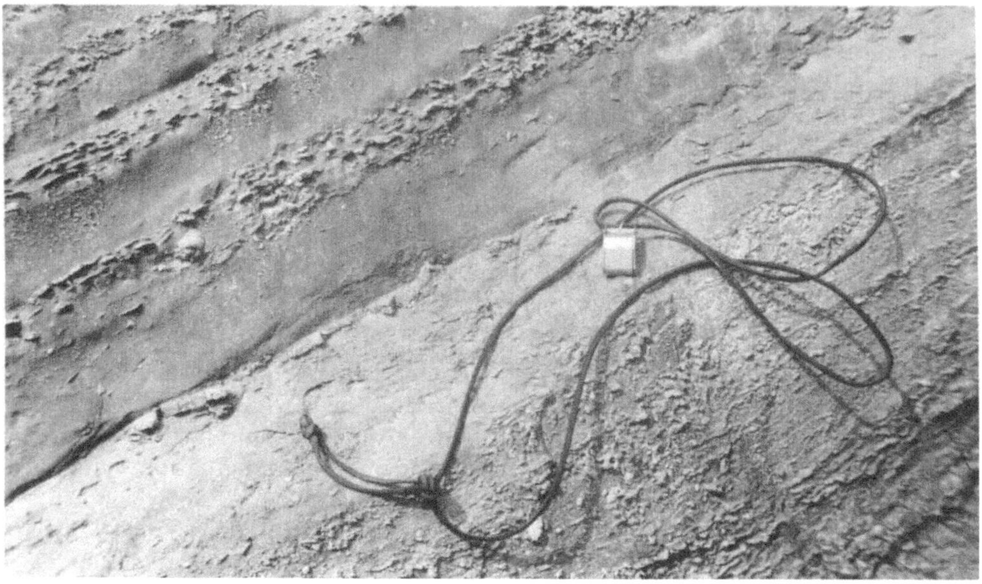

Figure 7-6*a*. Progressive upward thinning of paired clay-barite layers. Note the constant thickness ratio of clay to succeeding barite layer from pair to pair. Visible are nodular, lenticular, and layered forms of barite.

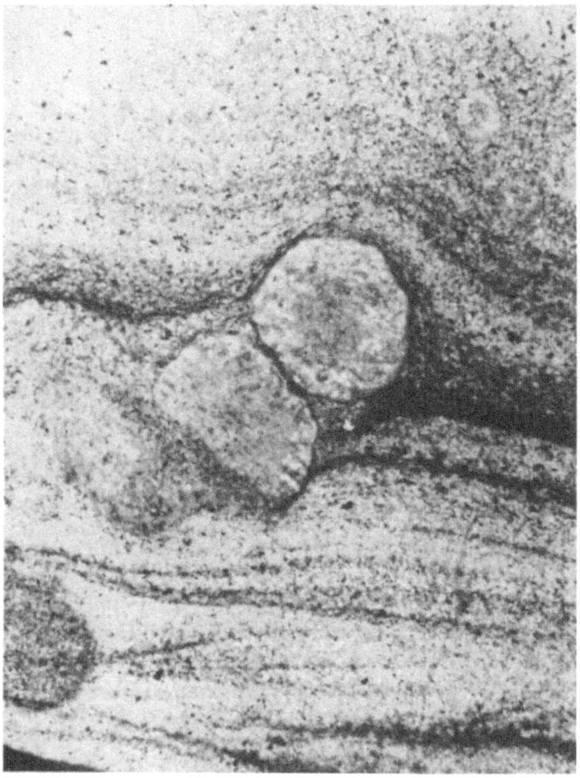

Figure 7-6*b*. Barite nodules in silty shale. Wedges of dark sediment along wide upper nodule as well as downward bending of underlying layers indicate that the nodules are syndiagenetic in age. Breakage of the nodule in center of figure likewise took place diagenetically as the dark layers above the left broken half form a hump due to compaction. The two nodules in the lower left half of the figure crosscut cross beds and fill pore spaces and formed during post-compaction diagenesis. Chamberlain Creek syncline.

of a bed to the scale of a nodule, one observes that the same relationship between barite nodules and shale, that of curved shale layers bending around nodules, is present as with the bending of shale layers with barite layers (Figs. 7-3 and 7-6*b*). Note the unmistakable cross-bedding in the lower half of the thin section, revealing, with certainty, ocean depths where the sediment was still affected by currents and perhaps wave energy (Fig. 7-6*b*).

In one unweathered outcrop deeper in the syncline where the rhythmicity has been interrupted by hard siltstone beds, which may vary from 6 cm to 10 cm thick, sharp contacts with the barite beds are seen. Barite nodules occur in places and are oriented parallel to the bedding. The nodules range from minute spheres up to 1 cm in size. The average size in the hand specimen is about 2 mm or 3 mm in diameter. Similar sizes and shapes occur in the same layer. Shapes and sizes will vary from layer to layer, but the changes in shape and size of the nodules is always transitional from layer to layer. The larger nodules occur in wider layers than do the smaller nodules. Nodules measuring in the 2 mm range may occur in a section 1 cm wide. Nodules measuring about 0.25 mm occur in a section about 3 mm wide.

A thin section along the contact between one of the layers of baritic siltstone of specific gravity 2.85 and an overlying barite bed of specific gravity 3.82 revealed the barite siltstone to consist of alternating thick light layers and thinner dark layers, with the light layers ranging up to 1 mm thick and the darker layers about half as thick. Both layers contain angular detrital quartz grains, mica, angular plagioclase feldspar fragments, barite, and disseminated calcite and pyrite grains. Calcite is more abundant in the light layers. The barite occurs generally as lenticular to spherical masses and ranges between 0.02 mm and 0.07 mm in diameter. Most of the masses consist of granular barite, and only a few of the largest masses consist of nodules of radiating barite. Most of the barite masses and nodules have fine-grained interiors. Many show calcite intermixed with barite in the interiors. The smaller masses occur in the dark layers associated with the fine-grained pyrite (pyrite grains range from 3 μm to 4 μm in diameter). The larger barite masses occur between the dark and light layers or within both layers near their contact. The case varies from layer to layer and laterally along a given layer. The inner parts of the lighter layers consist of the small 3–4 μm pyrite grains and larger pyrite grains of 30–40 μm. The quartz grains average about 20–30 μm in diameter. Within the darker layers, the small pyrite grains may be so closely spaced locally along a very thin discontinuous clay or shaly horizon as to form a lens of pyrite. Such a layer will often curve gently above or below barite masses that occur at the same horizon. Pyrite grains often occur in or with the barite and may occupy the central parts of the granular masses, or may form a concentric layer at some distance from the center of the mass, or may be randomly distributed.

The contact between the baritic siltstone and overlying barite begins at the top of the uppermost light silty layer of the baritic siltstone. A thin layer of fine-grained pyrite occurs at the top of the light, baritic siltstone layer in places. Above this is a thick layer of dense fine-grained barite that ranges up to 2 mm in thickness (see Fig. 7-7). The lower surface of this layer is nearly flat, whereas the upper surface is very uneven. This layer is thinnest where radiating barite nodules occur just above the contact of the underlying baritic siltstone (see Fig. 7-7). The nature of the upper surface of this barite layer shows that the presence of barite nodules close to the contact affects the thick-

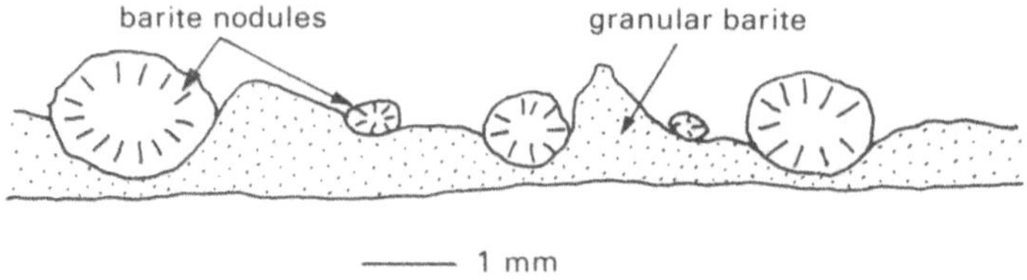

Figure 7–7. Layer of granular barite (dotted) depressed in places by loading of barite nodules, and underlain by rhythmic laminations of (a) pyrite, calcite, and minute barite lenses in siltstone, and (b) carbonate-bearing siltstone.

ness of the granular barite layer. This thinning in places beneath the barite nodules may perhaps be due to sedimentary or diagenetic settling and compaction of hard nodular barite onto a soft mud or jell-like layer of colloidal barite. These relationships suggest that the fabric is due to gravity, and the features produced may, therefore, be considered to be geopetal features. Nodules are abundantly distributed throughout the upper barite bed and range up to 3.5 mm in diameter. Most nodules consist of barite radiating entirely through the nodules, with some disseminated carbonate, pyrite, and detrital quartz grains. One nodule, measuring 3 mm in diameter, consists of about 50 percent detrital quartz and 50 percent barite, without showing any sign of radiating structure. Throughout the upper barite bed layering is discontinuous and only locally present. Besides the barite nodules, other material in this bed consists of locally accumulated, discontinuous layered masses of fine-grained barite similar to the layer present at the contact. Also, lenses and discontinuous layered masses of detrital quartz with minor barite occur in places. In general, flat, horizontal layering does not prevail as in the underlying siltstone. Layering is crudely developed, as either thin, local, discontinuous, generally inclined curved planes, or on a large scale as a crude layering of a particular type of material, such as barite nodules, siltstone, or fine-grained barite. The crude development of the layered fabric, together with the high abundance of barite relative to the underlying siltstone, would tend to indicate a change in normal detrital sedimentation conditions that existed while the underlying siltstone bed was being deposited. The barite may have been precipitated very rapidly in contrast to the normal sedimentation of detrital fragments. In the description of the breccia bed (see above), which occurs higher in the stratigraphic section, it is observed that the same fabric prevails, together with large "floating," that is, disconnected, fragments of novaculite. In other areas where the novaculite breccia bed is present outside the barite horizons, as at Gap Mountain (Fig. 7–8), it is observed that the fabric consists of closely spaced novaculite and other detrital fragments.

PRIMARY, IRREGULARLY BANDED, AND LENTICULAR BARITE (ALGAL TYPE LAYERING) OF FANCY HILL

Of the three main ore types at Fancy Hill (Figs. 7–9*a* and 7–9*b*) including massive barite, beds with mostly barite nodules, or nodules and lenses, and irregularly banded lenticular and layered barite, the last mentioned is most

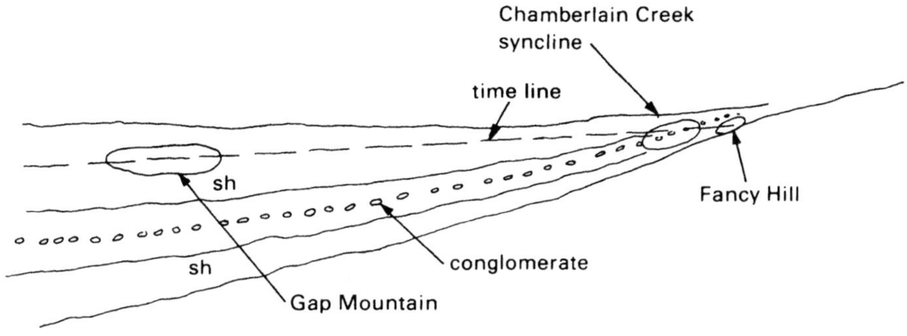

Figure 7-8. Stratigraphic position of the conglomerate and barite beds of the Chamberlain Creek syncline, Gap Mountain, and Fancy Hill deposits (after Zimmermann, 1965). The Gap Mountain barite lacks the primary rhythmically banded barite of the not too distant deposits of Fancy Hill, but, instead, forms dense, blocky beds of fine-grained barite, possibly of reworked origin and deposited under deeper marine conditions.

provocative as to the origin of the layered, lenticular, and nodularly formed bodies; for a layered or lenticular form must have arisen as some body on the ocean floor, assuming a definite shape and size. Now, the irregular layers of barite at Fancy Hill all have the same or similar thickness, as though the same origin were involved in each layer or lens across the barite section. Some features, given below, that appear to have formed so early as to have been a kind of fossil material on the ocean floor, such as algae, deserve explanation. The only other feasible explanation would be a primary precipitate (if such precipitates can first be proven to actually form on the ocean bottom).

The lenses and layers, with the exception of shape, display close similarities in composition, zoning, thickness, and tension breaks. These lenses and layers are separated by equally thick layers (or matrix) of black fine-grained quartz. Observed under the hand lens, tension breaks are seen to have formed very early and point therefore to a primary origin of the layered barite (see Fig. 7-10). These particular breaks did not form by later tectonic activity; breaks of such relatively early age require the existence of a relatively hard to brittle fossil material, characteristics which only algae fossil matter can provide. From existing morphological features present, it is concluded that algae should be considered as being the original bodies later diagenetically filled with and replaced by barite. (Primary features preserved after diagenetic barite replacement have been documented [Martin et al., 1984] even over the scale of an entire stromatolite belt in the Granada Basin of southeastern Spain). In the Fancy Hill barite, the largest lenses display breaks of the earliest origin involving only one layer and therefore must have involved forces acting only on one layer occurring at the sediment-water interface, or close to this interface where an individual layer could be disturbed and opened up over a very thin space of one layer only.

Occurring in close association are the irregular nodular forms several times the thickness of the layers. Later tectonic squeezing again appears to have played a minor role, if any at all, to form these masses and many examples do display that curling (with often incomplete closure) probably was in play

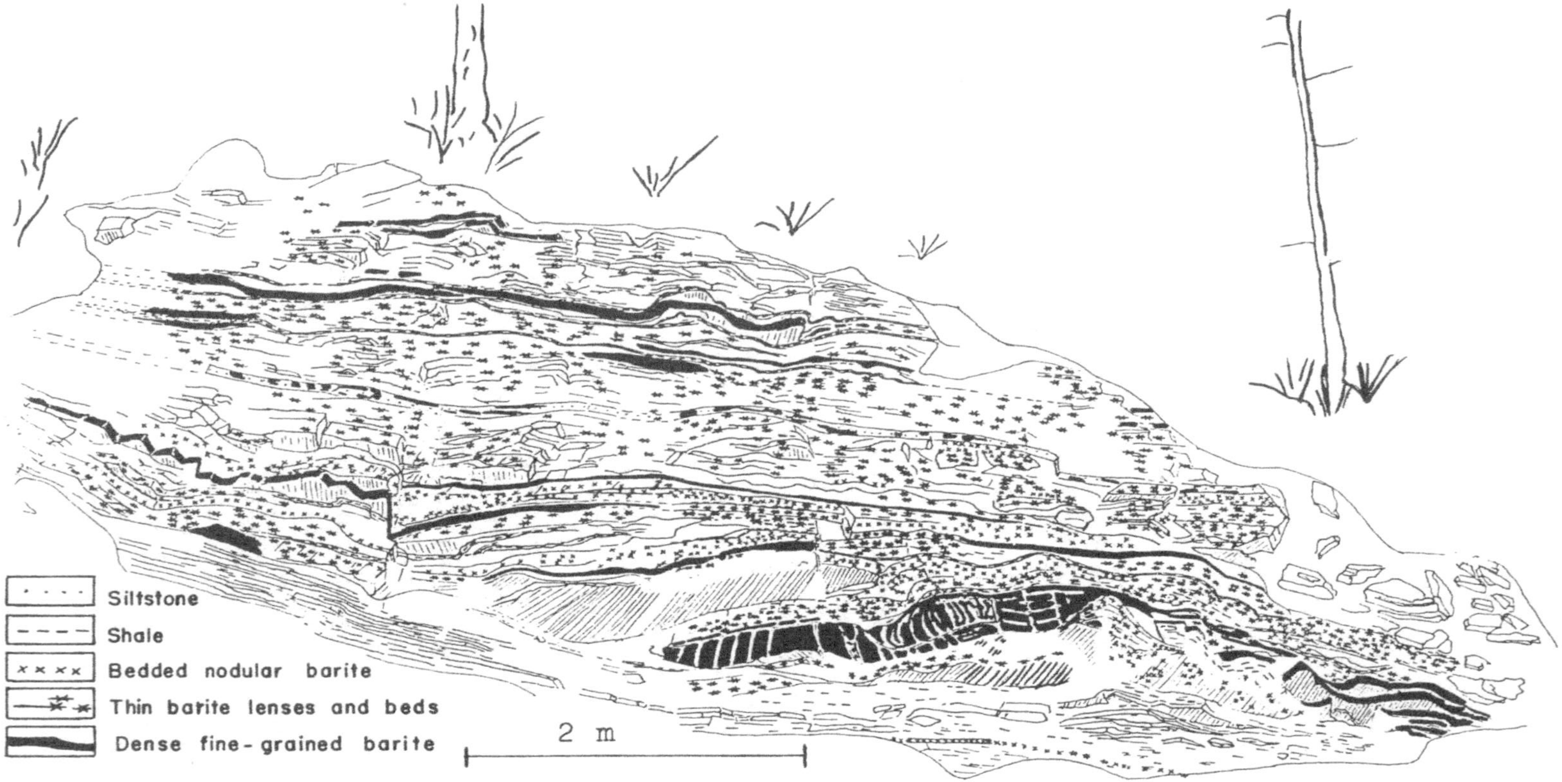

Figure 7–9*a*. Overturned barite section (upper part) in lower part of the Stanley Shale (Mississippian) at Fancy Hill displaying three main barite types: (a) dense, fine-grained barite, (b) nodular barite in shale, and (c) thin barite lenses and layers in chert matrix forming hard, blocky barite of algal(?) origin. Not shown are 3 m of interbedded shale and stratabound barite nodules forming top of section (bottom of sketch). Section measured is from exploratory pit. Henderson deposits (1961). Vegetation shown in background. See Figure 7–9*b* for view in field.

Figure 7-9*b*. Fancy Hill barite beds, overturned and dipping away from viewer so that top of section lies at bottom of figure. Bed below hammer (for scale, center of figure) is thick dense, fine-grained barite bed in lower right half of the section drawn in Figure 7-9*a*.

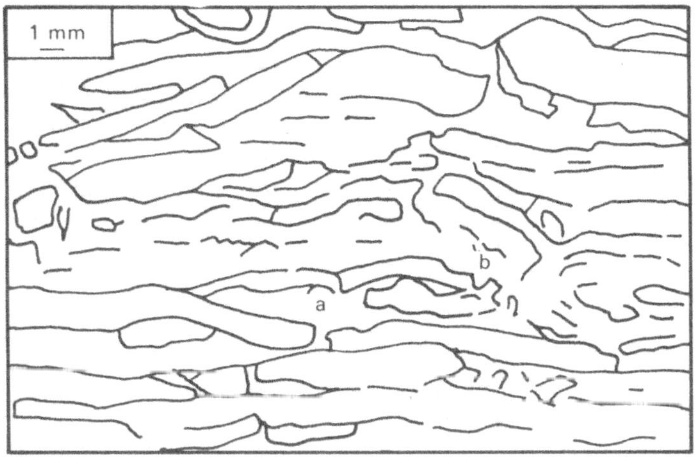

Figure 7-10*a*. Layers of the algal-like type barite, with local breaks in the layers (a) filled now with black chert (see Fig. 7-10*b*), and V-shaped cracks in curled layers (b). Fancy Hill, Arkansas.

Figure 7-10*b*. Fancy Hill layered, lenticular, and nodular ore. The barite-chert section, as illustrated in Figure 7-10*a*, is located here in the center of the upper-right quarter of the photograph.

to develop the irregular nodular form (some small well-rounded nodules in cross section are actually tubular in the third dimension and over five times the diameter of the nodules in length).

Many layers also show 180° bends upon themselves, such as can be commonly observed as having formed early in algal mats.

The breaks (Fig. 7-10) that extend neither above nor below the affected layer are filled by local black quartz matrix squeezed into the opening. The process causing the breakage affects only one layer where conditions below the broken layer remain unaffected by the breakage process because of plasticity of the intermediate black siliceous matrix sediment. Layers above the broken layer are unaffected because no upper layers had yet been deposited; the only reasonable process must be one taking place in soft unconsolidated sediment at the interface of the algae layer and the upper sea water. Abrupt physical changes in conditions met with here occurred in moist or water-soaked sediment undergoing subaqueous or subaerial desiccation.

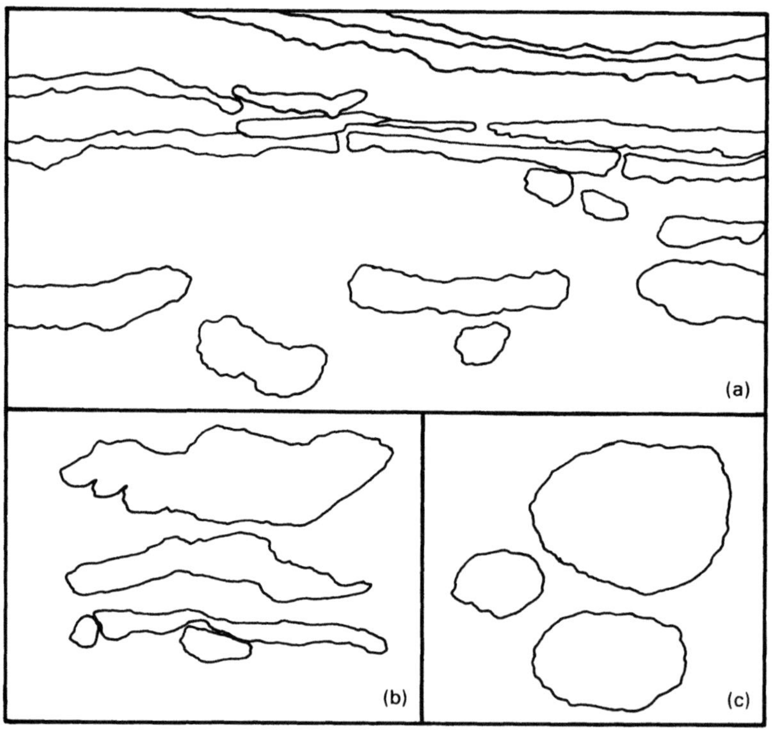

Figure 7-11. Forms of layers (a) from Fancy Hill, lenses (b), and nodules (c) from the Chamberlain Creek syncline deposit. The lower parts of nodules, layers, and lenses commonly show a downward sag. The lateral margins of lenses, layers, and nodules are often steep and show a sharp juncture with the upper surface. It is suggested that the shapes are the result of settling and compaction. The thick layers in the lower half of (a) are about 0.5 mm thick; lengths of lenses in (b), top (6 mm), middle (1 cm), bottom (1.5 cm); large nodule at top of (c) is 1 cm in diameter.

FORM OF NODULES, LAYERS, AND LENSES

In the foregoing materials, asymmetric shapes have been observed at the terminations of nodules, lenses, and layers of barite. A greater degree of rounding at the ends and lower surfaces of the barite bodies are geopetal features in the nodules, layers, and lenses. The tapered lower surfaces and nearly flat upper surfaces are the result of sagging during sedimentation and diagenesis. Several examples are given in Figure 7–11.

FEATURES OF DISRUPTION

DISRUPTION IN MASSIVE BARITE

Disruption took place in massive barite (Fig. 7–9) shortly after deposition so that the resulting features actually go unnoticed on the face of a broken or even cut surface. Upon wetting a smoothly cut surface of massive dense barite, fractures do appear that are bounded by fine breaks, which are filled with the same or nearly similar barite type (grain size, etc.) as the main rock itself. Coupled with these fractures are flowage features in the massive rock revealing mobilization of dense barite sometime during diagenesis but still shortly after deposition. Figure 7–12 shows the texture resulting from the end part of a partly consolidated barite layer (white) of one type of barite floating and slowly disintegrating within the darker, more fluid barite matrix, forming a flaring pattern. The darker barite contains somewhat more quartz, and other fragments of one type of barite floating in another are often rounded, indicating movement and flowage before consolidation, or with the material still in a plastic state (Figure 7–13).

SLUMP STRUCTURES IN BANDED BARITE

Slump structures are not uncommon in stratabound deposits—the nicest example we have seen occurs in the Sullivan mine of British Columbia. Geologists were so impressed by it that the section was not only left unmined, but a block was carefully cut out and put on exhibition worldwide. We photographed and sketched a slump structure of similar scale from the Chamberlain Creek syncline (Fig. 7–14). Of course slumps of smaller size are also common, and beautiful examples are pictured in Amstutz et al. (1964) from Mount Isa, Australia.

CROSSCUTTING RELATIONS WITH IGNEOUS INTRUSIONS

Igneous intrusions crosscut barite nodules and other materials of the mudflow breccia bed described above. Barite nodules are clearly broken at the

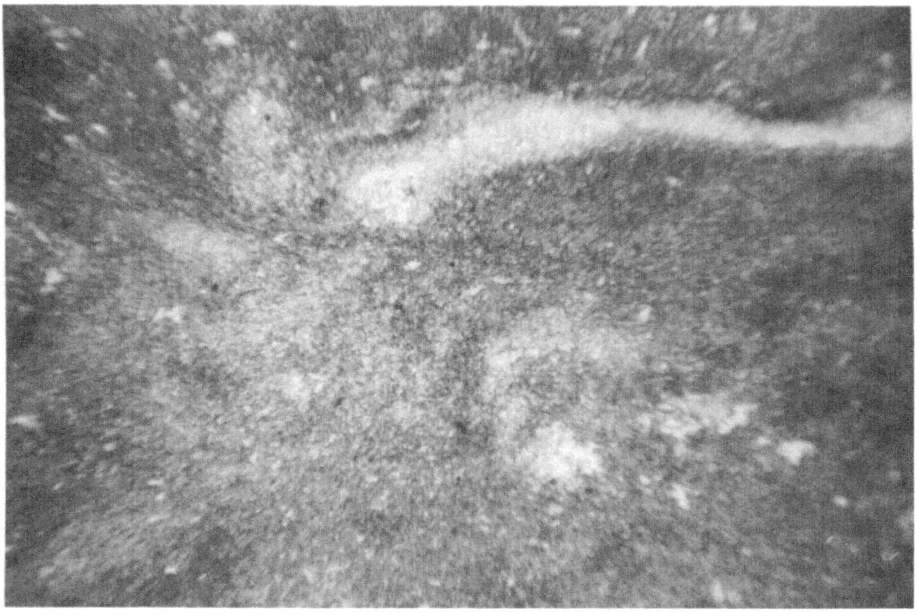

Figure 7-12. Internal features of disruption common in thick bed of dense, nearly pure barite consist here of a partly consolidated barite layer (white), the end of which displays a flaring pattern within a darker barite matrix and is the result of disintegration of the white layer in more fluid darker barite very early in diagenesis. Lower frame is 3.5 mm in length. Top-bottom as in nature. Fancy Hill deposit. See Figure 7-13 for overall rock texture.

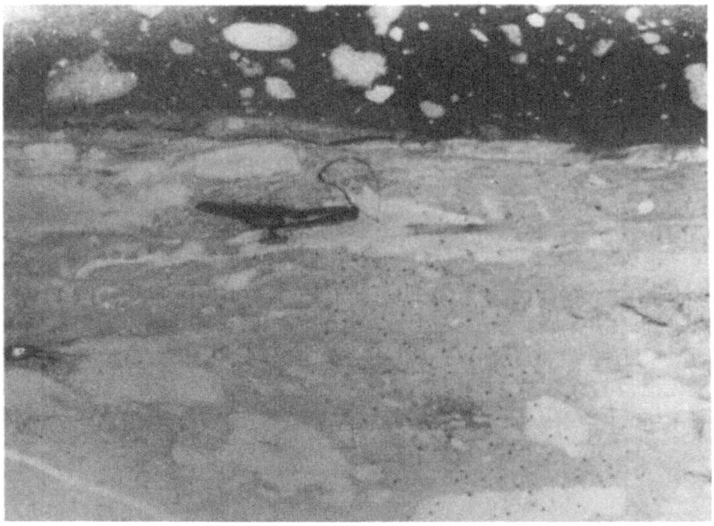

Figure 7-13. The flowage feature in Figure 7-12 is located left of center in the above picture. Abundant, minute, egg-shaped bodies, consisting of barite and quartz, probably radiolarians (not visible here), are located in top, right of center. A limonite stain formed by weathering occurs in the dark layer in the upper part of the figure (also top in nature).

Figure 7–14. Sedimentary slump structure in the form of a giant "S" and made up of interlayered barite and silty shale with minor folds of irregular distribution and size. Section is overturned with strike of upper and lower beds parallel to upper and lower frames of the figure, which parallel the axis of the syncline. South flank, Chamberlain Creek syncline (Fig. 7–2*a*).

contacts of such intrusions, leaving little doubt as to which came first. The lack of symmetrical distribution of barite around the intrusions is evidence that the barite and intrusions are unrelated.

BRECCIA

Conglomerate beds are known from the lower part of the Stanley Shale (Miser and Purdue, 1929). The same units designated as conglomerates by Miser and Purdue contain abundant sharply angular novaculite fragments in these beds in the barite sections of the various deposits (see Fig. 7–8) and are therefore referred to as breccia beds. In the Chamberlain Creek syncline deposit, one such breccia bed is over 1 m thick and lies in the upper part of the section. Another breccia bed was observed many meters up in the Stanley Shale above the tuffite section in the Chamberlain Creek syncline. In other barite deposits, such as the Gap Mountain deposit (Fig. 7–1), the breccia bed lies below the barite section (Fig. 7–8). The breccia bed in the Chamberlain Creek syncline deposit is without internal layering. Strewn together are, as already mentioned, novaculite fragments, fragments of sandstone, barite nodules (apparently not entirely lithified on arrival in the breccia bed as they are squeezed by compaction in places), shale fragments, pyrite fragments, and so forth (see Zimmermann, 1982). Obviously the material contained here was derived from already sedimented materials; still undigested diagenetically were the barite and accompanying sandstone and shale components. Already lithified was the chert (or novaculite), further evidence that the Arkansas Novaculite was at the emergent source undergoing erosion. Detrital fragments, mostly novaculite, were then released and transported as a mudflow or turbidite and disrupted locally still unlithified shale, sandstone, and barite of sedimentary barite deposits. As the barite components, especially barite nodules, are still in rather unbroken condition, the mudflow could not have moved a very great distance.

CHERT MUD VOLCANO CONCRETIONS OF THE STANLEY SHALE

The Arkansas Novaculite and Stanley Shale are given extra attention in this chapter because they locally make up the country rocks of the barite deposits (only the Stanley Shale in all of the deposits of the Chamberlain Creek syncline and Fancy Hill district). In the small deposits of the Hatfield district in the westernmost part of the barite belt near the Arkansas–Oklahoma state line, chert (novaculite) of the Arkansas Novaculite or lower part of the Stanley Shale are interbedded with shale and barite.

In the Chamberlain Creek syncline (easternmost known part of the barite belt), the lowermost shale beds of the Stanley Shale are intruded by black chert giving rise to concretions made up of alternating chert and shale shells around chert–shale interiors (Fig. 7–15). Where the concretions have been removed by quarrying, chert necks protrude in places. The chert blanket beneath these concretions is depressed and broken along cracks due in part to subsidence on extrusion of the chert and shale mud. These features are fossil mud volcanos formed in a subaqueous environment where water depths were quite shallow; otherwise hydraulic pressure in deep water would have been too great to allow for extrusion of mud. To date, at least one occurrence of sandstone concretions has been identified to have formed by extrusion of mud. These are the sandstone concretions in the Carboniferous of County Clare, Ireland, investigated by Gill and Kuenen (1957). One crack pattern in the chert blanket referred to above is made up of (secondary) curved cracks terminating at larger cracks as displayed in a desiccation crack pattern (Fig. 7–16). For this to happen, some contraction or shrinkage must have taken place in unconsolidated sediment.

LITHOLOGIC, MINERALOGIC, AND CHEMICAL NATURE OF THE BARITIC SHALE AND BARITIC SANDSTONE STRATA

The Stanley Shale is a siliceous (quartzose) shale (Goldstein and Hendricks, 1953, in Dapples, 1979) with calcite, pyrite, sericite, clay, barite, and benstonite in the area of the deposits. The quartz in the quartz–barite layers is made

Figure 7-15. Alternating layers or shells of black chert and cherty quartzite in concretions at the base of the Stanley Shale. Chert necks, some with black shale interiors and lying along the same horizon, point to an origin by extrusion.

Figure 7-16. Quarry wall of the Chamberlain Creek syncline barite pit parallels the south flank of the syncline exposing the black chert blanket and concretions and lowermost black shale seams of the Stanley Shale. In the lower right portion of the illustration are chert necks of extruded black chert and shale. Some of the crack patterns, especially the area to the left of the flat 1 m in diameter concretion, appear to be contraction cracks similar to dessication cracks.

up of very fine grains, often forming, when abundant, a well-bedded network with fine-grained barite. Detrital quartz of silty and finer grains is most common in shaly portions (where sericite and clay and often pyrite predominate; see also Zimmermann, 1976).

The extremely siliceous (and calcareous) nature of the shale places the depositional setting of the shale in a setting similar to novaculite and perhaps the Big Fork Chert, also in the Lower Paleozoic section. The Devonian–Mississippian Arkansas Novaculite that underlies the Stanley Shale is essentially a recrystallized chert with abundant microfossils, and, in the upper part, disseminated carbonate, but containing detrital quartz and shale beds. Keller et al. (1977) regard the novaculite, including parts of the Big Fork Chert and Arkansas Novaculite, as thermally metamorphosed siliceous rock, originally a chert of biogenic origin or another siliceous rock, on the basis of isotope and petrographic data, a siliceous biogenic ooze (Jones and Knauth, 1979). According to Wise and Weaver (in Burgess, 1979, p. 422), chert will apparently form wherever there is sufficient biogenic opal to act as a source rock. Barite likewise is known to occur interbedded with shale and novaculite of the Arkansas Novaculite near the town of Hatfield.

The origin of the barite appears to be linked with the silica and carbonates in the same strata. Concentration of the ocean water and sediments with barium and strontium and other metallic salts may have affected the biogenic activity to where enough silica was available to form only the very siliceous shales and a later diagenetic environment able to hold both silica and barite. With the end of continuous chert or novaculite formation of the Arkansas Novaculite, shale would be the normal deposit, especially when sandstone deposition would be absent (see the Lower Paleozoic section in Table 7–1). The organic carbonate fraction would likewise be reduced in comparison to its content in the Arkansas Novaculite but still account for a considerable amount of disseminated carbonate in the Stanley Shale. Further, with not enough iron present, in a somewhat reducing environment, much of the carbonate formed is dissolved by oxidation of reduced sulfate and reaction with any carbonate formed (for diagenetic reactions and processes the reader is referred to Berner, 1980).

Of course, barite does not occur everywhere above the Arkansas Novaculite, but this does not preclude its absence in sea water and depositing sediments during Late Mississippian time. The changing conditions brought about by the beginning of the Ouachita orogeny certainly had an effect on the depositional environment during deposition of the upper part of the Arkansas Novaculite. A relationship existing between silica and barite is indicated near Hatfield, where the presence of barite between the novaculite beds is accompanied likewise by a considerable amount of shale in the section. The seas during the time of Stanley Shale deposition are considered to have been carbonate deficient (save for the few limestone beds in the barite section of the Chamberlain Creek syncline) and high in silica, such that carbonate organisms were rare or absent (a lime-poor habitat, according to Goldstein and Hendricks (1953, in Dapples, 1979). According to Goldstein and Hendricks (1953, in Dapples, 1979, p. 122), silica in the Big Fork Chert was derived from buried volcanic debris because of its chemical instability. Goldstein and Hendricks, also state that the abrupt lower contacts and gradational upper contacts of the siliceous sediments resulted from a sudden beginning and spasmodic diminuation of volcanic activity. At any rate, although there is a lack of any local spatial congruency with the barite deposits, the alkali rock trend of Upper Cretaceous age does follow a somewhat east-west or east-northeast-west-southwest course across Arkansas, and these rocks are probably the coarse-grained equivalents of the Late Mississippian volcanic tuffs.

SANDSTONE–SHALE RHYTHMITES (TURBIDITES) AND TUFFITE OF THE STANLEY SHALE AS RELATED TO THE BARITE (IN THE CHAMBERLAIN CREEK SYNCLINE)

SANDSTONE–SHALE RHYTHMITES

The anomalously great thickness of the Stanley Shale relative to the only moderately thick units of Lower Paleozoic or pre-Late Mississippian shale, chert, and sandstone below is to a considerable extent due to influx of sandstone (Figs. 7-2*b* and 7–17) into more slowly accumulating fine clastic shale deposits.

The sandstone exposed in the barite open pits shows features that indicate that they were deposited either as mudflows or from turbidity currents. They are rhythmically interbedded with shale. Most evident is the sharp transition

Figure 7–17. Siltstone-sandstone/shale rhythmite (subsection 9 and overlying beds in Fig. 7-2*b*) illustrating the thickness relationships. Upper part of figure (above topmost black shale band) forms lowermost beds of tuffite. West wall of Chamberlain Creek syncline open pit. Magnet Cove, Arkansas.

of underlying black shale to overlying sandstone, showing up as an abrupt contact (in contrast to the upper contacts of the sandstone units, which grade to black shale), large shale fragments in the basal parts of the sandstone units, and the presence of rapidly accumulated plant fragments in the upper parts of the sandstone beds and worm burrows (slow deposition of the overlying clay allowed for this worm habitat), as well as the absence of cross-bedding or ripple marks.

The sandstone units form rhythmic sets (Fig. 7–17). The sandstones are to an extent rhythmically interbedded with black shale, although the bed thicknesses vary, and these features indicate that normal fine detrital black shale deposits were interrupted by rapid and short influxes of sand.

Two relationships in the sandstone–black shale sequence (in the section below the tuffite) emerge: (1) a progressive increase in sandstone to shale ratio upward, when the sequence is uniformly subdivided, as in sections about 20 m in thickness, and (2) a rhythmic pattern appears such that the upper part of the entire section up to the tuffite contains notably thinner sandstone and shale beds.

Tuffite

The tuffite (Fig. 7–2 and 7–17) follows a rather thick shale bed, which in turn follows a sandstone–shale sequence with progressively thinning sandstone beds upward. Although the sandstone–shale beds are relatively thin near the upper part of the section, the reason for the rhythmicity could not be taken to mean seasonal, or even longer, climatic causes to cause the sandstone deposits since the intermediate shale beds are, in general, too thick to prove such a relation. Thinning of the shale beds upward indicates more frequent influxes of sand, due perhaps to more frequently occurring earthquakes. The final thick shale bed below the tuffite then points to a longer period of quiescence. Considering the entire sandstone–shale section below the thick shale bed, one sees that over each thickness of section of about 20 m, a steadily increasing sandstone-to-shale ratio occurs. The point to be made here is that probable earthquakes and finally volcanism relate the tuffite in its particular position in the section to the alternating sequence of sandstone–shale beds, ending in a longer period of quiescence (the thick shale) and, in turn, interrupted and buried by a thick tuffite unit containing not only volcanic tuff but also abundant sandy quartz grains of detrital origin, the components of a sand released into the basin and mixed with the tuffite.

Paleogeography and Depositional Environment

Quite notable, as well as coincidental in the Ouachita barite deposits, is their position at the base of the thick sequence of detrital upper Paleozoic sedimentary rocks at the base of the Stanley Shale. At that stratigraphic level, a major change in the Ouachita stratigraphy occurs where a return to conditions similar to those occurring during the deposition of the Missouri Mountain Slate (see Table 7–1) seems to have been prevalent during deposition of the lower part of the Stanley Shale. The correspondence of barite with the shale, however, seems to indicate that the shale may have been related to the changing paleogeography. Although the Stanley Shale lithology shows continuity of sandstone and shale upward in the section, barite deposits recur first several thousand meters higher in the section just above another major change in the stratigraphy, the Paleozoic–Lower Cretaceous unconformity near Dierks, Arkansas. The deposits at Dierks, Arkansas, are of minor interest and the barite cements sand and gravel of the Lower Cretaceous Trinity Sand. The deposits form stratabound lenses and are of the Laisvall or Mississippi Valley type.

Supratidal conditions are believed to have existed during deposition of the Arkansas Novaculite to the west in Oklahoma (Jones and Knauth, 1979), and shallow water to open marine conditions to the east in Arkansas. The barite deposits all occur parallel to the northern rim of the Ouachita trough, which places them also north of the deepest part of the trough. According to Glick (1979), broad areas of northern Arkansas were probably exposed to low tide at the end of interval C Mississippian time when the Upper Arkansas Novaculite was deposited. Barite, in such an environment, thus formed syndiagenetically under conditions conducive to algae and sulfate-reducing bacterial growth, which could account for the prevalence of irregularly banded to lenticular textures in the Fancy Hill deposits, given fuller treatment above.

At Gap Mountain, between layers of fine-grained, dark gray to black siliceous shaly barite, and entirely within the section of barite beds, orbiculoid brachiopods were observed and collected about 35 m above the contact of the Stanley Shale and underlying Arkansas Novaculite. The locality is about 5 miles north of Hopper (about 10 miles west of Glenwood), Montgomery County, Arkansas. To our knowledge, orbiculoid brachiopods are not included in any of the descriptions listed by Chamberlain (1978) and Chamberlain and Basan (1978) in their summaries of trace fossils of the Ouachita geosyncline. It appears that these are the only orbiculoid brachiopods ever to have been collected from this horizon and geographic location in the Stanley Shale. As the shells from the weak-shelled brachiopods are quite delicate and poorly preserved, identification is difficult. However, they appear complete and unbroken and therefore were quite likely not transported to their present site. However, although the greater percentage of living brachiopods can live in shallow water, no generalizations as to bathymetric conditions can be made for extinct brachiopods.

COMPARISONS WITH RELATED DEPOSIT TYPES

MASSIVE SULFIDE-TYPE DEPOSITS

Massive sulfide deposits the world over are rather conspicuously associated with barite in amounts up to 50 percent, or, in places, in bodies almost entirely composed of barite (for example, the Grauerz orebody of Rammelsberg), so that barite is not just considered as gangue or part of the wall rock, but also as a mineral genetically related to the sought after sulfide ore. A number of such massive sulfide deposits are known around the world, most of them having been extensively developed only in recent years. The Meggen (Lennestadt) mine located in a hilly region of the Rhine Slate Mountains of West Germany is one of these; the deposit not only shows similarities with the Arkansas deposits, but also has been the object of attention by numerous ore geologists through the years. The most recent view is that volcanic exhalations are no longer considered as the origin of the formation of the ores.

Sedimentary and diagenetic structures exist including geopetal, rhythmic, septarian, and diagenetic differentiation–crystallization features (Zimmermann, 1970). Microfossils are replaced by pyrite and sphalerite and appear to make up the bulk of the ore layers. The variation in grain size, and therefore variation in brassy hue of the pyrite, allows geopetal deformations and other features to be readily recognized. Load casts of sphalerite in pyrite occur in the nodular pyrite and sphalerite at the foot of the deposit. The topmost sulfide bed and layers of the deposit contain in places tiny folds and slumps between the two minerals, probably the result of a combination of rapid deposition and differential diagenetic settling. Most striking in regard to the barite are the lens shapes, outlined by very thin seams of bituminous material and evidently of shallow water origin, as the textures are quite similar to features observed in present-day tidal channel and tidal flat deposits of sand and clay. Stylolitization in the barite has led to some lens forms and outline lenses, but for the most part, lenses appear primary. Likewise, there is abundant additional evidence pointing to a geographic high during the Givetian Stage of the Middle Devonian in the Meggen area. The similarity of the stratabound nature of the Meggen and Arkansas barite deposits tends to indicate some similarities in origin such as paleogeographic environment. Yet, the Stanley Shale is probably not a good stratigraphic target for massive sulfide deposits as indicated by recent chemical analyses on drill core in and near the Fancy Hill District by Howard and Hanor (1987).

MISSISSIPPI VALLEY-TYPE DEPOSITS

Barite in the Ouachita Mountain Stanley Shale deposits lies in siliceous shale (calcareous in the Chamberlain Creek syncline); the major difference between the bedded Arkansas barite and the Mississippi Valley barite occurrences (see also Zimmermann and Amstutz, 1983) is that the barite of the Ouachita Mountains is found in a mobile belt but is not associated with lead and zinc sulfides or fluorite. Therefore, a distinction between the two types may be made on the basis of (1) depositional environment, (2) country rocks, (3) deposit type, and (4) mineral associations. Alone, as a criterion, the presence or absence of the Pb–Zn–F mineral association or of later tectonic movements of thick geosynclinal sections may not be decisive in separately classifying the Arkansas type from the Mississippi Valley type. For example, evidence was given above pointing to the shallow water origin of the lower part of the Arkansas Stanley Shale section.

The barite-bearing Lower Cretaceous Trinity Sand of Arkansas is of the Laisvall or Mississippi Valley type with the barite-bearing sandstone and cherty conglomerate lying above the Paleozoic–Mesozoic unconformity. The SrO content of one sample of 0.5 percent SrO (Zimmermann, 1964) is similar to the arithmetic mean of 0.6 percent for Sr contents determined by Clark (1970, in Leach, 1980) for central and southeast Missouri barite.

Barite crystals from central and southeast Missouri show an increase in Sr concentration toward the younger parts of the crystal (Leach, 1980); barite nodules from the Chamberlain Creek syncline deposit and Gap Mountain deposit of Arkansas likewise show an increase in Sr from the centers of the nodules outwards, followed, however, by a progressive decrease in Sr over most of the nodules' inner parts so that the lowest Sr concentrations are in the youngest shells. A progressive decrease in Sr has been determined from older to younger beds in the Gap Mountain deposit (Zimmermann, 1964, 1976). The change in the Sr contents of the Arkansas barite nodules is due in part to the concentrations of pore solutions at different horizons during dewatering of the sediments, in addition to an overall change from shallow to deeper water depositional environment as revealed by facies indicators. Similar fluctuations in depth are also known from Missouri (Zimmermann and Spreng, 1984).

VEIN DEPOSITS

Veins are commonly but not always symmetrically banded inward from the wall rocks. For stratabound layers, even those that have been partly recrystallized, this symmetry test expected for veins is not met because the ages of formation are different.

Higher contents of Sr appear to occur in vein deposits as compared to stratabound types as determined by Krimmel and Emmermann (1980) for the sedimentary, stratiform, and vein deposits of the Rheinland-Pfalz and Meggen areas. However, according to Church (1970, in Church, 1979), barites formed in different environments seem indistinguishable on the basis of Sr content alone.

SrO content in Stanley Shale barite ranges from 0.4 percent to 0.6 percent in most analyses made. A decrease in Sr due to remobilization of barite by surface waters is noted by Tufar (1962, 1963, in Tufar, 1965) for Austrian barite deposits.

Fluorite is a frequent gangue mineral accompanying barite in veins, and minerals of Bi, Co, Ni, As, Ag, and U are often detected, whereas, in massive sulfides, as well as in karst fillings of Mississippi Valley type deposits, minerals of the base metals of Cu, Pb, and Zn, including mainly chalcopyrite, galena, and sphalerite, are common. However, due to their inclusion in organic matter, some trace metals occur in anomalously high concentrations in banded rocks of algae and chert, including deposits of the Arkansas barite type and Mississippi Valley type. We are presently engaged in geochemical investigations of the Arkansas barite deposits for various trace metals.

REFERENCES

Amstutz, G. C., and W. C. Park. 1971. The paragenetic position of sulfides in the diagenetic crystallization sequence. In *International Association of the Genesis of*

Ore Deposits, Tokyo-Kyoto, Papers and Proceedings, ed. Y. Takéuchi, pp. 280–282. Tokyo: Society of Mining Geologists of Japan.

Amstutz, G. C., P. Ramdohr, F. El Baz, and W. C. Park. 1964. Diagenetic behavior of sulfides, in *Sedimentology and Ore Genesis,* ed. G. C. Amstutz, pp. 65–90. Amsterdam: Elsevier.

Amstutz, G. C., R. A. Zimmermann, and E. H. Schot. 1971. The Devonian mineral belt of Western Germany (the mines of Meggen, Ramsbeck and Rammelsberg). In *Sedimentology of Parts of Central Europe,* ed. G. Mueller, pp. 253–272. Frankfurt: Verlag Waldemar Kramer.

Berner, R.A. 1980. *Early Diagenesis,* Princeton, N.J.: Princeton University Press, 241p.

Brobst, D. A. 1958. Barite resources of the United States, *U.S. Geol. Surv. Bull.* **1072B:** 67–130.

Brobst, D. A. 1973. The association of barium in the Stanley Shale of the southern Ouachita region, Arkansas, *Geol. Soc. America Abstracts with Programs* **5:**249.

Burgess, C. J. 1979. The development of a Lower Jurassic carbonate tidal flat, Central High Atlas, Morocco, 2: Diagenetic history, *Jour. Sed. Petrology* **49**(2): 413–428.

Chamberlain, C. K. 1978. Trace fossil ichnofacies of an American flysch. In *A Guidebook to the Trace Fossils and Paleoecology of the Ouachita Geosyncline,* ed. C. K. Chamberlain, pp. 23–37. Tulsa: Society of Economic Paleontologists and Mineralogists.

Chamberlain, C. K., and P. B. Basan. 1978. Field guide to the trace fossils and paleoecology of the Ouachita geosyncline. In *A Guidebook to the Trace Fossils and Paleoecology of the Ouachita Geosyncline,* ed. C. K. Chamberlain, pp. 38–68. Tulsa: Society of Economic Paleontologists and Mineralogists.

Church, T. M. 1979. Marine barite. In *Marine Minerals,* vol. 6, *Reviews in Mineralogy,* ed. R. G. Burns, pp. 175–209. Washington, D.C.: Mineralogical Society of America.

Dapples, E. C. 1979. Silica as an agent in diagenesis. In *Diagenesis in Sediments and Sedimentary Rocks,* eds. G. Larsen et al., pp. 99–141. New York: Elsevier.

Davies, D. K., and E. A. Williamson. 1976. Shallow-marine origin for Ordovician rocks of the Ouachita Mountains, Arkansas. *Geology* **4**(6):340–344.

DeHairs, F., R. Chesselet, and J. Jedwab. 1980. Discrete suspended particles of barite and the barium cycle in the open ocean, *Earth and Planetary Sci. Letters* **49:**528–550.

Fontbote, L., and G. C. Amstutz. 1983. Diagenetic crystallization rhythmites in Mississippi Valley-type ore deposits. In *International Conference on Mississippi Valley-type Lead-Zinc Deposits,* eds. G. Kisvarsaniy, S. K. Grant, W. P. Pratt, and J. W. Koenig, pp. 328–337. Rolla, Mo.: University of Missouri-Rolla.

Frezon, S. E., and G. H. Dixon. 1975. Paleotectonic investigations of the Pennsylvania System in the United States, Part I: Introduction and regional analyses of the Pennsylvania System. Texas Panhandle and Oklahoma. *U.S. Geol. Survey Professional Paper 853-J,* pp. 177–195.

Gill, W. D., and P. H. Kuenen. 1957. Sand volcanoes on slumps in the Carboniferous of County Clare, Ireland, *Geol. Soc. London Quart. Jour.* **113:** 441–460.

Glick, E. E. 1979 Arkansas. In *Paleotectonic Investigations of the Mississippian System in the United States, Part I, Introduction and Regional Analyses of the Mississippian System,* L. C. Craig and C. W. Connor, coordinators pp. 125–145. U.S. Geological Survey Professional Paper 1010-H.

Graham, S. A., R. V. Ingersoll, and W. R. Dickinson. 1976. Common provenance for lithic grains in Carboniferous sandstones from Ouachita Mountains and Black Warrior basin. *Jour. Sed. Petrology* **46:**620–632.

Hanford, C. 1979. Late Mississippian paleogeography along the Ouachita-Southern Appalachian belt. *Geol. Soc. America Abstracts with Programs* **11:**148.

Hanor, J. S., and L. R. Baria. 1977. Controls of the distribution of barite in Arkansas. *Arkansas Geol. Comm Bull.* **2**:42–49.

Hass, W. H. 1950. Age of Lower part of Stanley Shale. *Am. Assoc. Petrol. Geol. Bull.* **34**:1578–1584.

Howard, K. W., and J. S. Hanor. 1987. Compositional zoning in the Fancy Hill stratiform barite deposit, Ouachita Mountains, Arkansas, and evidence for the lack of associated massive sulfides. *Econ. Geol.* **82**:1377–1385.

Jones, D. L., and L. P. Knauth. 1979. Oxygen isotopic and petrographic evidence relevant to the origin of the Arkansas Novaculite. *Jour. Sed. Petrology* **49**:581–598.

Jones, T. A. 1948. Barite deposits in the Ouachita Mountains, Montgomery, Polk, and Pike Counties, Arkansas. *U.S. Bur. Mines Rept. Inv. 4348,* 15p.

Keller, G. R., and S. E. Cebull. 1973. Plate tectonics and the Ouachita system in Texas, Oklahoma, and Arkansas, *Geol. Soc. America Bull.* **84**:1659–1666.

Keller, W. D., G. W. Viele, and C. H. Johnson. 1977. Texture of Arkansas Novaculite indicates thermally induced metamorphism, *J. Sed. Petrology* **47**(2):834–843.

Keroher, G. C., et al. 1966. Lexicon of geologic names of the United States for 1936–1960, Pt. 1, A–F; Pt. 2, G–O; Pt. 3, P–Z. *U.S. Geol. Surv. Bull.* **1200**, 4341p.

Krimmel, M., and K. H. Emmermann, 1980. Geochemische Untersuchungen an Baryten. Ein Beitrag Zur Genese der Baryte in Rheinland-Pfalz. *Mainzer geowiss. Mitt.* **9**:127–166.

Leach, D. L. 1980. Nature of mineralizing fluids in the barite deposits of Central and Southeast Missouri. *Econ. Geol.* **75**:1168–1180.

Ligasacchi, A., and G. C. Amstutz. 1960. Origin of the barite deposits of Missouri (abs.). *Geol. Soc. America Bull.* **71**(12):1916.

Lowe, D. R. 1975. Regional controls on silica sedimentation in the Ouachita system. *Geol. Soc. America Bull.* **86**:1123–1127.

Martin, J. M., M. Ortega-Huertas, and J. Torres-Ruiz. 1984. Genesis and evolution of strontium deposits of the Granada basin (southeastern Spain): Evidence of diagenetic replacement of a stromatolite belt. *Sedimentary Geology* **39**:281–298.

McElwaine, R. B. 1946*a*. Exploration for barite in Hot Spring County, Arkansas. *U.S. Bur. Mines Rept. Inv. 3963,* 21p.

McElwaine, R. B. 1946*b*. Exploration of barite deposits in Montgomery County, Arkansas. *U.S. Bur. Mines Rept. Inv. 3971,* 24p.

Miser, H. D., and A. H. Purdue. 1929. Geology of the DeQueen and Caddo Gap quadrangles, Arkansas. *U.S. Geol. Surv. Bull. 808,* 195p.

Niem, A. R. 1976. Patterns of flysch deposition and deep-sea fans in the Lower Stanley Group (Mississippian), Ouachita Mountains, Oklahoma and Arkansas. *Jour. Sed. Petrology* **46**:633–646.

Parks, B., and G. C. Branner. 1932. A barite deposit in Hot Spring County, Arkansas. *Arkansas Geol. Surv., Inf. Circ. 1,* 52p.

Pettijohn, F. J. 1975. *Sedimentary Rocks,* 3rd ed., New York: Harper & Row, 628p.

Scull, B. J. 1958. Origin and occurrence of barite in Arkansas. *Arkansas Geol. and Conserv. Comm. Inf. Circ. 18,* 101p.

Tufar, W. 1965. Geochemische Untersuchungen an oesterreichischen Baryten. *Tschermaks Mineralog. Petrog. Mitt.* **3**(9): 242–251.

Wickham, J., D. Roeder, and G. Briggs. 1976. Plate tectonics models for the Ouachita fold belt. *Geology* **4**(3):173–176.

Zimmermann, R. A. 1964. The origin of the bedded Arkansas barite deposits (with special reference to the genetic value of sedimentary features in the ore). Ph.D. Dissertation. Rolla, Mo.: University of Missouri at Rolla.

Zimmermann, R. A. 1967. Sedimentary features in the layered barite deposits of Nevada, Wisconsin and Meggen (Germany). In *7th Int. Sediment. Congress, Great Britain, 1967,* Preprint.

Zimmermann, R. A. 1969. Strata-bound barite deposits in Nevada: Rhythmic layering, diagenetic features, and a comparison with similar deposits of Arkansas. *Mineralium Deposita* **4**:401–409.

Zimmermann, R. A. 1970. Sedimentary features in the Meggen barite–pyrite–sphalerite deposit and a comparison with the Arkansas barite deposits. *Neues Jahrb. Mineral. Abh.* **113**:179–214.

Zimmermann, R. A. 1976. Rhythmicity of barite-shale and of Sr in strata-bound deposits of Arkansas. In *Handbook of Strata-bound and Stratiform Ore Deposits,* vol. 3, ed. K. H. Wolf, pp. 339–353. Amsterdam: Elsevier.

Zimmermann, R. A., 1982, Turbidite barite in the Ouachita Mountains of Arkansas, U.S.A. In *Ore Genesis: The State of the Art,* eds. G. C. Amstutz, A. El Goresy, G. Frenzel, C. Kluth, G. Moh, A. Wauschkuhn, and R. A. Zimmermann, pp. 66–72. Berlin: Springer.

Zimmermann, R. A., and G. C. Amstutz. 1961. Sedimentary structures in the Arkansas barite belt. *Geol. Soc. America Special Paper (Abstracts)* **68**:306–307.

Zimmermann, R. A., and G. C. Amstutz. 1964*a*. Small scale sedimentary features in the Arkansas barite district. In *Sedimentology and Ore Genesis, Developments in Sedimentology,* vol. 2, ed. G. C. Amstutz, pp. 157–163. Amsterdam: Elsevier.

Zimmermann, R. A., and G. C. Amstutz. 1964*b*. Die Arkansas-Schwerspatzone, neue sediment-petrographische Beobachtungen und genetische Umdeutung. *Zeitschr. Erzbergbau Metal.* **17**:365–371.

Zimmermann, R. A., and G. C. Amstutz. 1964*c*. Genesis of layered barite deposits in Arkansas and Germany. *Geological Society of America Special Paper 82,* p. 234.

Zimmermann, R. A., and G. C. Amstutz. 1983. Barite, its place and role in Mississippi Valley-type deposits. In *International Conference on Mississippi Valley-type Lead-Zinc Deposits,* eds. G. Kisvarsanyi, S. K. Grant, W. P. Pratt, and J.W. Koenig, pp. 279–288. Rolla, Mo.: University of Missouri-Rolla.

Zimmermann, R. A., and A. C. Spreng. 1984. Sedimentary and diagenetic features in the sulfide-bearing sedimentary dikes and strata of Lower Ordovician dolomites, Decaturville, Missouri, U.S.A. In *Syngenesis and Epigenesis in the Formation of Mineral Deposits,* eds. A. Wauschkuhn, C. Kluth, and R. A. Zimmermann, pp. 350–372. Berlin: Springer-Verlag.

8

Stratiform Barite Ore Fields in Western Europe

Yves A. Fuchs

Barite deposits and occurrences are particularly abundant in Western Europe. Vein type deposits, in the German Federal Republic (Dreislar) and in the French Massif Central (Lacan, etc.) are mined intensively. Stratiform barite deposits, however, are economically important in the German Federal Republic, Belgium, Austria, France, Italy, and so forth. These stratiform or strata-bound barite deposits belong to three different types:

- Stratiform lenses associated, or not, with massive sulfide ores, located in strata of the pre-Hercynian age of Europe. These deposits are reported to be of the exhalative sedimentary type (Rammelsberg, Meggen, Eisen in the Middle Devonian of the German Federal Republic, Magcobar in the Lower Carboniferous of Ireland).
- Stratiform lenses hosted in Lower Jurassic carbonate rocks or sandstones unconformably lying on the Hercynian crystalline basement. Pessens and Chaillac, both in France, belong to this type.
- Barite deposits with or without fluorite and sulfides in orebodies located below unconformities and presenting some characteristics more or less consistent with karstification processes. Examples include Chaudfontaines and Fleurus in Belgium, some deposits of minor economic interest on the northern flank of the Pyrenées Mountains, and Sulcis and Barega in Sardinia, Italy.

MASSIVE BARITE AND BARITE SULFIDE DEPOSITS OF NORTHERN EUROPE

DESCRIPTION

Important economic barite deposits are known in the Middle Devonian of Germany and in the Lower Carboniferous of Ireland: the Magcobar deposit in Ireland (2.5 millions tons), and in Western Germany, Rammelsberg (0.2 millions tons), Meggen (10 millions tons), Eisen, and Lorheim.

The ore body of Eisen (Gwodz et al., 1974) is located in the southwestern part of the Rheinisches Schiefergebirge in a tectonically disturbed zone where a window of red and violet Gedinnian schists appears through Middle Devonian (Hoffman, 1966). Near Eisen some mafic volcanics (Middle Devonian age?) are known. A slight metamorphism induced the formation of quartz veins, parallel to the schistosity. At a later stage, the formation of the grabenlike Sarre Nahe Permian basin leads to a 50° southeast rotation.

The ore deposit consists of a barite lens 140 m in length, an average 10 m thickness (locally up to 22 m), and an extension along dipping of approximately 90 m. The ore is a gray, fine- to coarse-grained barite with minor pyrite. The pyrite forms fine stripes interbedded with black argillaceous laminae. In thin sections, rare sphalerite can be observed within the pyrite. The barite was subjected to postdepositional tectonism and low-grade metamorphism. Metamorphism induced grain growth and some remobilization of barite, pyrite, and quartz in the form of fracture-filling, crosscutting veinlets in the host rocks.

Other deposits and occurrences are known in the Rhine region, such as Günterod and Lorheim (Stoppel, 1979). A stratiform barite deposit also exists in Fortuna, an iron mine of the Lahn Dill district where iron ores are related to an important spilitic volcanism (Gwodz et al., 1974).

Meggen is the most important deposit (Fig. 8–1). The Meggen region is located at the boundary of a shelf facies in the northwest and a trough in the southeast, these structures being active during the Middle Devonian. During this period, sedimentation on the shelf was characterized by deposition of thick sequences. This thickness is related to subsidence in a shallow water mobile environment. During the same period of time, in the southeasterly located trough, dark shales, silt and sand layers, and pelagic limestones were deposited. The sedimentation rate on the platform was higher than in the basin, due to the more mobile environment. The development of reef facies (Attendorn reef and Meggen reef) (Krebs, 1972*a*, 1972*b*) along the border of the shelf is significant. The Meggen ore deposit (pyrite, sphalerite, barite) is a lenticular-shaped orebody located between dark shales at the footwall and a pelagic limestone (Lagerkalk) of Upper Givetian age at the hanging wall. The existence of a tuff horizon above the deposit and underneath the Lagerkalk is significant and indicates contemporaneous volcanic activity.

The ore lens (Fig. 8–2) is subdivided by a ridge. The barite deposit is located along the margins of the two basins.

Underneath the ore deposit, the presence of disseminated sulfides and barite is indicated. The barite of the orebody itself is a fine-grained dark one. The

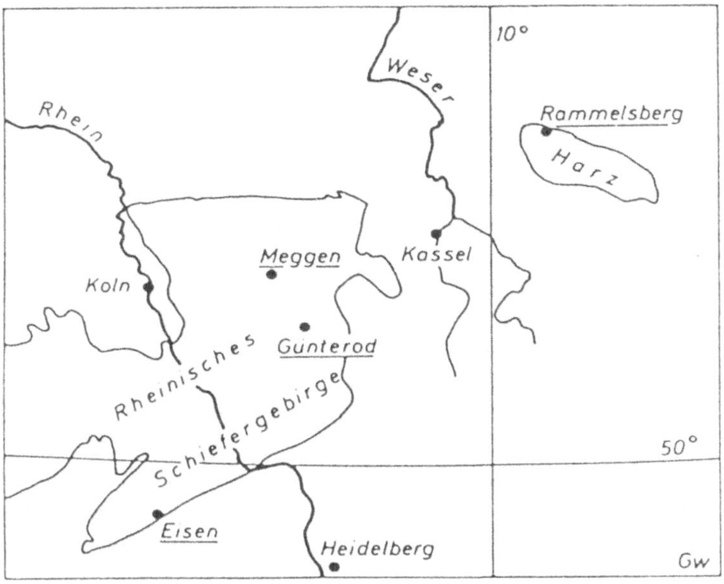

Figure 8–1. Location of the stratiform sulfide-barite deposits in the Harz Mountains and the Rhenish Schiefergebirge (after Gwodz et al., 1974).

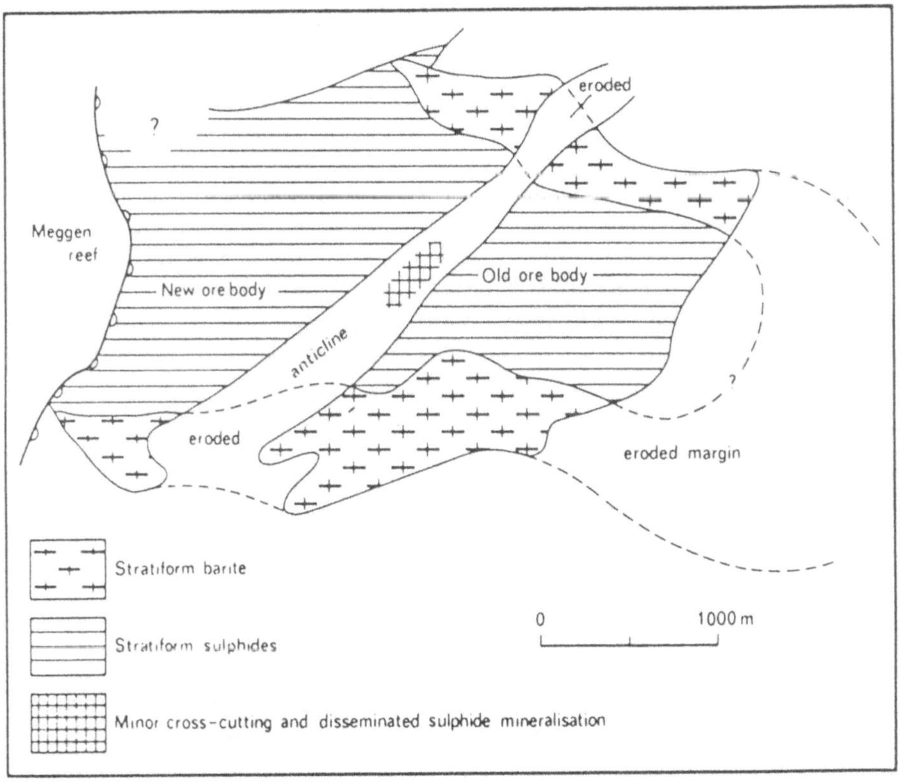

Figure 8–2. Palinspastic reconstruction of the original shape of the Meggen deposit (after Weisser, 1972, in Large, 1980).

barite beds are separated by thin layers of argillaceous material. Some pyrite layers and pyrite nodules are present, particularly near the sulfide ore lenses.

Farther east in the Harz Mountains, the Rammelsberg deposit, mined without interruption since the tenth century, is an important sulfide ore deposit. Barite ores are associated with these sulfidic ores. In the central part of the deposit, the massive ore is characterized by its high metal content, with up to 7 percent lead, 18 percent zinc, 1 percent copper, 12 percent iron, 20 percent barite, 120 g/t of silver, and 0.6 g/t of gold.

In the marginal parts of the orebody, however, one can observe interbedded layers of sulfides and sediments (mainly shales), the so-called banded ore. The shales below the deposit are of considerable thickness. During deposition (Eifelian), differences in subsidence rates induced a differentiated basin and ridge paleogeography. The Rammelsberg deposit itself is located within a depression, about a kilometer in diameter and showing important and rapid changes in thickness.

Tuff beds and diabases, those particularly well developed in the overlying Givetian, are significant of important volcanic activity. Beneath the orebody, the shales contain disseminated sulfide (pyrite) and some "stringers." This zone, called "kniest," shows an important enrichment in silica and is reported to be the result of hydrothermal circulation. Above the main orebody, there is a dominant baritic body (Grauerz).

In the northwestern part of Europe, another important barite deposit of this type is Magcobar in Ireland. This deposit, a part of the Silvermines sulfide deposit, is located in sediments of Lower Carboniferous age (Fig. 8-3). Paleogeographic reconstitution indicates a position between shelf and trough facies.

The Silvermines district is composed of several orebodies, some of which are sulfide stratiform bodies, some crosscutting sulfide bodies, and Magcobar is a stratiform barite lens (Taylor and Andrew, 1978). The footwall of the stratiform orebodies consists of bioclastic limestones, whereas their hanging wall shows a dominant reef facies. Magcobar is a barite lens with a thickness up to 18 m. The barite is recrystallized. The cap of the barite deposit consists of pyrite with hematite interlaminae and nodules.

CHARACTERISTICS OF THIS TYPE OF DEPOSIT

GEOLOGICAL PARAMETERS

The stratiform barite deposits of Middle Devonian age in Germany and Lower Carboniferous age in Ireland show some common characteristics. They all belong to the two northern branches of the Hercynian orogeny (Fig. 8-4). The southern branch (South Iberic belt) is particularly poor in barite (Routhier et al., 1980). In these northern branches, barite deposits are located in particular stratigraphic levels and in particular paleogeographic environments. They are always situated at the boundary of shelf and basinal areas, with significant

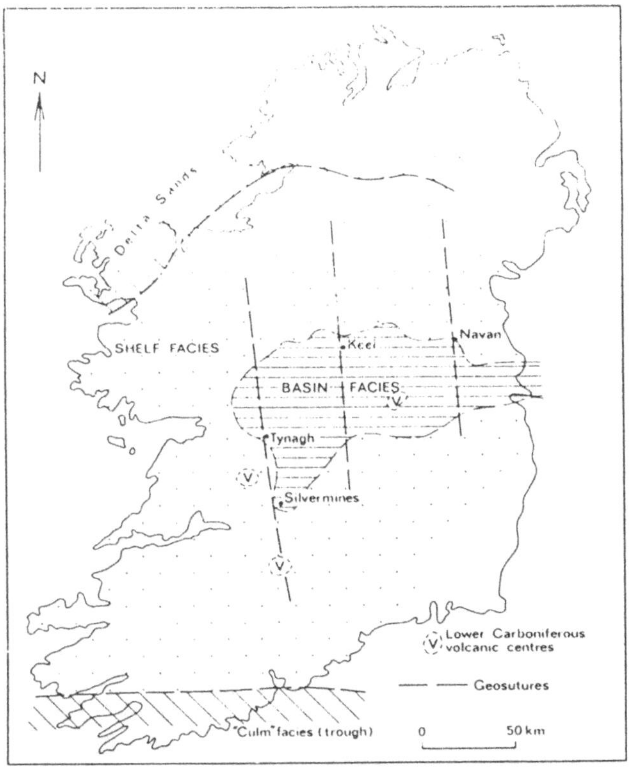

Figure 8-3. Lower Carboniferous paleogeography of Ireland showing the location of some lead–zinc deposits hosted by Lower Carboniferous carbonates (from Large, 1980).

variations of thickness and facies. Reefs are obviously frequent at the vicinity of the deposits (Meggen, Magcobar).

All deposits occur in stratigraphic levels characterized by tectonic instability and volcanic activity. Nevertheless, the barite deposits are not directly linked with the volcanic activity as in Kuroko-type deposits. Only some occurrences in the Lahn Dill iron deposits district seem directly connected with volcanic centers where spilitic lavas, pillow breccias, and glass-rich tuffs were vented. However, volcanic activity is obvious in most deposits or in their vicinity (Fig. 8-5). In Meggen, for example, the tuffite level directly above the ore lenses is not the only one; there are at least four tuffite levels and the northern flank of the older Upper Emsian keratophyre eruptive center is supposed to be at a depth of 1000 m below the Meggen deposit. The volcanic activity was more important in Rammelsberg, where about 20 tuffite layers, some up to 3 m thick, are quoted as well as some diabase sills.

In Magcobar the presence of tuffites is not clearly proven, but some chloritic marl layers could be related to volcanic influence and one can observe the existence of contemporaneous volcanics (Pallas Green Volcanics) (Evans, 1976).

Figure 8–4. Structural schematic map of the Hercynian system in Western Europe (after Arthaud and Matte, 1975) and location of the stratiform barite deposits.

(1) overthrust line; (2) transverse fault; (3) probable suture line; (4) distribution of basic and ultrabasic rocks, sometimes high-pressure granulite; (5) eclogites; (6) glancophane rocks; (A, B, C) location of the continental plates; (black stars and dots) stratiform barite deposits of the pure orogenic stage.

SULFUR ISOTOPES

Concerning the sulfur isotope ratios and oxygen isotope ratios of these deposits, two models are proposed by Large (1980): a restricted basinal one and an open basinal one.

A small restricted basin had no replenishment of sea water with normal sea water sulfate. Under such conditions the isotopic ^{34}S ratio will theoretically increase in biogenic pyrite upward in the sequence.

The open basinal model is preferred for Silvermines, Rammelsberg, and Meggen (Large, 1980). Sea water is proposed to be the source for the sulfate of barite in Magcobar. In Rammelsberg (Anger et al., 1966), the ^{34}S ratio of pyrite is dispersed and significant for a biogenic origin of most of the sulfur,

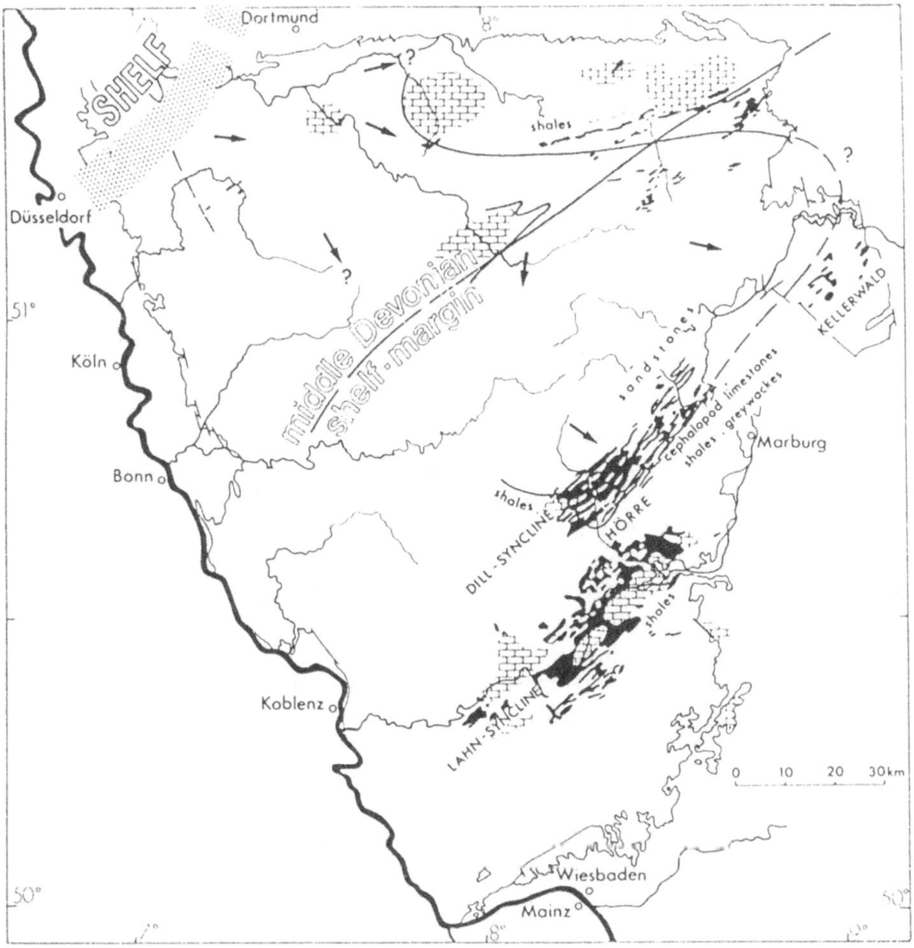

Figure 8-5. Paleogeography of the Rhenish Schiefergebirge east of the Rhine: Nehden Strufe (lower Famennian). Black areas: basaltic volcanics, Givelian to Visean; brick signature, Givetian/Fransmian reefs (from Franke et al., 1978).

whereas the isotopic ratios of other sulfides (galena, sphalerite) are in a narrow domain, which could indicate a supply from a hydrothermal source. The ^{34}S ratios of barite in the Rammelsberg main orebody are the same as those of Middle Devonian sea water sulfate sulfur. Nevertheless, in the overlying barite orebodies (Grauerzkörper and Liegendes Grauerz), the sulfur isotopic ratios are higher, and this could be isotopic evidence for an evolution to a restricted basin at the end of the process. Such an evolution would be consistent with our knowledge of the local paleogeography.

The ^{34}S isotopic ratios of the Meggen, Günterod, and Eisen deposits also occur within a narrow range, which is consistent with a sea water origin for the sulfate. However, the Lorheim deposit shows higher ratios (Krimmel and Emmerman, 1980), which could also be related to a restricted basinal evolution.

Concluding Remarks for this Type of Deposit

This type of barite deposit is associated with massive sulfide or iron hematitic deposits as a part of a sedimentary exhalative type. These barite deposits occur during the initial phase of the orogenic evolution. They were formed at the boundary of the continental plate (shelf limit) in zones where important fracturing induced thickness changes and restricted basinal evolution and permitted the upward movement of important heat flows along the cratonic margin. The presence of igneous activity is also related to this geodynamic evolution. Always present, this magmatic activity seems to be more distal from ore deposits, as in the Kuroko type, for example. With the exception of the Lahn Dill district, the deposits are not directly connected to volcanic centers even where vented volcanics, mainly tuffs, are present, as in Rammelsberg. Volcanism seems to be more an associated phenomenon as a source for the genesis of this type of deposit. The heat flow could initiate a hydrothermal system. The presence of zones with silica and sulfides below the stratiform orebodies in Meggen and Rammelsberg can be related to feeders. At the same time, the tectonic activity was able to induce a trough and graben paleogeography. Local restricted conditions in third-order basins with slow energy sedimentation and reducing conditions favored the precipitation of sulfides and barite from the hydrothermal fluids. The existence of a dual source for the sulfur of sulfides, whereas sulfur from sulfate is of marine origin, is in full agreement with such a model.

Postdepositional Transformation

Related to their geodynamic position, the host rocks of this type of deposit are often subjected to metamorphism during the orogeny. Middle Devonian strata in Germany and Lower Carboniferous strata in Ireland underwent only low-grade metamorphism during the Hercynian orogeny. Some recrystallization and remobilization can be observed.

Older deposits in Norway and Scotland were transformed during metamorphism. In Norway, near Oslo, Lower Ordovician shales were subjected to low-grade metamorphism. In these shales, barite pseudomorphs can be observed. Barite is then replaced by an assemblage of pyrite, quartz, calcite, and barium-potassium feldspars (Bjørlykke and Griffin, 1973).

The feldspars contain up to 33 percent mole of the $Ba\ Al_2\ Si_2\ O_8$ (celsian) molecule. According to Deer et al. (1966), the name *celsian* should be reserved for feldspars with more than 90 percent of the $Ba\ Al_2\ Si_2\ O_8$ molecule. Feldspars containing 5 percent to 30 percent of this molecule should be called *hyalophane*. Barite is stable in high metamorphic conditions, but in the presence of reducing agents, barium silicates can be formed (Segnit and Gelb, 1970).

The Aberfeldy Ba–Zn mineralization is located in the Dalradian (Lower Cambrian) metamorphics of Scotland. The barite and zinc mineralization is stratabound, hosted by graphitic quartz–muscovite schists. Coats et al. (1980) consider the Aberfeldy mineralization to be of sedimentary-exhalative origin. The region was subjected to amphibolite grade metamorphism. The barite lenses themselves are not transformed and the remobilization effect seems to be very weak. In the host rock, however, celsian and/or hyalophane, Ba-rich muscovite, are present (Fortey and Deddoe-Stephens, 1982). It is suggested that Ba-rich muscovite could have formed from clay minerals (illite?), which had incorporated important amounts of barite during diagenesis. Celsian/hyalophane, on the other hand, could have originated via the transformation of authigenic cymrite (Ba Al_2 Si_2 O_8 H_2O).

THE STRATIFORM TYPE IN STABLE SHELF ENVIRONMENTS

The stratiform type is hosted in shallow water deposits in stable platform environments. It can be divided into two different subtypes: carbonate and sandstone hosted.

STRATIFORM BARITE DEPOSITS IN CARBONATE ROCKS

Stratabound occurrences and deposits are known in different regions and at different stratigraphic levels. In Sardinia, for example, barite and sulfide ores (Pb–Zn) are located within the zebra dolomite (dolomia rigata) of Lower Cambrian age (Boni, 1985). Whereas sulfide ores (without barite) are associated with breccias and erosion surfaces, barite is related to facies with zebra and chickenwire structures (Gandin et al., 1974). Baritic orebodies are to be found in local evaporitic basins with high pH, in tidal to supratidal platform environments. The same conditions can be observed in the barite deposit of Pessens (France) in the southern part of the Massif Central where Lower Jurassic limestones and dolomites lie unconformably on metamorphic basement and continental sedimentary rocks of Permo-Carboniferous age.

This deposit, situated 5 km east of the city of Rodez, from 1928 to 1983 produced about 1.2 million tons of barite until the mine was closed in 1983, in relation to the economic conjuncture and despite some 0.5 million tons of potential reserves. The region itself is rich in diverse barite occurrences and deposits. A vein-type deposit at Lacan, some 8 km farther east, hosted in the Palanges gneisses and continental Stephanian (Upper Carboniferous sediments), is reported to contain about 1 million tons of barite between the surface and the deepest explored level (125 m below surface).

The deposit in Pessens is located in dolomites of Hettangian (Lower Jurassic) age, 25 m above the unconformity surface (Fig. 8–6). The Hettangian dolomites and limestones, interbedded with black and green centimeters-thick clay layers, are typical tidal flat deposits. North of the orebody, freshwater sediments, with root remnants and mud cracks, are typical of a palustrial environment. The host rock of the orebody forms on supratidal to intertidal flats with strong development of stromatolites. A north–south profile shows a progressive change in facies from freshwater swamp to tidal flat with stromatolites and rill marks (Fig. 8–7). Farther south, infratidal facies with more reducing conditions (black shales intercalated in carbonate rocks) becomes dominant.

Figure 8–6. Geological sketch map of Pessens.

(P) barite deposit; (1) crystalline basement; (2) Stephanian; (3) Autunian; (4) Saxonian; (5) Hettangian + Sinemurian S.S.; (6) Lotharingian; (7) Middle and Upper Lias; (8) Dogger; (9) Quaternary fluvial deposits.

Locally, however, in the southern part, evaporitic facies (anhydrite) can be observed. Changes in the nature of clay minerals also reflect these paleogeographic variations. The boron content of clay layers varies from less than 30 ppm in the freshwater environment up to 400 ppm in layers associated with anhydrite. The ^{13}C ratios in the carbonate rocks show a decrease of continental water influence from north to south. These important changes in paleogeography and sediment geochemistry can be related to the existence of a 110° east basement structure, which is assumed to have been active during the Jurassic. This is expressed by the development of early microfracturing and slumping in the stromatolites.

The Pessens deposit is a stratabound layer whose thickness varies from about 0.5 m up to 4 m. The paragenesis is very simple. In the main part of the deposit, barite is the only mineral present. Rare fluorite crystals can be observed at the northern end of the orebody, and small cinnabar crystals have been observed but are very rare and restricted to a small area.

The barite layer exhibits two different facies:

- A first generation (80 percent to 90 percent of the ore) is of fine- to medium-grained barite with interbedded and partially brecciated clay layers and sparse coal lenses (up to 15 cm). In this facies, replacement structures of former dolomicrite, locally with cross-bedding, and anhydrite can be observed.
- A second coarse-grained (with crystals up to 15 cm) type develops as a layer within the orebody but can be slightly discordant with the stratification. In this layer, barite megacrystals show centripetal growth toward the center where voids partially filled with calcite and manganese oxides occur. Spatially, this facies has a larger extension than the orebody itself. In the outer part of the deposit, this type of barite occurs as a coarse-grained cement in a breccia, whose elements are angular undisplaced or weakly displaced dolomicrite.

The limit of the second barite with the first generation facies and the element of breccias is always very sharp, showing no trace of replacement.

It is concluded (Fuchs, 1978, 1983) that the Pessens barite orebody was formed in a tidal flat environment between freshwater swamps and the open sea. Evaporitic conditions were present in local patchy basins. (See Fig. 8–7.)

The first generation seems to have been formed through remobilization during the diagenesis from a barium geochemical anomaly located in the sediments at the limit of the fresh and salt waters (Puchelt, 1967). This remobilization and concentration of barium could have been induced through wandering of connate water during diagenesis. The presence of reducing waters derived from the deeper parts of the basin and of sulfate rich water (evaporitic facies) could explain the removal of barium from the sediments under reducing conditions and the precipitation of barite by sulfate-rich water. Sulfur isotope ratios of barite from the first generation are in agreement with a marine origin from the evaporites. They show a strong genetic relationship with the host rock (Fig. 8-8). The second generation, however, shows sulfur isotope ratios grouped in a narrow range, which suggests a more distal origin of the fluids that initiated a hydraulic brecciation of the neighboring rocks.

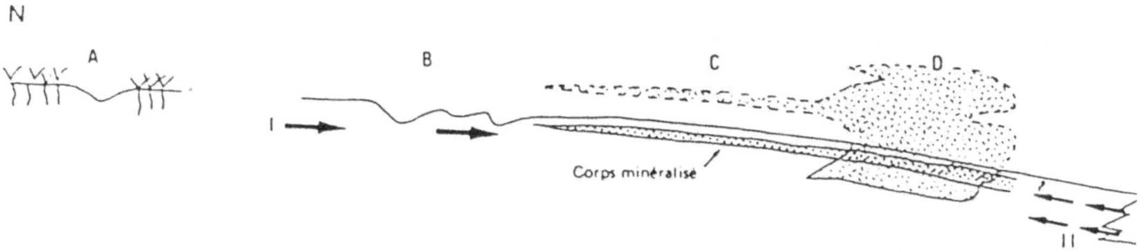

Figure 8–7. Paleographic reconstitution profile in the Pessens barite deposit.

(I) freshwater circulation; (II) SO_4-rich connate water; (A) swamp facies with soils; (B) supratidal zone with cut-and-fill structures and silicification; (C) zone with mud cracks, rill marks, and red clay layers; (D) stromatolites; (E) patchy evaporitic facies.

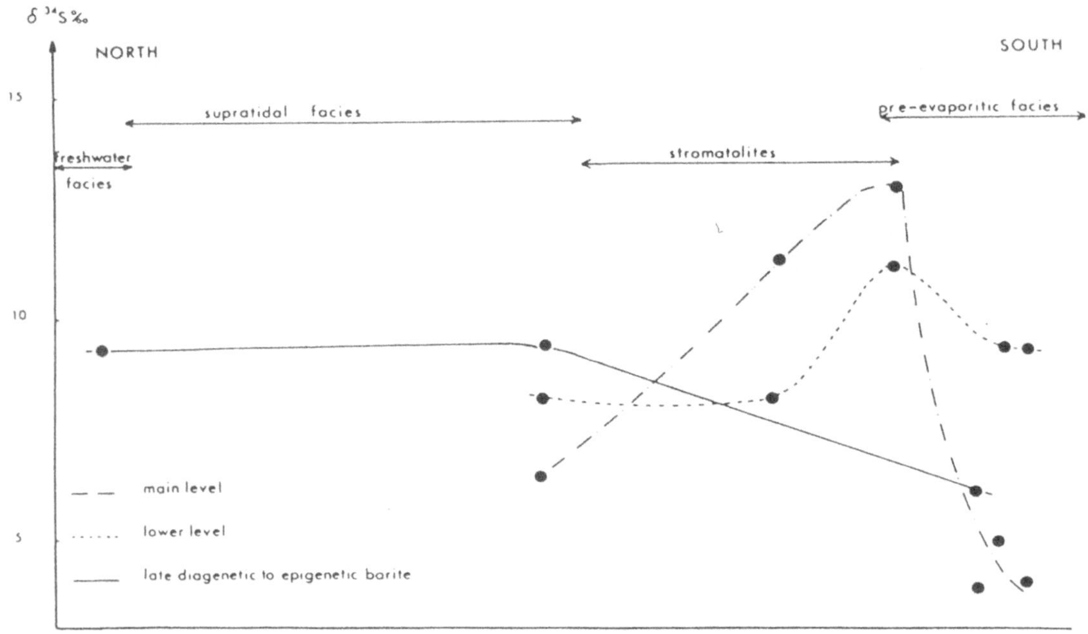

Figure 8–8. Evolution of $\delta\ ^{34}S$ from freshwater environment to pre-evaporitic environment in the barite deposit of Pessens (France).

STRATIFORM BARITE DEPOSITS IN SANDSTONES

The sandstones, which occur as peripheral facies of the Mesozoic transgression on the Massif Central, include numerous occurrences of barite. There is a large amount of barite trapped as low-grade ore in sandstones (5 percent to 20 percent $BaSO_4$). This was described by Samama (1969) for the Triassic sandstones of Largentière on the eastern border of the Massif Central. At Chaillac, in sandstones of Lower Jurassic age, on the northwestern border of the crystalline massif, a large barite deposit is actually mined. It consists of a stratiform orebody and a vein type, the second being hosted in basement gneisses. The total barite reserve is estimated at 1.8 million tons. There are also 0.4 million tons of fluorite. The deposit consists of a fluorite vein at Le Rossignol and the barite stratiform orebody of Les Redoutières, both associated with a regional north 10° east lineament. The Mesozoic transgressive stratas are discordant on the locally deeply weathered crystalline basement. The existence of dolomite crust (dolocrete) on the top of weathering profiles indicates pedological evolution under predominantly semiarid conditions. The base of the transgression (Hettangian) consists mainly of detrital strata (sandstone, clay), above which carbonates indicate a typical platform environment. The barite stratiform orebody is located in the basal detrital layers. From bottom to top it is possible to observe the following profile:

- Fine-grained, deeply weathered gneisses with a paleosurface and dolocrete locally at the top.
- Greenish to grayish, fine layers that are more or less continuous and do not contain barite.
- Coarse-grained basal sandstones are the host rock of a first barite deposit and at the periphery of the barite orebody contain fluorite. The thickness of these sandstones changes from 0 m to 5 m. These variations do not seem to be related to the paleotopography of the paleosurface. Green to red colored clays are lying on the sandstones. These so-called clays do not contain clay minerals but mainly a very fine association of quartz, feldspars, and micas.
- Lying on the basal sandstones and clays is the formation intercalaire (interpolated formation). It consists of clayey ochreous sands and meter-thick levels of dark sands (quartz + goethite).
- The iron–barite complex is up to 10 m thick and has a barite content of 70 percent. In the past, it was also exploited as iron ore. It shows a characteristic "striped" structure with alternating iron oxides and rose barite ribbons. The barite crystals show a fan-shaped disposition. This iron–barite layer disappears laterally very sharply. This change seems to be related to minor faults, which were active during the formation of the layer.
- The iron-barite formation grades progressively upward into the ferruginous member with an upward increasing iron content. The highest layer is an iron crust with a lateritic composition.

The Chaillac–Redoutières deposit is connected to the Rossignol vein. This consists of a near surface stratiform fluorite deposit located between basal sandstones and basement or, locally, between sandstone and clay units. This "plateure" (an old miner's word for horizontally lying orebodies) is connected to the vein itself. The vein is explored to a 180 m depth and about 800 m along the strike.

ISOTOPE GEOCHEMISTRY

Isotope data geochemistry was obtained by Sy (1983) on samples of barite from Chaillac. Strontium isotope ratios show results from 0.70976 $\pm$0.00009 to 0.71132 $\pm$0.00006, which are not very different from those of Pessens' 0.711782 $\pm$0.000016 and of the vein-type deposit of Lacan, near Pessens, of 0.709739 $\pm$0.000041. The $\delta^{34}S$‰ (C.D., Canyon Diablo meteorite) ratios of the sulfur in barite range from 13.8 to 16.7 in Chaillac and are conformed with an origin linked to leaching of evaporite as well as a direct origin from marine water. The $\delta^{18}O$ (SMOW, standard medium oceanic water) ratios range from 9.12 ‰ to 10.65 ‰ and do not permit an easy interpretation. They suggest that there was some mixing of water of continental origin with marine water, but they can also be interpreted as the result of successive oxido-reduction processes. The negative correlation shown by the $\delta^{34}S$ ratios versus $\delta^{18}O$ in different stratiform barite deposits of this period as well in carbonate rock environments as in sandstones could also be related to a redox type process (Fuchs, 1984).

The Karstic-type Deposit

The most important karstic-type barite deposits in Western Europe are those from Sardinia described in Chapter 9. Some occurrences are also known on the northern side of the Pyrénées (France). Other important occurrences are known in Belgium, including the discovery in three boreholes near Chaudfontaines described by Dejonghe (1985).

The karstic-type barite deposit of Chaudfontaines consists of a lead–zinc–barite mineralization with predominant barite. Paleogeographic reconstruction shows that the early diagenetic barite occurrences are located within the Upper Frasnian strata in limestones deposited in a tidal environment. The $\delta^{34}S$ ratio of sulfur from barite shows that this element is probably of marine origin, whereas strontium may not be of marine origin according to its isotopic composition. Lead isotopic composition of the galena accompanying the barite is similar to those of Middle Devonian deposits.

Mineralization seems to have developed in karstic cavities formed during emersion of the tidal platform. This deposit could also be an exogenic one formed by reworking of early massive sulfide–sulfate (barite) deposits or disseminated ore of Middle Devonian age.

Another complex barite deposit in Fleurus, Belgium, is quoted by De Magnee and Doyen (1981). It is the most important barite deposit in Belgium and produced about 700,000 tons of barite from 1890 to 1928.

The ore deposit is located in a karstic depression formed in the Visean limestones covered by Eocene sands. The main orebody is massive barite, whereas in the outer parts of the deposit it grades to an unconsolidated breccia of barite, rock fragments of Namurian age, quartz, and clay minerals. The Visean host rock is locally silicified and pyritized at the contact. The Eocene sands, lying unconformably on the deposit, reworked barite as well as quartz and Namurian phtanites in their basic layers.

Outcrops are very scarce and the absence of recent mining activity does not permit a genetic interpretation of the Fleurus barite deposit, whose morphology seems to be that of karstic deposits.

Conclusions

Three main types of stratiform barite orebodies are known in Western Europe. Each type is linked in time and space to a particular stage of the geodynamic evolution of plates.

The first type is represented by the stratiform barite deposits of northern Germany (Middle Devonian) and Ireland (Lower Carboniferous). Closely associated to a massive sulfide deposit in most cases (Meggen, Rammelsberg, Silvermines), the barite deposits were located at the margin of a cratonic area during the pre-orogenic period. They are linked to hydrothermal activity and are also influenced by sedimentologic and local paleogeographic conditions that are controlled by the instability of the shelf border lineaments. They can be characterized as exhalative-sedimentary deposits.

On the other hand, stratiform barite orebodies were formed during the Cambrian in Sardinia (Boni, 1985) and Lower Jurassic around the Massif Central (France). These deposits are hosted in carbonate rocks or in sandstones.

Sedimentologic studies and paleogeographic reconstitutions indicate a tidal flat environment with evaporitic or pre-evaporitic conditions. Migration of connate water favored barium mobilization and reconcentration. These deposits appear to be diagenetic and sedimentary postorogenic, formed on a stable platform affected only by light epeirogenic movements. Barbier (1976) has attempted to replace the formation of some vein-type deposits linked to stratiform deposits, such as Le Rossignol in Chaillac. He concluded that the fluorite–barite mineralization formed from $CaSO_4$ with F- and Ba-rich continental water. This mixing could occur in paleosabkha environments (fluorite and barite in sandstones in Chaillac, barite in Pessens) and/or within fractures of the basement (vein-type like Le Rossignol). However, one may assume that most of the barite and fluorite vein-type deposits in the Massif Central are controlled by north 10° east lineaments of Stephanian age, with which hydrothermal systems and late Hercynian, predominantly acid, volcanism are associated.

Karstic morphology is not rare being particularly abundant in Sardinia (see Chapter 9). Other deposits of this type are known in Belgium where they were formed partly during short emersion of tidal flat facies (Chaudfontaines) or later during the continental evolution of the plate under particular weathering conditions (Fleurus). Concerning their location, deposits with karstic morphology are located in districts (Sardinia, Belgium, Germany) that are rich in other barite deposits. They represent the last remobilization stage in the continental plate evolution in areas characterized by high barium potential.

REFERENCES

Anger, G., H. Niclsen, H. Puchelt, and W. Ricke. 1966. Sulphur isotopes in the Rammelsberg ore deposit (Germany). *Econ. Geology* **G1**:511–563.

Arthaud, F. and Ph. Matte, 1975. Les décrochements tardihercyniens du Sud-Ouest de l'Europe. Geométrie et Essai de reconstitution des conditions de la déformation. *Tectonophysics* **25**:139–171.

Barbier, J. 1976. Sur la signification paleogeographique de certaines minéralisations filoniennes à fluorine-barytine in "Metalogenie et geologie minière," *Mem. h. ser. soc. geol. Fr.* **7**:85–94.

Bjørlykke, K. O., and W. L. Griffin. 1973. Barium felspars in Ordovician sediments, Oslo region, Norway. *Jour. Sed. Petrology* **43**(2):461–465.

Boni, M. 1985. Les gisements de type Mississippi Valley du Sud-Ouest de la Sardaigne (Italie): une synthèse. *Chron. rech. min.* **479**:7–34.

Coats, J. S., C. G. Smith, N. J. Fortey, M. J. Callaghan, F. May, and W. J. Mac Court. 1980. Stratabound barium-zinc mineralization in Dalradian schist near Aberfeldy, Scotland. *Inst. Mining. Metal. Trans.* **89B**:110–122.

Deer, W. A., R. A. Howie, and J. Zussman. 1966. *Rock Forming Minerals,* vol. 4. London: Longmans, 435p.

Dejonghe, L. 1985. Constitution à l'étude metallogénique du synclinorium de Verviers (Belgique). Ph.D. dissertation. Paris: Université Pierre et Marie Curie.

De Magnee, J., and L. Doyen. 1982. Le gisement de barytine de Fleurus. *Symp. gites filomiens basse temperature.* Orléans Bureau Recherches Geol. et Minieres.

Evans, A. M. 1976. Genesis of Irish base metal deposits. In *Handbook of Strata-bound and Stratiform Ore Deposits,* vol. 5, ed., K. H. Wolf, pp. 231–256. Amsterdam: Elsevier.

Fortey, N. J., and B. Deddoe-Stephens. 1982. Barium silicates in strata-bound Ba-Zn mineralization in the Scottish Dalradian. *Min. Mag.* **46:**63–72.

Franke, W., W. Eder, W. Engel, and F. Langenstrassen. 1978. Main aspects of geosynclinal sedimentation in the Rhenohercynian zone. *Zeitschr. Deutsche Geol. Gesell.* **129**(1):201–216.

Fuchs, Y. 1978. Sur un exemple de relation entre une minéralisation barytine et un milieu à evaporites. Le gite de Pessens (Aveyron). *Sci. Terre* **22**(2):127–146.

Fuchs, Y. 1984. Migration of fluids during diagenesis: An ore-forming process in carbonate rocks. In *Syngenesis and Epigenesis in the Formation of Mineral Deposits,* eds. A. Wauschkuhn, C. Kluth, and R. A. Zimmermann, pp. 287–293. Berlin: Springer-Verlag.

Gandin, A., G. Padalino, and M. Violo. 1974. Correlation between sedimentation environment and ore-prospecting. Sedimentological and ore genesis studies of "Cambrian arenarie" and "dolomic rigate" formations (Sardinia, Italy): deposition and concentration of barite in an evaporitic environment. *Rend. Soc. It. Min. Petrogr.* **30:**251–303.

Gwodsz, W., H. Kruger, D. Paul, and A. Bauman. 1974. Die Liegendschichten der devonischen Pyrit and Schwerspat-Lager von Eisen (Saarland), Meggen und des Rammelberges. *Geol. Rundschau* **63:**74–93.

Hofmann, R. 1966. Lagerstattenkundliche Untersuchungen in Bereich der Schwerspatgrube Eisen (S. W. Hunsrück). *Neues Jahrb. Geol. Paläontol Monatsh.,* pp. 24–26.

Krebs, W. 1972*a.* Zum geologischen Rahmen schichtgebundener Sulfid-Baryt-Lagerstätten. nr 1. Die Paläogeographisalfazieller Aussagen zur Position des Meggener Layers. *Schr. Ges. Metal. Bergleute* **24:**187–196.

Krebs, W. 1972*b.* Zum geologischen Rahmen schichtgebundener Sulfid-Baryt-Lagerstätten—nr 3. Facies and development of the Meggen Reef (Devonian, West Germany). *Geol. Rundschau.* **61:**647–671.

Krimmel, M., and K. H. Emmerman. 1980. Geochemische Untersuchungen an Baryten Ein Beitrag zur Genese der Baryte in Rheinland-Pfalz. *Mainzer geowiss. Mitt.* **9:**127–166.

Large, D. E. 1980. Geological parameters associated with sediment-hosted submarine exhalative Pb-Zn deposits: An empirical model for mineral exploration. *Geol. Jahrb.* **40D:**59–130.

Puchelt, H. 1967. Zur Geochemie des Bariums im exogenen Zyklus. *Sitzungber. Heidelb. Akad. Wiss. Math. Nat Kl.* **4:**1–205.

Routhier, P., F. Aye, C. Boyer, M. Lecolle, P. Moliere, P. Picot, and G. Roger. 1980. La ceinture sud iberique à amas sulfures dans sa partie espagnole médiane. Tableau geologique et metallogenique. Synthèse sur le type amas sulfures volcano-sédimentaires. *Mem. Bur. Recherches Geol. Min. 94,* 265p.

Samama, J. C. 1969. Contribution à l'étude des gisements de type Red Beds: Etude et interprétation de la géochimie et de la métallogénie du plomb en milieu continental. Cas du Trias Ardéchois ed du gisement de Largentière. Ph.D. dissertation. Université Nancy.

Segnit, E. R., and T. Gelb. 1970. Reactions of Kaolinite with Barium carbonate and Barium sulphate. *Australian Jour. Ceramic Soc.* **1:**12–18.

Stoppel, D. 1979. Das Schwerspatlager von Schönheid bei Günterod. *Geol. Jahrb. Hessen* **107:**93–104.

Sy, D. 1983. Géologie et géochimie du gisement métallifère de Chaillac (Indre). Contribution à l'exploitation et à la valorisation d'un gisement de barytine. Ph.D. dissertation. Paris: Université Pierre et Marie Curie.

Taylor, S., and C. J. Andrew. 1978. Silvermines orebodies, County Tipperary, Ireland. *Inst. Min. Metal. Trans.* **87B:** 111–124.

Yaman, S., A. Ziserman, and J. C. Touray. 1978. Un modèle de filon lié à une paléosurface d'après l'étude des gites de fluorine et de barytine de la Marche Occidentale (France). *Chron. Rech. Min.* **441:** 45–48.

9

Stratabound and Karstic Deposits of Barite in Sardinia

Giorgio Padalino, Sandro Tocco, and Maurizio Violo

Sardinia, the largest island in the western Mediterranean along with Sicily, contains Italy's most important mining district for nonferrous metals, barite, and fluorite. Sardinian mining at present provides about 70 percent of Italy's domestic barite production, mainly from Cambrian terrains outcropping in the southwestern part of the island (Fig. 9–1). Three main barite occurrences are found in the Cambrian sequence, namely:

1. Stratiform barite, interbedded in the Lower-Middle Cambrian dolomites;
2. Stratabound bodies of silica (quartz and chalcedony) with barite and argentiferous galena. These bodies are located along the Cambrian–Ordovician unconformity;
3. Funnel-shaped bodies, sometimes very rich in barite, in karstic cavities in the Cambrian limestone. From an economic point of view, this is the most important occurrence. For instance, the Barega mine in Cambrian limestone to the south of Iglesias (Tamburrini and Violo, 1965) produced about one-eighth of Europe's barite in the 1960s from this type of deposit.

Veins and lodes, in which the barite occurs along with quartz, fluorite, and ankerite as gangue of sulfide mineralization, are not dealt with in this chapter.

Barite mineralization models have evolved greatly of late, not least because of an improvement in knowledge of the relevant general geology, sedimentology, tectogenesis, geochemistry, and so forth, but also as a result of the actual mining of known bodies. Until the 1960s, the barite, fluorite, and sulfide min-

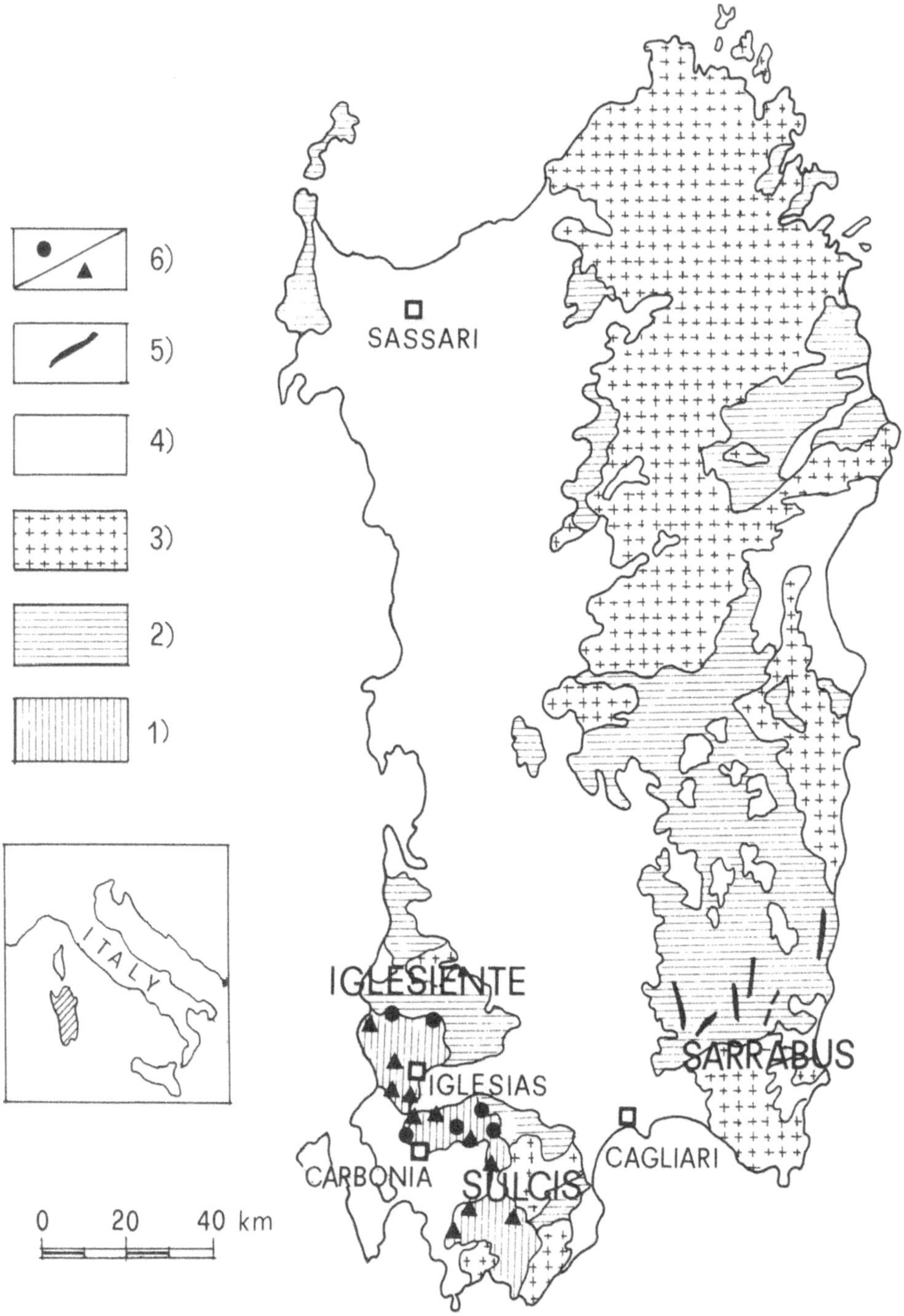

Figure 9-1. Schematic geological map and barite occurrences of Sardinia (Italy).

(1) Cambrian terrains; (2) Ordovician, Silurian, and Devonian terrains; (3) Hercynian granite; (4) post-Paleozoic terrains; (5) barite mineralization: veins; (6) barite mineralization: circle = stratiform, triangle = karstic.

eralizations were held to be of hydrothermal origin, linked to the "granitoid" stocks of the Hercynian orogenesis. The barite ore deposits of Sulcis–Iglesiente (southwest Sardinia), and particularly those at Barega mine, were also believed to be hydrothermal, even though it was thought that some bodies had been deposited "per descensum" in karstic cavities (Tamburrini and Violo, 1965).

Acquired data and geological investigations on the Sardinian Cambrian in the 1970s (Carannante et al., 1974; Cocozza et al., 1974; Cocozza, 1979; Boni et al., 1981) improved the models of barite deposits, singling out several mineralizing stages, involving different genetic environments and ages (Tamburrini, 1968; Marcello, 1969; Tamburrini and Zuffardi, 1969; Gandin et al., 1973, 1974; Padalino et al., 1973, 1980; Boni, 1978; Pretti et al., 1978), while two recent papers provide even more detailed knowledge on the deposition, remobilization, and concentration aspects. The first of these papers (Marcello et al., 1983) furnishes a detailed classification of all Sardinian barite occurrences, covering sedimentary, reworked, and hydrothermal deposits. The second (Garbarino et al., 1984) outlines the paleogeographic and geodynamic evolution of the Cambrian basin, with primary depositions of galena, sphalerite, pyrite, and barite in the early metallogenic epoch (the Cambrian).

The subsequent metallogenic epochs, namely, that in the Hercynian, with mesoepithermal veins of barite–fluorite, and the Mesozoic, with barite bodies in karstic cavities, represent the results of different endogenous and exogenous processes. The latter, along with sedimentary barite, are described in this chapter.

GENERAL GEOLOGY

It may be useful to outline the main occurrences of barite mineralization within the framework of the sedimentologic, tectonic, and geomorphologic evolution of the host terrains, since knowledge of barite depositional parameters can furnish guidelines for ore prospecting in the metallogenic provinces, as described later.

1. The most important barite mineralization, as regards volume and content, occurs in southwest Sardinia, particularly in the Lower-Middle Cambrian terrains. This mineralization differs somewhat in genesis and age; in the Cambrian carbonatic sediments in this area there are stratiform bodies, syngenetic with Cambrian dolomites, stratabound horizons of barite and quartz along the Cambrian-Ordovician unconformity, and epigenetic veins and bodies in karstic cavities, as well as recent alluvial deposits. This barite mineralization, although varying widely in occurrence, age of deposition, and volume (and hence in economic importance), is closely linked to the great sedimentologic cycle of the Cambrian platform, its erosion, and post-Hercynian peneplanations.
2. The barite deposits in central Sardinia occur mainly as veins (Fig. 9-1). These veins—in which barite, along with fluorite, sometimes represents the gangue mineral of mixed sulfide mineralization—exhibit epigenetic deposition, widely connected with emplacement of the Hercynian granitoid stocks.

Thin beds of barite, interstratified in the Silurian slates and volcanites, occur in the internal trough of central Sardinia (Garbarino et al., 1984). This barite is frequently closely associated with stratabound sulfide mineralization (Zn, Fe, Pb, Cu). Though this type of barite mineralization is of little importance, it is of interest because in the western Mediterranean the Silurian and Devonian periods represent a metallogenic epoch for nonferrous metals, barite, and fluorite; this type of mineralization in fact occurs in Spain and in the Central Pyrenees (Pouit et al., 1979), and it is believed to be of volcano-sedimentary genesis (Garbarino et al., 1984; Ferrer et al., 1984).

Certain authors (Calvino, 1961; Marcello et al., 1983) dubiously ascribe some barite and fluorite veins to Alpine orogeny volcanism.

Geologic Setting of Sardinia

As far as the Paleozoic sequences are concerned, the Sardinian microplate may be divided into three main structural areas (Cocozza et al., 1974; Carmignani et al., 1982).

1. *Southwestern Sardinia,* characterized by the sedimentation of the carbonate platform (Fig. 9–2). This area, in which the Cambrian sequence ourcrops, may represent the foreland of the Hercynian orogene (Carmignani et al., 1979). Important ore deposits with Pb, Zn, and Ba occur in the Cambrian dolomites and limestones, exhibiting interesting diagenetic concentration and tectonic and supergene remobilization of metals. In this sedimentary basin, the post-Cambrian sequence is relatively thin but along the Cambrian–Ordovician unconformity stratabound barite mineralization, hosted in a silica horizon, is to be found.
2. *Central Sardinia* ("internal trough," Cocozza et al., 1974) area, from Sarrabus to Barbagia and Nurra, includes the thickest Paleozoic sequence of the Sardinian microplate. Apart from some basal laminated porphyries ("Porfiroidi") whose age is still obscure (Precambrian?), sediments and volcanites from the Upper Cambrian up to the Lower Carboniferous are present and can attain a thickness of 10,000 m. The volcanites and tuffites are rather more widespread than in the southwestern region and stratabound mixed-sulfide mineralizations are present. Barite and fluorite occur mainly as veins and/or epigenetic features.

Figure 9–2. Schematic geological map and barite occurrences of southwest Sardinia.

(1) Lower Cambrian: sandstone and siltstone (Nebida formation); (2) Lower Cambrian: dolomite and limestone (Gonnesa formation); (3) Middle Cambrian: slate (Cabitza formation); (4) Ordovician–Silurian: conglomerate, sandstone, and siltstone; (5) Hercynian granite; (6) post-Paleozoic terrains; (7) Silica horizon along the Cambrian–Orovician unconformity, stratabound occurrences of barite; (8) occurrences of barite: circle = stratiform, triangle = karstic bodies.

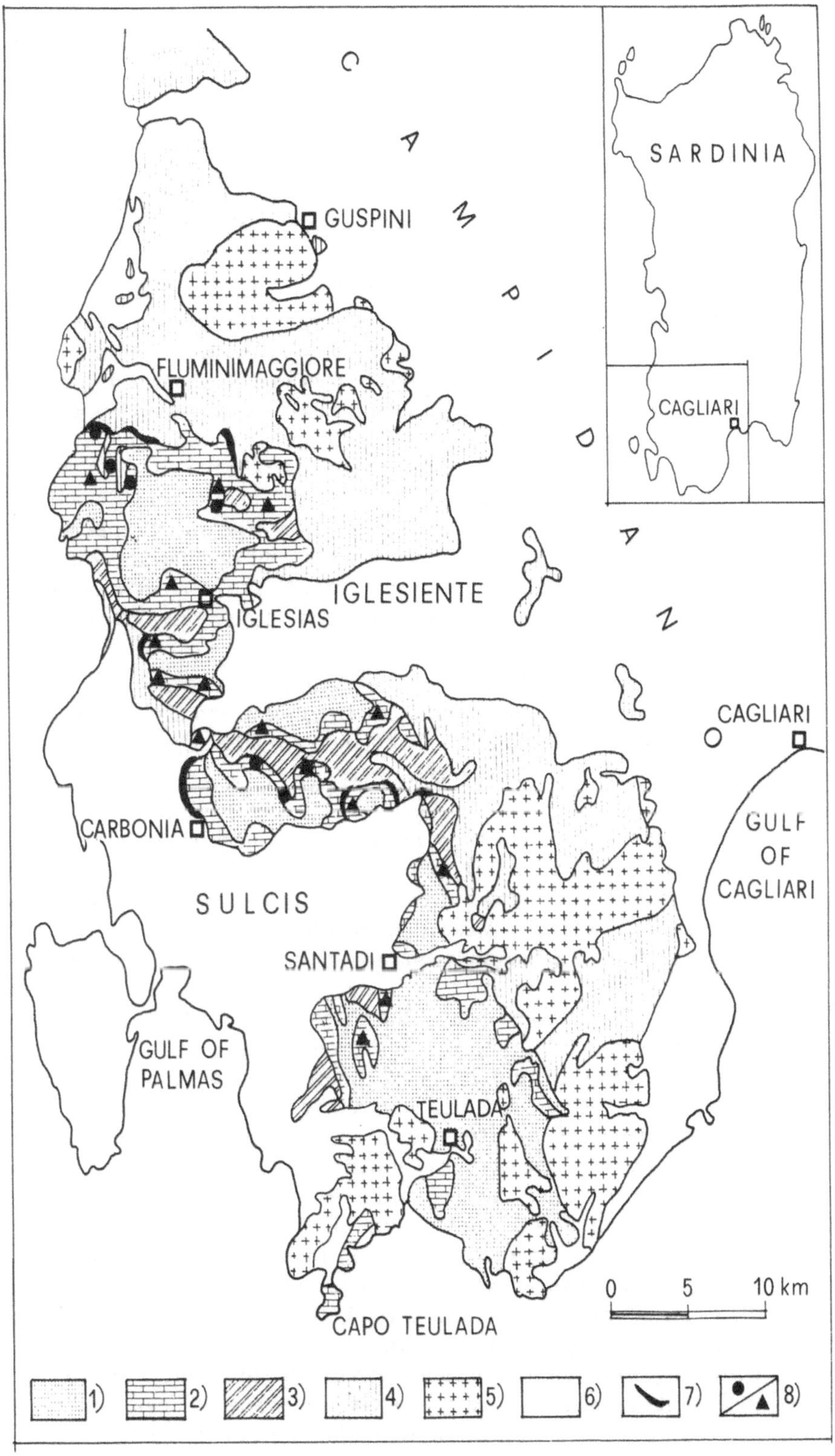
CAMPIDANO
SARDINIA
CAGLIARI
GUSPINI
FLUMINIMAGGIORE
IGLESIENTE
IGLESIAS
CAGLIARI
CARBONIA
GULF OF CAGLIARI
SULCIS
SANTADI
GULF OF PALMAS
TEULADA
0 5 10 km
CAPO TEULADA
1)
2)
3)
4)
5)
6)
7)
8)

3. *Northern Sardinia and Corsica* area is believed by some authors (Carmignani et al., 1979) to be the root of the Hercynian belt. No outcrops of barite mineralization occur in this area.

This Paleozoic basement is overlain by a Mesozoic and Cenozoic cover, affected by different types of paleoenvironments and mineralizations.

These three main structural belts are characterized by the following metamorphic and structural zoning (Carmignani et al., 1982; Ghezzo et al., 1982).

1. *Northeastern Sardinia:* The highest metamorphism occurs in this belt, which has migmatites with eclogite relicts. No barite mineralization is found in this area (Di Simplicio et al., 1974; Carannante et al., 1981).
2. *Central Sardinia:* This belt is characterized by polyphase tectonics and green schist metamorphism. Strainslip cleavage, shortening, and overthrusts are very common in this area. Important veins of barite and fluorite occur in the southern part of the central belt (Fig. 9–1). Minor occurrences of stratiform barite, along with Zn, Pb, and Cu sulfides may be found, as mentioned, in the Silurian–Devonian sequences.
3. *Southwestern Sardinia:* This belt, where Cambrian terrains outcrop, is characterized by low metamorphic grade (epizonal) metamorphism, thus the primary sedimentologic characters are readily recognizable.

The schematic structural frame of the Cambrian outcrops may be represented by a system of folds and faults due to at least two different main orogenetic events; the Caledonian and the Hercynian. The Alpine orogenesis exhibits mainly wrench fault characters.

1. The oldest tectonic structures are represented by an early "Caledonian phase," which, acting in a north–south stress direction, caused east–west folds and a series of faults striking north 30° west and north 30° east. The north 30° west folds may be considered to have a typically Caledonian direction owing to the great number and development.
2. The subsequent tectonic events, the Caledonian and Hercynian, at the beginning still exhibit a north–south stress direction (Poll and Zwart, 1964; Poll 1966). During this phase the early Caledonian folds shrunk and the north 30° west faults were rejuvenated.
3. The persistence of a north–south trend enabled the development of a second stress phase acting in northwest-southeast and northeast–southwest directions. These could be responsible for the wide dispersion of the fold axes (Carmignani et al., 1982). In the Cambrian terrains, the folds whose axis strikes north–south may be imputed to these phases. In the post-Cambrian terrains, the typical Hercynian direction is north 70° east. The final structure of the Cambrian sequence, caused by the above tectonic events, is characterized by folds with axes running from east–west to north–south and a set of fractures striking north 30° west, north–south, north 30° east, and north 70° east. Several papers are now available on the tectonic structures of the Sardinian microplate; the reader is referred to these for more

detailed information (Arthaud, 1963; Poll, 1966; Mc Moore, 1969; Carannante et al., 1981; Ghezzo et al., 1982).

DEPOSITIONAL HISTORY OF THE CAMBRIAN BASIN

As already indicated, the geology of southwest Sardinia is characterized by a sedimentary Paleozoic sequence, consisting essentially of a considerable thickness of Cambrian carbonatic platform lithofacies (Fig. 9–2; Fanni et al., 1981).

This Cambrian sequence may be schematically subdivided into three main formations (Fig. 9–3), the bottom one involving prevalently clastic sedimentation (Nebida formation sandstone) followed upward by carbonate sedimentation (Gonnesa formation dolomites and limestones), while the deposition of slates and siltstones (Cabitza formation), lying conformably on the preceding carbonates, closes the Lower-Middle Cambrian sedimentologic cycle.

This sequence lies on what is assumed to be the Precambrian (Bithia formation; Junker and Schneider, 1979), outcropping in the southern part of the island and is covered by the clastic, transgressive Ordovician deposition.

The Cambrian formations as a whole may reach 1500 m in thickness; they testify to a sedimentologic evolution starting from a clastic shelf, which became a carbonate platform during the Lower Cambrian and was subsequently drowned during the Middle Cambrian.

The depositional history of the Cambrian basin may be outlined as follows (Figs. 9–2 and 9–3);

1. *Lower Cambrian:* Clastic sedimentation with sandstone of variable grain size and siltstone evolving upward to an alternation of clastic and carbonatic sediments. This sequence, which is locally known as the Nebida formation, is about 650 m thick and has two members: Matoppa member (lower) and Punta Manna member (upper).

   The lower member is characterized by the presence of carbonatic lenses, mainly in the upper part, in the siltstone. Depending on shape and faunal association (algae, archaeocyatha), these lenses may be classified as bioherms or mounds (Debrenne, 1964; Bechstadt et al., 1985).

   The upper member has carbonatic horizons, interbedded in the sandstone–siltstone, with oolitic and/or algal levels. Upward the carbonatic horizons exhibit early dolomitization, announcing a modification of the sedimentologic and diagenetic environment.
2. The Gonnesa formation lying conformably on the preceding formation is characterized by thick, continuous carbonatic sedimentation. It is about 600 m thick and subdivided into two members: Laminated dolomite ("Dolomia rigata" Auct.) at the bottom and Ceroide limestone at the top.

   The Laminated dolomite member consists of dolomite, with a large number of stromalites, sometimes alternating with thinly laminated muds. These muds frequently exhibit oncolites and drying phenomena, together with levels and nodules of chert, with phantoms and pseudomorphs of evaporites.

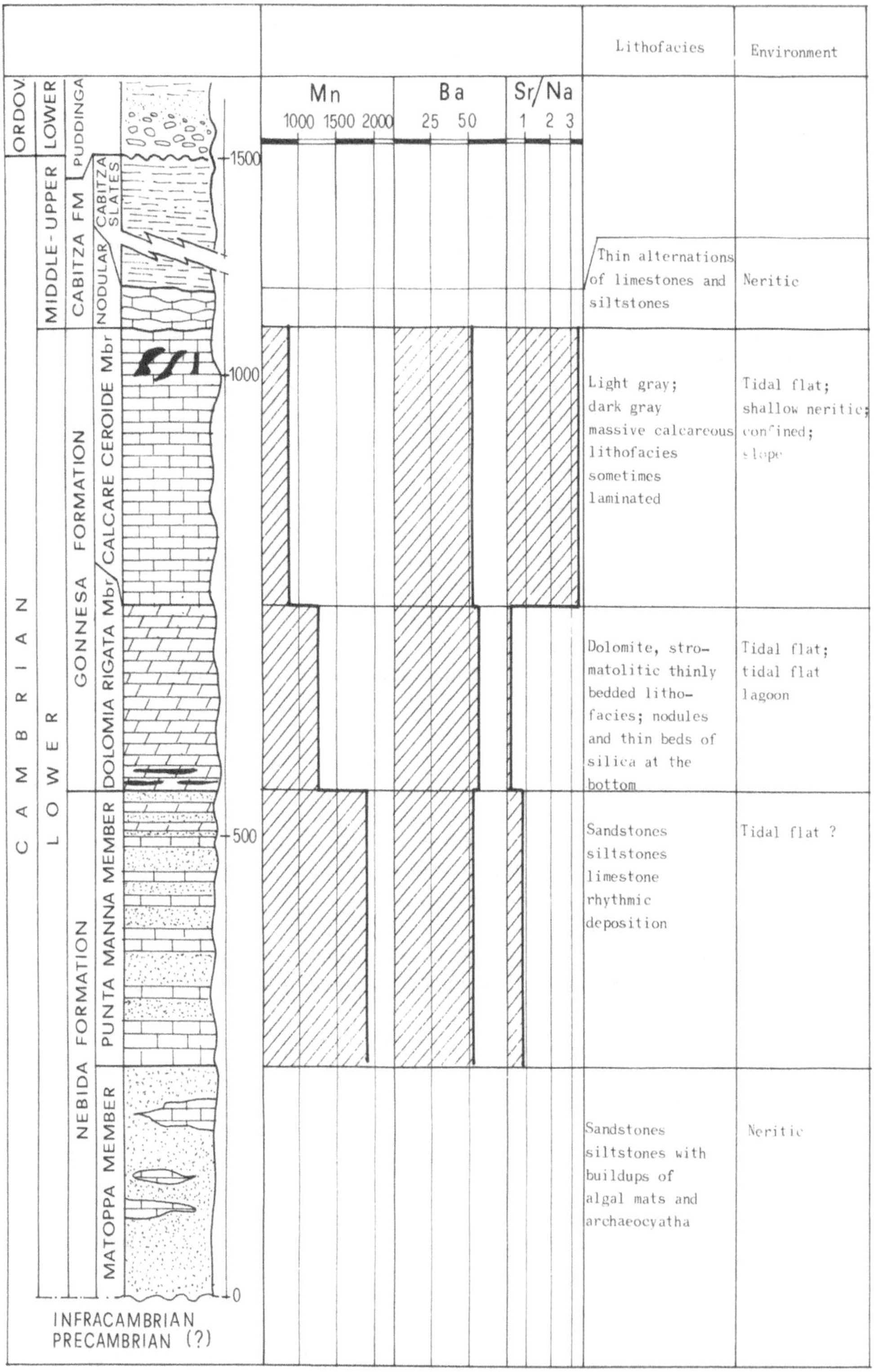

Figure 9-3. Cambrian-Ordovician sequence in southwest Sardinia.

Carbonate microfacies	Structure	Biota	Energy	Diagenesis	Mineralization
Bioclasts wacke-pakstone	Early diagenetic nodular fabric	Trilobites echinoderms lingulids etc.	Low to Medium	Marine Phreatic	
Mudstone; algal bindstone; wackestone; fenestral mudstone	Fenestral bio-clasts; detrital quartz bioturbation; stylolites	Algae; trilobites; echinoderms; lingulids; hyolites; chanceloria	Low	Marine phreatic	Barite and sulphides in nodules, dissemination and veinlets
Bindstone; mudstone; vadose pisolite ooids, oncoids; wackestone; grainstone	Algal mat planar lamination bioclasts tepee sulphate pseudom. length slow chalcedony	Trilobites echinoderms	Low	Schizohaline to marine vadose	Stratiform barite
Mudstone bindstone ooid, oncoid, peloid, intraclast bioclast, wackestone, packestone, grainstone	Planar lamination flaser bedding algal mat bioclast cross bedding cut and fill birdseye stylolite detrital quartz	Trilobites echinoderms, lingulids, archaeocyatha girvanella	Low to High	Marine phreatic; marine vadose schizohaline	
Bindstone wackstone mudstone	Cut and fill breccia burrowing antigene quartz stylolite	Archaeocyatha epiphyton trilobites chanceloria hyolites chabakovia	Low to High	Phreatic marine	

The Ceroide limestone member, following the preceding one conformably, is characterized by a very fine-grained limestone, sometimes finely laminated, due to muddy sedimentation, but with algal events.

3. The slates and siltstones of the Cabitza formation testify to the drowning of the Cambrian platform. The evolution from carbonatic to clastic sedimentation is relatively sudden and is marked by the presence of the Nodular limestone formation. This formation, a few meters thick, consists of alternating clayey and carbonatic beds sometimes very rich in fossils. The Cabitza formation develops upward to a thick sequence of clastic sediments, testifying to the establishment of an epicontinental basin.

The evolution from a platform environment to a clastic basin is probably connected with rifting phenomena, during a period of extensional tectonics. The subsequent Ordovician sedimentation, in fact, starts with conglomeratic facies, transgressive on Cambrian terrains; this transgression covers the Cambrian sequence, gently folded by the early Caledonian orogenetic phase.

PALEOGEOGRAPHIC EVOLUTION OF THE CAMBRIAN PLATFORM

The Cambrian lithofacies described seem to indicate an early regressive phase (Laminated dolomite) followed by a relative deepening of the sea (Ceroide limestone deposition).

Development of the paleogeographic and diagenetic environments, particularly as regards the carbonatic facies, which is the most important for barite mineralization, may be summarized as follows:

1. Carbonatic sedimentation becomes established on clastic terrains, formed under high energy conditions in a deltaic environment.
2. Detrital supply tends to decrease, allowing deposition of carbonate horizons. A lagoon, between a detrital beach and open sea, begins to take shape; sedimentation is characterized by rhythmic deposition of clasts and carbonates, the latter with oolites, oncolites, algae, archaeocyatha, trilobites, and so forth. The trend toward emergence is marked by the presence of inter- and supratidal facies and drying phenomena; the environment seems to be a tidal-flat lagoon, with diagenesis evolving from marine-phreatic to marine-vadose, schizohaline. Geochemical investigations (Table 9–1) on carbonatic facies have revealed the high Mn content (about 1800 ppm on average). This Mn content seems to indicate the proximity of a paleocontinent and an oxidizing environment; the value of the Sr/Na ratio (0.9) may also indicate rather high salinity tending to evaporitic (Land and Hoops, 1973). The average Ba content in the carbonate horizons is 53 ppm.
3. Deposition on Laminated dolomite marks the establishment of a new balance in the basin. With peneplanation of the paleocontinent, the detrital supply is completely absent and thick, continuous dolomite deposition develops on a tidal flat. The most common sediments con-

Table 9–1. Geochemical Data on the Carbonatic Facies

		Ba	*Mn*	*Sr*	*Na*	*Sr/Na*
Ceroide limestone member	$\bar{x}$	51	512	152	44	3.3
	σ	19	169	43	24	
	n	97	95	98	94	
Laminated dolomite member	$\bar{x}$	46	1176	82	172	0.5
	σ	32	509	39	69	
	n	76	79	73	71	
Punta Manna member	$\bar{x}$	53	1866	126	131	0.9
	σ	26	687	49	48	
	n	72	57	68	67	

Notes: $\bar{x}$ = average content in ppm; σ = standard deviation; and n = number of samples. The number of samples stems from x^2 test, which shows a different distribution for each element; only samples inside the $\pm 3\sigma$ mean have been utilized.

sist of finely laminated muds, shadowing vertically or laterally to stromatolitic and algal dolomite, sometimes largely dried. The occurrence of chert levels, sulfate pseudomorphs, length-slow chalcedony, and precocious dolomitization mark the evaporitic trend on the environment.

The climate seems to be warm and dry on a tidal flat with reduced connection with the open sea; the energy is low, and diagenesis is marine-vadose and schizohaline. The Sr/Na ratio (0.5) is higher than the preceding one (Table 9–1), being indicative of considerable salinity, while the Mn content (about 1200 ppm) confirms an oxidizing environment. The average barium content is 46 ppm.

4. Ceroide limestone deposition marks an important variation in basin characteristics. The regressive trend seems to become exhausted and slight deepening of the basin appears; the climate becomes temperate and wet.

   The most common facies are represented by mudstones, peloidal mudstones, and algal bindstones. Subordinately laminated facies are found, indicative of the presence of a gentle slope in the central and southwestern area of the platform (Fig. 9–4). Different deposition environments occur on the platform such as tidal flats, vanishing toward a shallow neritic, locally confined environment and the slope just mentioned. Energy is relatively low, while diagenesis is marine-phreatic and sometimes marine-vadose. The Sr/Na ratio (3.3) indicates a sharp variation of salinity (Table 9–1), while the Mn content is reduced to 50 ppm, testifying to a less oxidizing environment than the preceding ones. The Ba content is 51 ppm; sulfide and barite mineralization occurs in the upper part of the Ceroide limestone.
5. The nodular limestone, followed by the deposition of Cabitza formation, marks a break in the sedimentologic balances. The carbonatic sedimentation tends to run out, while the detrital supply increases, owing to the sudden drowning of the platform.

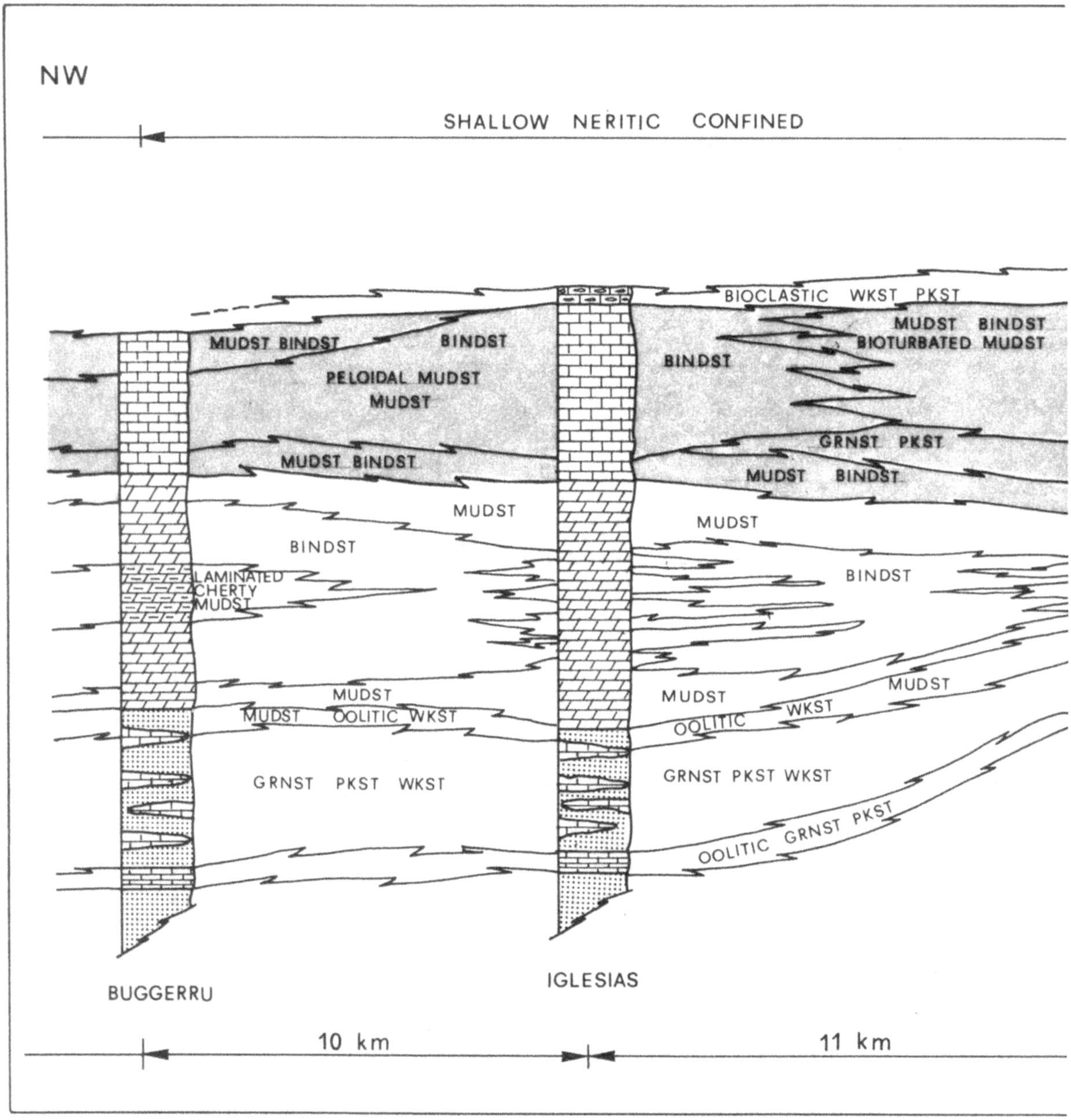

Figure 9–4. Different Cambrian sequences in southwest Sardinia from northwest to southeast, showing the microfacies and the assumed paleogeographic environment of the Cambrian platform.

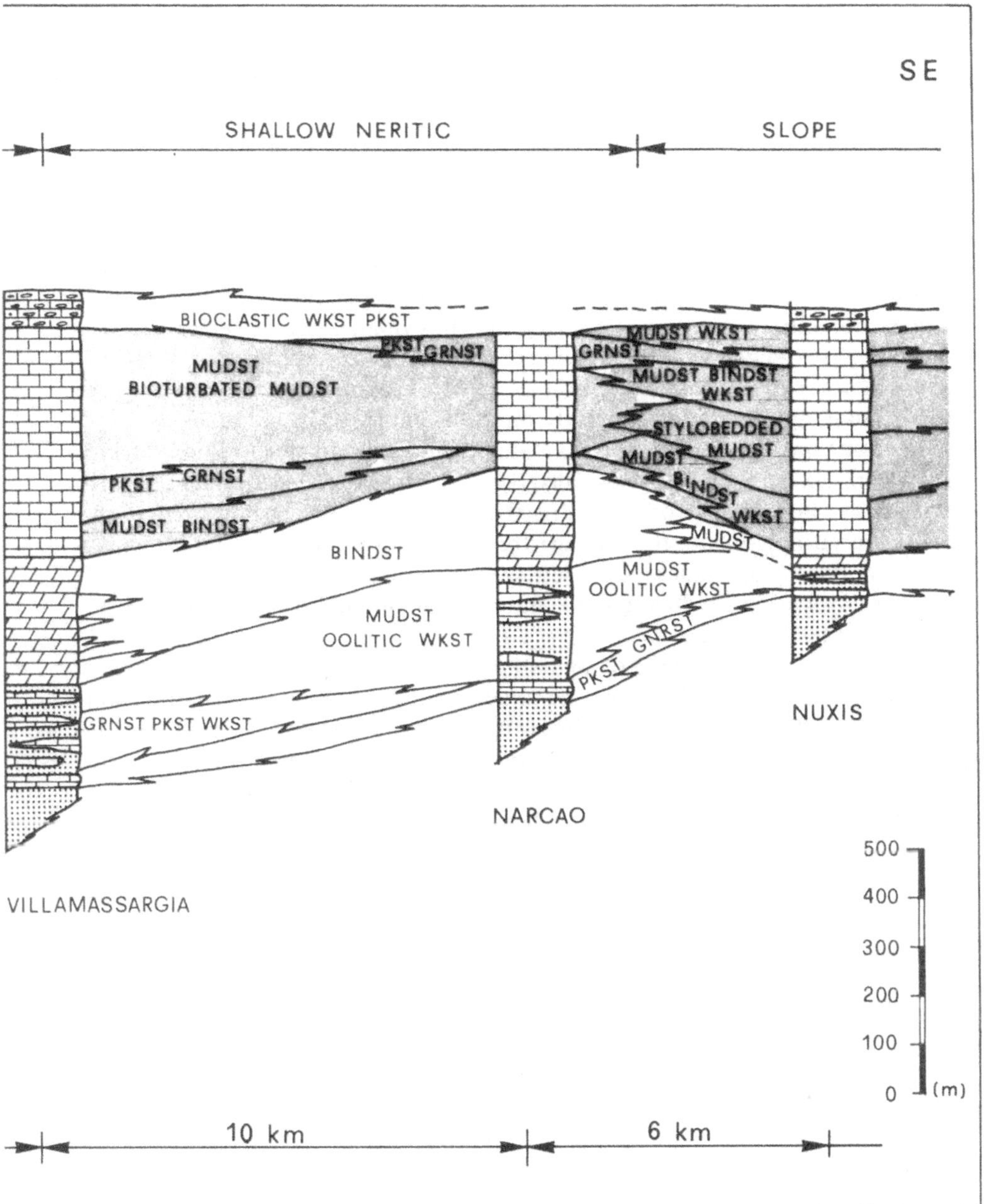
SE
SHALLOW NERITIC
SLOPE
BIOCLASTIC WKST PKST
PKST GRNST
MUDST
BIOTURBATED MUDST
PKST GRNST
MUDST BINDST
BINDST
MUDST WKST
GRNST
MUDST BINDST
WKST
STYLOBEDDED
MUDST
MUDST
BINDST
WKST
MUDST
MUDST
OOLITIC WKST
MUDST
OOLITIC WKST
PKST GNRST
GRNST PKST WKST
NUXIS
NARCAO
VILLAMASSARGIA
500
400
300
200
100
0
(m)
10 km
6 km

CAMBRIAN–ORDOVICIAN UNCONFORMITY

The Cabitza formation closes the Middle Cambrian, and perhaps the Upper Cambrian sedimentation (Barca et al., 1981). This member is covered unconformably by a polygenic conglomerate (''Puddinga'' Auct.) of Ordovician age (Fig. 9–5).

The conglomerate, with its arenaceous-clay matrix, reddish for hematite, is present with angular unconformity along the whole contact between the Cambrian and Ordovician terrains. Some authors (Del Bono, 1965; Brouwer, 1966, 1986) believe this ''Puddinga'' to be a sort of tectonic breccia, and not a transgressive conglomerate, produced during the Hercynian cleavage phases. However, whatever the origin of ''Puddinga,'' the unconformity represents an important marker from the metallogenic point of view, especially where the Ordovician conglomerate lies directly on the Cambrian limestone.

The Cambrian carbonate terrains, which emerged and were folded during an early orogenetic phase (the so-called Sardic phase of the caledonian orogenesis) have, in fact, been eroded and karstified, the processes being favored by climatic variations during the passage from the Cambrian to the Ordovician. Faunal analyses (Spjeldnaes, 1961) testify to a sudden change in climate, passing from hot and damp during the Cambrian to cold and dry with periglacial features in the Ordovician.

Furthermore, the presence of hematite in the matrix of the conglomerate could well confirm the erosion of the underlying limestone and the formation of karstic cavities, subsequently filled with barite, silica, and, sometimes, argentiferous galena at the beginning of the Ordovician sedimentation.

The unconformity is marked in the field by the presence of a thick silica horizon particularly well developed when the Ordovician conglomerate lies di-

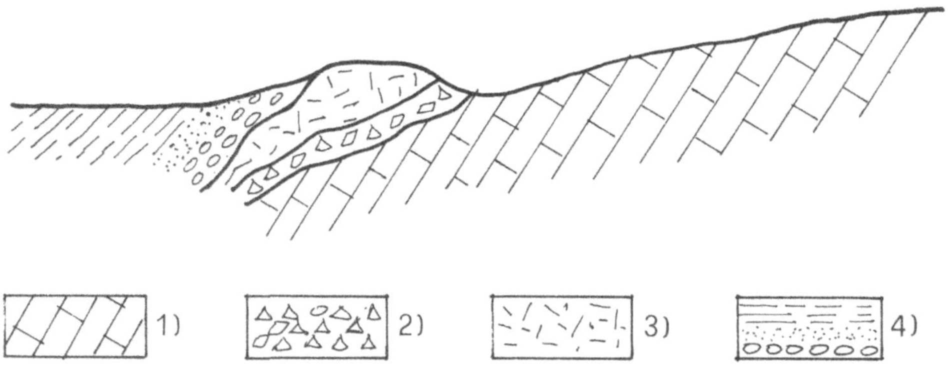

Figure 9–5. Cross section of the silica horizon along the Cambrian–Ordovician unconformity.

(1) Cambrian limestone (Gonnesa formation); (2) breccia made up of fragments of limestone and dolomite, more or less intensely silicified, cemented by quartz and barite; (3) Massive quartz and chalcedony, sometimes with barite and argentiferous galena; (4) Ordovician transgression with basal conglomerate shadowing upward to sandstone and siltstone.

rectly on the Cambrian limestone (Fig. 9–6). The map in Figure 9–2 shows the occurrence of this silica horizon in southwestern Sardinia.

Two hypotheses may be put forward to explain the origin of silica along the unconformity surface:

1. Change of paleoenvironment and climate between the Cambrian and the Ordovician. The silica, removed from the weakly deformed Cambrian terrains, may have circulated on the Cambrian carbonates, owing to the high solubility of silica at high pH. Then in the Ordovician, in the small paleobasins of the eroded Cambrian limestone, the silica may have been deposited owing to a colder climate and, probably, a lower pH. This idea is summarized in Table 9–2.
2. Volcanic supply of silica. The end of the Cambrian and the whole of the Ordovician are characterized by intense volcanic activity, represented nowadays by thick bodies of metamorphyries (''porfiroidi'' Auct.), rhyolites, and tuffs. Intensification of the role of volcanism during the Ordovician is also borne out by the higher fluorite contents of Ordovician slates than on those of the Cambrian sequence (Mazzella et al., 1979).

It is thus possible to postulate an increase in the silica content of the volcano-sedimentary fluids circulating along the more porous horizons, in this case the basal conglomerate of the Ordovician transgression.

Figure 9–6. Silica horizon along the Cambrian–Ordovician unconformity. Cambrian limestone can be seen on the right in the photo and Ordovician conglomerate and siltstones on the left.

Table 9–2. Barite–Silica Deposition along the Cambrian–Ordovician Unconformity

Age	*Geological Structure*	*Paleoenvironment*	*Climate*	*Possible pH*	*Mineral Deposition*
Middle-Upper Cambrian	Gentle folds (early Caledonian tectonic phase)	Weakly developed karstic cavities	Hot and wet	$\simeq 8$	Sulfates and locally sulfides
Middle Ordovician	Marine transgression	Delta and beach	Cold and dry	$\simeq 7$	Hematite, silica, and small veinlets of barite

POST-HERCYNIAN PENEPLANATION

The last stages of the Hercynian orogenesis were followed by a general rejuvenation of the Paleozoic terrains along with a weak phase of extensional tectonics leading to the creation of small lacustrine basins, subsequently filled by Lower Permian clastic sediments and rhyolitic ignimbrites.

During the Triassic, the horst-graben structure (Gaspari and Gelmini, 1979) favored an intense cycle of erosion on the highlands with consequent denudation both of Permian sediments and basement rocks, and deposition of clastic and chemical sediments in the basins. These sediments, a few meters thick and spatially rather limited, have barite as cement for the clasts and bedded barite, quartz, and carbonates in the chemical deposits. This type of mineralization appears to be of little economic importance.

Denudation of the subhorizontal Permian sediments, as well as part of the Ordovician and Cambrian sandstones and slates, permitted an intense cycle of weathering acting mainly on the Cambrian carbonates (Gonnesa formation) and giving rise to a large, branched karstic system.

Erosion and alteration were favored and speeded up by several factors, namely:

1. The presence of horsts as a consequence of extensional tectonics in early Permian times.
2. The Hercynian faults and fractures, which provided preferential pathways for percolating waters that dissolved the carbonates.
3. Change in climatic conditions, from hot and wet in the Permian to dry in the Triassic, giving rise to the formation of iron and barium-rich soils.
4. Occurrence of sulfides in the Gonnesa formation; oxidation, weathering, and leaching were thus favored by the presence of acids in the percolating waters.

As will be described below, the karst system provides the loci and traps for the most important supergene barite deposition in Italy.

During the Tertiary and Quaternary the block structures of the Alpine orogenesis led to reactivation of the old karst system, the cavities being filled by new materials such as Eocene brown coal, detrital galena, and barite.

The economic value of these ores is described in the next section devoted to economic geology.

ECONOMIC GEOLOGY

The history of barite deposition in the geological sequences of Sardinia is quite complex: several stages of mineralization and remobilization from the Cambrian to the present modified the primary structures and gave rise to different ore accumulations (Tamburrini, 1968, Tamburrini and Zuffardi, 1969, Zuffardi 1969, Padalino et al., 1980).

The phenomenon of heredity from the oldest barite occurrence to the most recent is evident from study of the geological evolution of the Paleozoic terrains. Geochemical and fluid inclusion research has also furnished some general data, namely:

1. Determination of the Sr content of several samples of barite (Barbieri et al., 1984) indicates that, apart from variations in Sr content among the sedimentary, karstic, and so-called hydrothermal barites, the origin of different barite occurrences seems closely connected with the remobilization of barium included in the basal Cambrian sandstones and of barite in the laminated dolomite.
2. Fluid inclusion composition and temperature of formation (Boni, 1986) are in good agreement with the idea of low-temperature barite deposition (karstic deposits). Part of the ore-forming fluids were warm and very saline; these were believed to be responsible for the so-called hydrothermal veins of barite.

Nevertheless, the high salinity could be derived from the evaporitic sediments of the Permo-Triassic, while the high temperature may have been the result of diagenetic–epigenetic exothermic processes, and of an increase in geothermal gradients due to granitic intrusions; the influence of magmatic intrusions concerns the barite deposits along the Cambro-Ordovician unconformity (Fig. 9–7).

Some ideas may be postulated, too, regarding the primary origin of barium. The lowest Cambrian terrains, namely sandstones and siltstones, testify to the

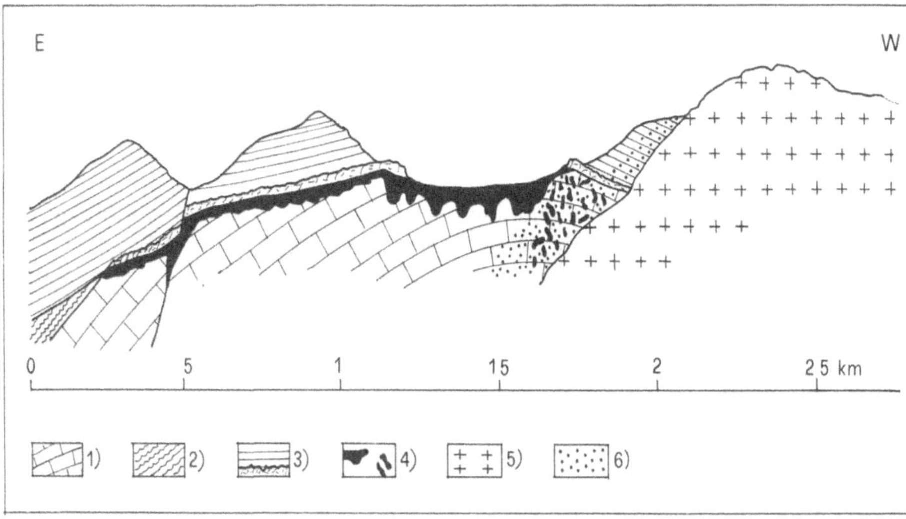

Figure 9–7. Metamorphosed mineralization in the silica horizon along the Cambrian–Ordovician unconformity in the Arenas deposit, northeast of Iglesias (after Padalino et al., 1973, pp. 213 and 215).

(1) Cambrian limestone (Gonnesa formation); (2) Cambrian slate (Cabitza formation); (3) Silica horizon (bottom) and Ordovician conglomerate and sandstone; (4) barite and galena mineralization; (5) Hercynian granite; (6) thermal metamorphic halo.

erosion of a Precambrian continent made up of granitoid stocks, and consequent transport to and sedimentation in the Lower Cambrian basin.

The freeing of metallic cations, along with the chemical alteration of micas and feldspars, might well have been the source of the mineralization "matter." The barium, leached from K-minerals, might have been removed, as bicarbonates and chlorites, depending on the composition and nature of the carrying waters.

The deposition of barite depends heavily on the shape of the sedimentary environment and the physicochemical parameters.

STRATIFORM BARITE IN THE LOWER CAMBRIAN

As has been remarked above, the end of detrital sedimentation on the shelf marks the establishment of the carbonatic platform, characterized by very shallow waters and an undulating floor.

The stratiform barite appears just at the beginning of dolomite sedimentation, and the lenses of sulfate are included, conformably, in the laminated dolomite; the dolomitic lithofacies points to the presence of small lagoons and supratidal flats. This platform morphology might well have permitted the establishment of evaporitic environments in which sulfates were deposited.

It is possible to think in terms of at least two sizes of paleolagoons: those located on the platform "highs" were small and characterized by oxidizing conditions, high pH, and positive Eh; thus providing the preferable locum for sulfate deposition.

Other lagoons instead played a regional role in southwest Sardinia. Maximum carbonate thickness occurs, in fact, in the Iglesiente area where sedimentation was confined and there was a reducing environment and perhaps negative Eh. Sulfide deposition occurred in this part of the Cambrian platform.

Figure 9–8 furnishes a paleoenvironmental hypothesis regarding sulfate–sulfide deposition. This is confirmed, at regional scale, by the presence of the clear sulfate–sulfide zoning.

The Iglesiente area has stratabound mineralization of mixed sulfides (pyrite–sphalerite–galena) surrounded, particularly toward the east and the southeast, by several occurrences of stratiform barite (Fig. 9–9). This barite outcrops as small bodies, perfectly interstratified between sandstones and dolomites, vanishing laterally, by pinch-out, to laminated dolomites.

The structure of the barite may be massive or laminated, the latter being made up of thin alternations of baritic and dolomitic beds (Fig. 9–10). Microscopically, the barite beds, in turn, consist of alternations of fine and coarse sulfate crystals (Fig. 9–11). The barite does not always occur as interbeds in the dolomite, however. Indeed, detailed microscopic and chemical analyses have revealed the presence of submicroscopic nodules of barium sulfate as inclusions in the dolomite (Fig. 9–12, on page 188).

From an economic point of view this type of mineralization does not seem to be of great importance, as the bodies are limited in horizontal continuity and thickness.

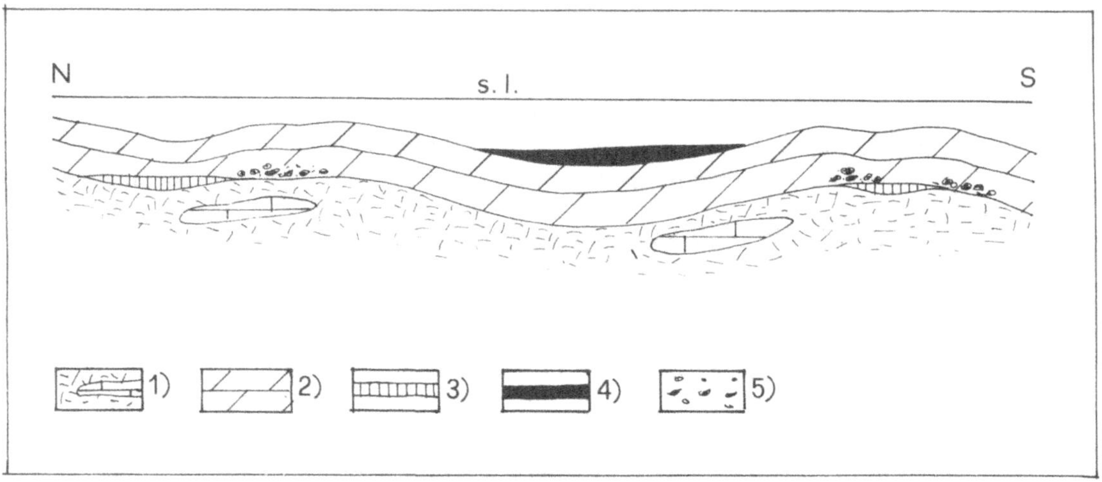

Figure 9–8. Possible Cambrian paleoenvironment with barite and sulfide deposition during the laminated dolomite sedimentation (base of Gonnesa formation).

(1) Cambrian sandstone with limestone lenses (Nebida formation); (2) Cambrian laminated dolomite (Gonnesa formation); (3) stratiform barite; (4) stratiform sulfides; (5) nodules and/or fine intergrowth of barium sulfate and dolomite.

Stratabound Barite along the Cambrian-Ordovician Unconformity

The mineralization located along the unconformity consists of lead–zinc oxides in the northern and central part of Cambrian platform, where primary sulfide occurrences are prevalent, and mainly barite in the southern and eastern part (Figs. 9–3 and 9–9).

This mineralization has been fossilized by intense silicification, now visible as a more or less continuous body of vacuolar, brecciated quartz and chalcedony. The silica horizon is present almost exclusively along the contact between the Ordovician conglomerate and the Cambrian limestones. Where the Ordovician lies directly on the Cambrian slates, this horizon is absent or reduced to a thin silicified crust.

The thickness of the silica horizon may range from a few centimeters to some tens of meters (southern Sardinia, Fig. 9–6).

The barite mineralization may occur in different shapes and structures:

1. Barite bodies in paleokarst below the silica horizon (Fig. 9–13). The vertical development of karstic cavities is not great; the barite occurrence has a flattened shape and some silica is present in the barite mineralization. This type of barite deposit seems to be connected not so much with the weathering processes leading to karstification, as with the sedimentation of the basal Ordovician (Leleu, 1966; Lagny and Rovier, 1976; Boni and Amstutz, 1982); the paleokarst cavities

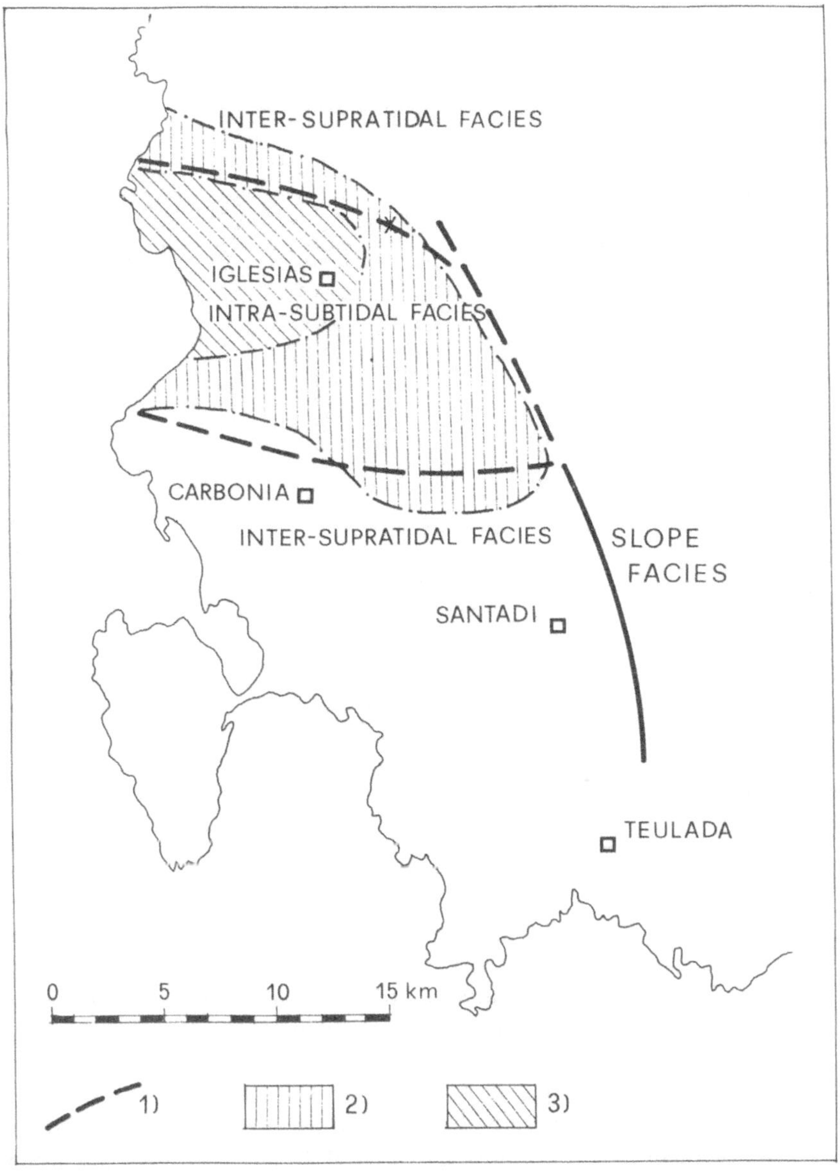

Figure 9–9. Regional sulfate–sulfide paleozoning in southwest Sardinia.

(1) Boundaries of different laminated dolomite facies; (2) area of mainly sulfate deposition; (3) area of mainly sulfide deposition.

acted as traps for barite deposition, giving rise to a rather discontinuous type of mineralization.

2. Barite layers, veinlets, nodules, and breccia within the silica horizon. The barite layers and veinlets exhibit no extensive, continuous spatial continuity, whereas brecciated structures, made up of fragments of limestone broadly cemented by barite and quartz (Fig. 9–14), are very frequent. Galena and framboidal pyrite are not uncommon (Marcello

Figure 9–10. Stratiform, laminated barite occurring at the bottom of laminated dolomite (Gonnesa formation; after Gandin et al., 1973, p. 337).

et al., 1983). The Hercynian granitoid intrusion might have caused some thermal metamorphism on the paleokarst located along the Cambrian–Ordovician boundary. In this case the ore-formation temperature was, of course, higher (Boni, 1986).

In some ore deposits (e.g., at Arenas, Fig. 9–7), the paleokarst has been transformed into skarn, made up essentially of garnets (Benz, 1964; Zuffardi, 1976) more or less mineralized with galena and barite. The dip of this type of mineralization generally ranges from about 30° to almost vertical; for this reason it has been explored and mined to a depth of only a few meters, since underground working increases costs and makes exploitation uneconomic.

BARITE IN PERMO-TRIASSIC PALEOKARST

Permian weathering, acting on the Cambrian carbonates, developed a massive system of karstic cavities and epigenetic dolomitization. These cavities slowly filled during the Triassic, and replacement by barite, quartz, and argentiferous galena occurred (Fig. 9–15). This filling was favored by the concomitance of several factors, namely:

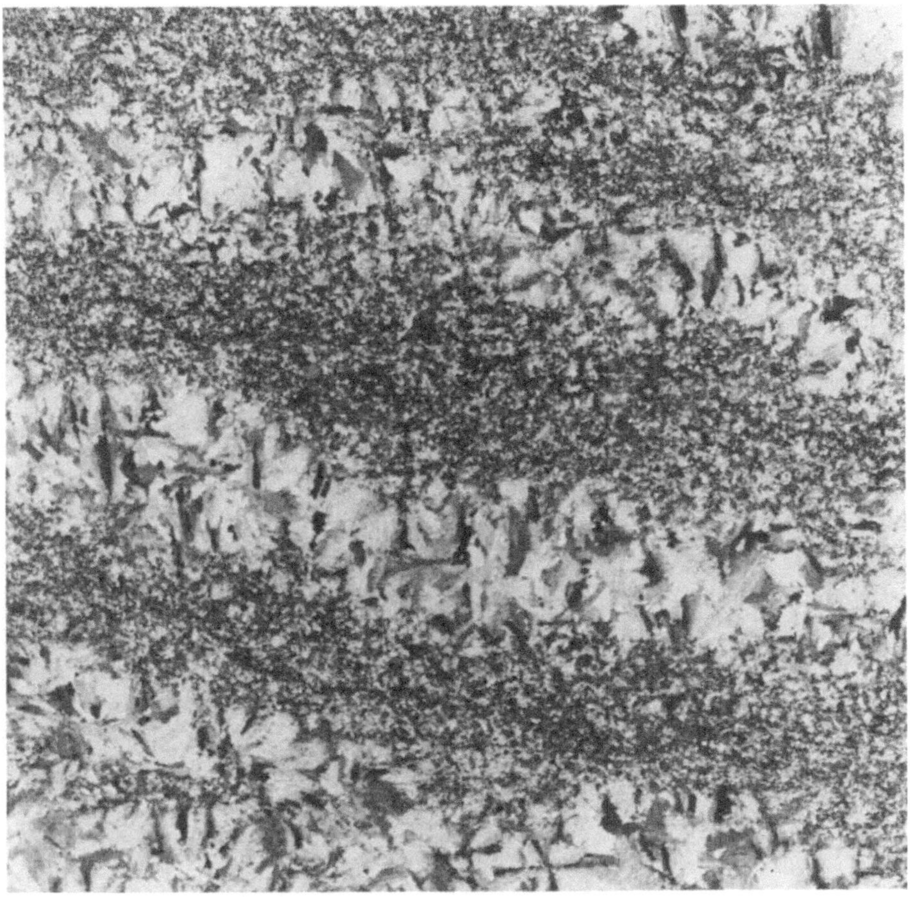

Figure 9–11. Stratiform, laminated barite made up of an alternation of fine and coarse barium sulfate crystals. Crossed Nichols; 100×.

1. Climatic changes; the Triassic was an arid period; rainfall became more sporadic and evaporation increased.
2. pH increased during the Triassic, causing high solubility of silica.
3. The percolating waters were loaded with salts, so the mobility of Ba and other ions became very high in a chloride-bearing solution.

The karstic cavities, acting as ore deposition traps, were thus slowly filled with quartz, sulfates (Fig. 9–16), and—in a senile stage of karst development—sulfides too (Bernard and Leleu, 1967; Bernard, 1971; Padalino et al., 1972). Collapse breccia, cemented by barite, are generally more common than fine-grained internal sediments. These sediments are of detrital or chemical origin and occur mainly at the bottom of karstic pockets (Fig. 9–17, on page 192). The prevalence of collapse breccia and vadose deposition over internal sediments and phreatic environments again testifies to the rather arid climate.

Barite mineralization may occur with different types of fabric, namely:

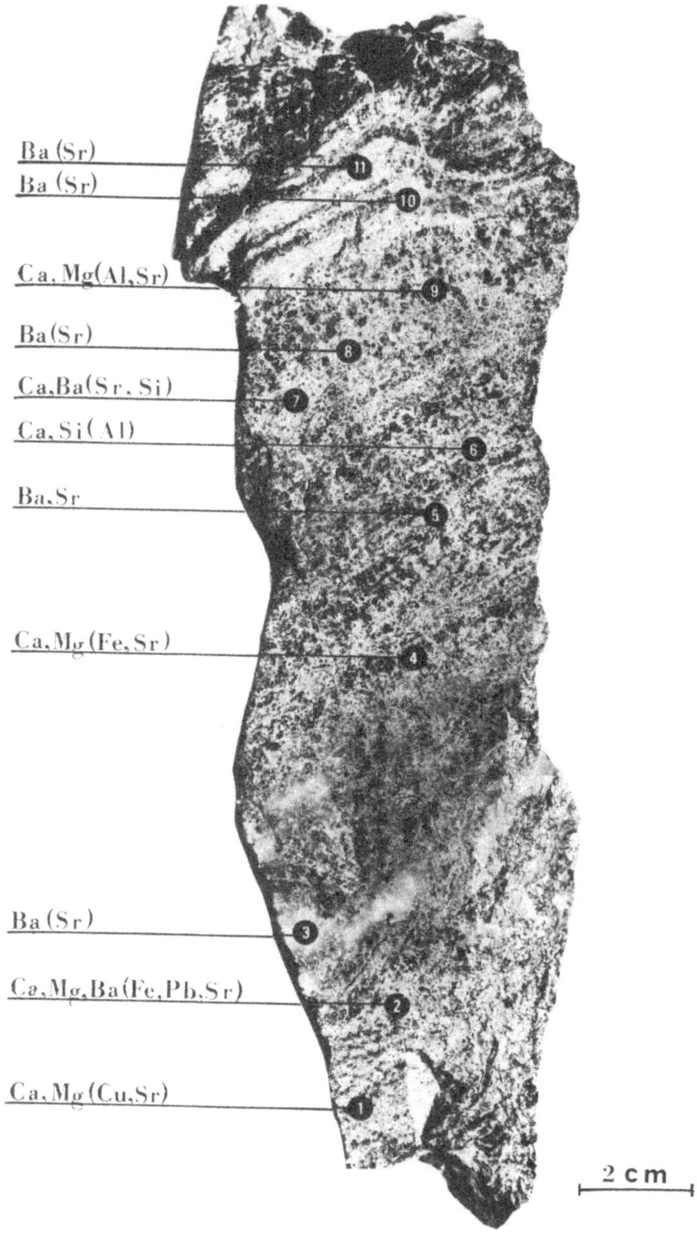

Figure 9-12. Laminated dolomite (Gonnesa formation. Nodules and small veinlets of barite are visible; sample analyzed by LASER Microspectral Analyzer LMA 1 (after Gandin et al., 1973, p. 336).

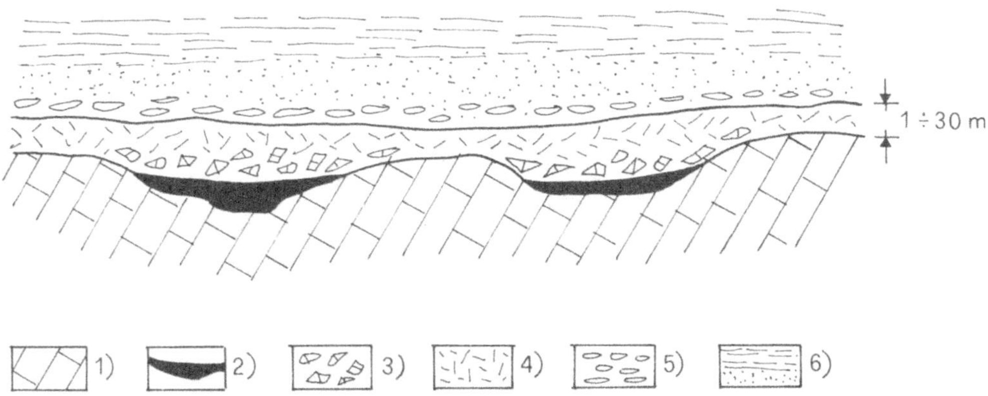

Figure 9-13. Barite mineralization in the silica horizon along the Cambrian-Ordovician unconformity.

(1) Cambrian limestone (Gonnesa formation); (2) massive barite; (3) brecciated limestone, intensely silicified; (4) quartz and chalcedony; (5) Ordovician transgressive conglomerate; (6) Ordovician sandstone and siltstone.

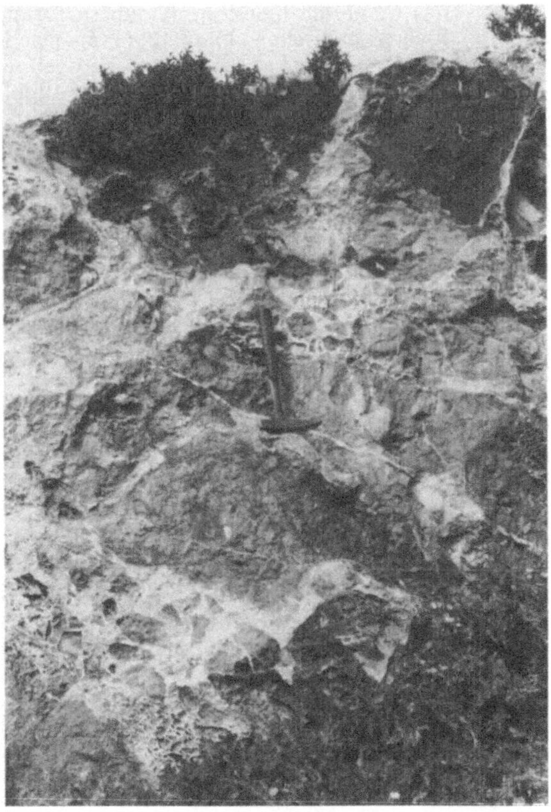

Figure 9-14. Barite mineralization in the silica horizon along the Cambrian-Ordovician unconformity. Breccia, consisting of fragments of intensely silicified limestone, cemented by barite (white).

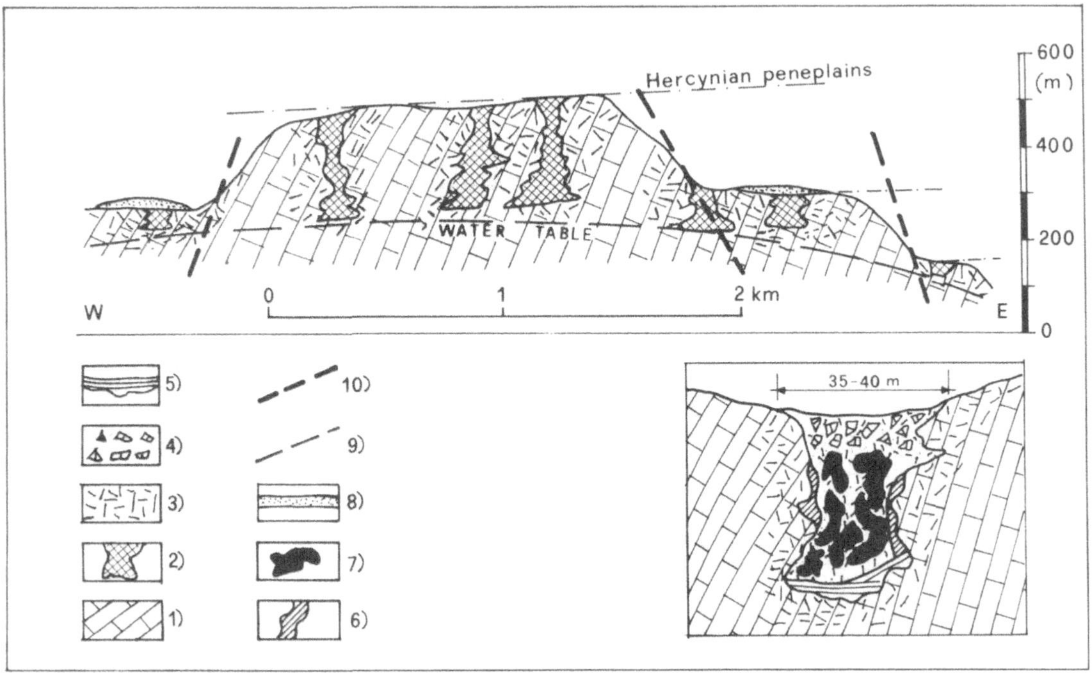

Figure 9–15. Triassic karst on the post-Hercynian peneplains.

(1) Cambrian limestone (Gonnesa formation); (2) karstic cavities filled with barite, quartz, and yellow dolomite; (3) epigenetic yellow dolomite; (4) collapse breccia, made up of fragments of limestone and yellow dolomite, generally silicified, cemented by barite; (5) Subhorizontal or gently dipping internal sediments; (6) coarse crystalline quartz; (7) massive barite replacing limestone and yellow dolomite; (8) subhorizontal Triassic sediments made up of sandstone or fine conglomerate with barite; (9) surface of old peneplains; (10) faults.

1. Cement of collapse breccia, made up of fragments of limestone and yellow dolomite, sometimes intensely silicified.
2. Massive, replacing limestone and yellow dolomite or bedded in the karst cavities (Fig. 9–18).
3. Fine needles of barite crystals in the cave sediments; these crystals are sometimes replaced by quartz.

The shape, content, and volume of barite bodies may be quite different; some karstic cavities may be a few tens of meters deep and contain several thousand tons of barite, but in other cases, as at the Barega mine, the volume amounts to around 1 million tons of ore, grading about 40 percent barium sulfate. Gangue minerals consist of coarse crystalline quartz, calcite, and dolomite; quartz is abundant also as the fine-grained constituent of the internal laminated sediments.

This great cycle of erosion and ore deposition was interrupted by the deposition of a subhorizontal cover made up of quartz, barite, and fragments laid down on the underlving slates mainly during the Triassic. These sediments,

Figure 9-16. Karstic barite on the Hercynian peneplain. Karstic cavities filled up by barite (white); the Cambrian limestone is intensely replaced by yellow, ankeritic dolomite.

generally have a brecciated structure, though thin lamination may be present. The cover played a very important role during the Alpine rejuvenation, preserving the karstic bodies from further erosion, and it now provides an important marker when prospecting for buried karstic bodies as described below.

Where the Alpine erosion removed this cover, the tops of the karstic mineralization were weathered, and finely powdered barite occurs along with quartz and red clay (terra rossa) on gently undulating surfaces (Fig. 9-19). The barium sulfate content of this residual clay is about 20-25 percent, and it can be extracted easily and economically, though it needs desliming before concentration in the mineral dressing plant.

Metallogenic Periods

According to the evolution of the geologic parameters and ore field development situation detailed above, four main periods of barite deposition may be outlined in Sardinia (Table 9-3), namely:

- Lower-Middle Cambrian,
- Cambrian-Ordovician unconformity,

Figure 9–17. Barite mineralization in karstic cavity. Hammer head lies on bedded barite; the bottom of the karst is filled with subhorizontal internal sediments consisting of fine grained quartz and needles of barium sulfate. In the top left of the photo, note the crusts of barite deposited under vadose conditions from percolating waters.

- Triassic peneplanation, and
- Cenozoic and recent alluvial and placer deposits.

Shapes, paragenesis, and economic importance of the barite deposits formed during the different metallogenic epochs are quite different. The Cambrian period produced limited stratiform bodies of barite (some thousands of tons), only sporadically exploited, depending on the market price. Sometimes the barite lenses have a high barium content and contain only a small quantity

Figure 9-18. Karstic-bedded barite of the post-Hercynian peneplains. The beds are made up of an alternation of barium sulfate and ankeritic dolomite with quartz.

Figure 9-19. Terra rossa and residual barite on the Hercynian peneplain. Barite is finely disseminated in the residual clay. Note erosional shape of Cambrian limestone.

Table 9–3. Schematic Breakdown of Barite Mineralization Stages by Age, Paleoenvironment, and Shape

Age	*Environment*	*Paragenesis*[a]	*Magmatic Trend*	*Orogenesis*	*Shape*
Cambrian	Platform	Ba (Pb, Zn, Fe, Ag)	—	Early Caledonian	Stratiform
Ordovician–Silurian	Margin	Ba (F, Cu, Fe, Pb, Zn, Sb, W)	Calc-alkaline	From Caledonian to early Hercynian	Stratabound and veins
Carboniferous Permian	Converging	Ba (F, Pb, Zn, Mo, Cu, Sn, V)	Calc-alkaline plutonites	Hercynian	Veins
Mesozoic		Ba (Pb, Ag)	—	—	Collapse breccia, massive bodies in karsts
Cenozoic	Basin and converging	Ba (F)	Calc-alkaline volcanites	Alpine	Veins, placers and alluvial

[a]Other ore associations occurring in the same metallogenic epoch are shown in parentheses.

of quartz; at other times, instead, the silica content increases, thus limiting the value of the mineralized bodies. Sulfides tend to be absent in this type of barite deposit, owing to the physicochemical sedimentation parameters.

The second period produced economic barite bodies, sometimes containing valuable quantities of argentiferous galena. These bodies have a stratabound structure and are found along the Cambrian–Ordovician unconformity.

This barite occurrence is certainly of greater economic importance than the preceding one. The mineralized horizon may be up to 10 m or so thick and run to several hundred meters in outcrop; it can thus be estimated to contain some million tons of resources, but it is economically exploitable as ore only locally.

The third metallogenic period represents the most important economic event as regards barite deposition. During the post-Hercynian erosion of the Cambrian carbonatic terrains, karstic fractures, cavities, and pockets were filled with several million tons of massive ore having a $BaSO_4$ content in excess of 35 percent.

The vertical development of the karstic cavities and the thickness of the barite bodies from the peneplains downward tends to be rather variable, ranging from a few meters to several hundred. The Scraper stope, for instance, at the Barega mine has been worked from Level 425 to Level 270.

The recent alluvial deposits of barite represent the products of post-Alpine erosion of the preceding mineralization. Their economic importance is rather limited, and they can be exploited economically only where they occur along with primary mineralization. The deposits have a $BaSO_4$ content of between 20 percent and 25 percent; the sulfate grain is fine and is intimately mixed with terra rossa.

It seems fairly clear that barite deposition in the different metallogenic periods is attributable to the reworking of primary barium sulfate resulting from the geodynamic evolution of the Cambrian–Ordovician basin. This reworking began just at the end of Cambrian sedimentation and is still proceeding today.

METALLOGENIC PROVINCES

Regarding the stratabound mineralization of barite, the location of the metallogenic provinces in Sardinia is relatively simple. The orebodies all occur in the southwestern part of the island, in the Cambrian terrains.

Barite deposition and remobilization took place in a fairly narrow area but over a very long period of geologic time; thus, the barium content of Cambrian limestone and dolomite migrated toward the geologic traps formed during the geodynamic evolution of the Cambrian platform. The close relationship among primary stratiform mineralization, deposition along the Cambrian–Ordovician unconformity, the filling of karstic cavities, and recent alluvial deposits all within the Cambrian provinces, is thus readily understandable.

Similar deposits of barite occur in southern Europe, particularly in Spain and France. The Cambrian sequence of the Cantabrian zone (northwest Spain), quite similar to that of Sardinia, consists of two members (Zamarreño, 1972). The lower one is made up of laminated dolomite indicating deposition environments analogous to present-day carbonate tidal flats, whereas the upper member consists of fossiliferous biomicrites, sometimes mineralized with Pb–Zn and barite.

The mineralization of barite in the Lower-Middle Cambrian exhibits remobilization phenomena "per descensum" not unlike those found in southwest Sardinia.

The difference in the level of economic importance of the kinds of deposits in the two countries (the filling of karstic cavities and the stratiform bodies of barite being apparently more important in Sardinia) may be due to differences in the supply of barium connected with local variations in sedimentologic parameters.

NEW ORE FIELD POTENTIAL

Hopes of finding new mineable resources of stratiform and karstic barite hinge closely on the occurrences in the Cambrian carbonatic terrains. Economically exploitable ore deposits, including Barega, Mont'Ega, and Punta Candiazzus, are beginning to show worrying signs of impending depletion.

The stratiform barite at the bottom of the Cambrian dolomite and the recent alluvial barite can be ignored, being less important from an economic point of view. Ore prospecting must thus be concentrated mainly on the mineralization present in the two most important occurrences, namely: (1) the silica horizon along the Cambrian–Ordovician unconformity, and (2) the karstic cavities of the post-Hercynian peneplains.

About 90 percent of Sardinian production comes from these occurrences. Only a few thousand tons per year are produced by small workings along the contact between the Cambrian sandstone and dolomite.

SILICA HORIZON ALONG THE UNCONFORMITY

In the northern part of the Cambrian outcrops, the silica horizon has been worked mainly for sulfides. It is only recently that small pits have been opened up to work barite that occurs both in massive form or as cement in the carbonatic breccia.

In the southern area (Sulcis) the sulfides are less important and the barite mineralization plays a bigger role; in this area the silica horizon may be up to 50 m thick.

Exploration for this type of mineralization may yield interesting results. Mining in the past was concentrated exclusively on the bodies nearest the surface, thus prospecting of the contact between Ordovician slates and Cambrian limestone now comes to the fore. However due account must be taken of two factors:

1. The thickness of the silica horizon, which may be as much as 30–50 m, though some mining operations appear to indicate that it thins in depth below the outcrop. This may be explained by assuming surficial thickening caused by silica remobilization during post-Cambrian stages of emergence.

2. The higher cost of underground mining operations. The rate at which the value of barite ore has risen in recent years, of course, has been lower than that of mining costs, so underground extraction is unlikely to be profitable, though open-pit operations may well prove a viable proposition.

KARSTIC CAVITIES OF THE POST-HERCYNIAN PENEPLAINS

The several stages of carbonate erosion, from early times (Cambrian) to present, have produced a network of cavities filled by different materials such as clay, spathic calcite, carbonatic collapse breccia, quartz, and barite and/or sulfides.

From the metallogenic point of view the most important period of mineralization-forming processes was the Permian–Triassic peneplanation (Padalino et al., 1973; Boni, 1980; Zuffardi, 1976). During this period the cavities were filled with barite of different textures; this type of mineralization is at present the main source of Sardinian production, and the discovery of new bodies could give a big boost to regional mining activities.

Several parameters must be taken into account when searching for karstic barite.

Geomorphology. The old Triassic peneplains were faulted during the Alpine orogenesis and different blocks of Cambrian terrains now occur at various levels (Fig. 9–15). Prospecting must thus be performed bearing in mind the presumed form of the old peneplains and their spatial continuity before the Alpine tectonics.

Mesozoic Sediments. As described earlier, the Triassic karsts were fossilized and preserved from erosion by clastic horizontal sediments that acted as cover. These sediments represent an important marker for ore prospecting; they consist of fine-grained quartz cementing fragments of dolomite and limestone (and occasionally slates) more or less silicified.

A small amount of blasting and ripping with a crawler tractor should easily and economically reveal any ore that occurs beneath the Triassic cover.

Paleogeography and Lithology. The Cambrian sequence exhibits sulfide-sulfate zoning, subsequently modified, at least in part, by Alpine tectonics. Mineralization in the Iglesias Valley area is mainly of the sulfide type (Fig. 9–9). Sulfate mineralization occurs instead in the outer part of the Iglesias carbonatic ring and toward the Sulcis area as a result of paleogeographic environmental conditions. Prospecting must therefore be concentrated where primary barite mineralization and/or higher geochemical barium contents occur. The lithology too is quite important when selecting the areas to be prospected; the most important karstic phenomena, in fact, are concentrated in the Cambrian limestone, very frequently replaced by yellow ankeritic dolomite. The primary, basal dolomite was unaffected by karstification.

Tectonics. At the end of the Hercynian orogenesis the southwestern part of Sardinia was split up into horst-graben structures by north–south and northeast-southwest striking fractures and faults. Then during the Mesozoic, the uplifted blocks and faulted zones were subjected to intense erosion and provided the most favorable loci for karstic deposits.

To conclude, therefore, at regional scale, the areas to be prospected for stratiform and karstic barite mineralization are relatively restricted. Nevertheless, the occurrence of numerous mineralized bodies and the presence of mineral-dressing plants in this restricted area may support the search for new bodies, albeit of limited size.

ACKNOWLEDGMENTS

The authors wish to thank colleagues Prof. Carlo Garbarino, who furnished the X-ray fluorescence analyses of the carbonates (Philips PW 1400), and Dr. Silvana Grillo for the sedimentologic data on the Cambrian sequence.

REFERENCES

Arthaud, F. 1963. Un exemple de tectoniques superposées dans le Paléozoíque de l'Iglesiente (Sardaigne). *C. R. Somm. Soc. Geol. de France* **3**:302.

Barbieri, M., V. Masi, and L. Tolomeo. 1984. Strontium geochemical evidence for the origin of the barite deposits from Sardinia, Italy. *Econ. Geology* **79**:1360–1365.

Barca, S., M. Del Rio, and P. Pittau Demelia. 1981. Acrintarchs in the "Arenarie di S. Vito" of south-west Sardinia: Stratigraphical and geological implications. *Soc. Geol. Ital. Boll.* **100**: 369–375.

Bechstadt, Th., M. Boni, and M. Selg. 1985. The Lower Cambrian of SW Sardinia: From a clastic tidal shelf to an isolated carbonate platform. *Facies* **12**:113–140.

Benz, J. P. 1964. *Le gisement plombo-zincifère d'Arenas (Sardaigne).* Travaux du Lab. Sciences de la Terre, Nancy, 126p.

Bernard, A. J. 1971. Metallogenic processes of intra-karstic sedimentation. Ore in sediments, *8th International Sed. Congress, Heidelberg,* pp. 43–57.

Bernard, A. J., and M. S. Leleu. 1967. A propos de la concentration résiduelle de la blende et la galène. *Acad. Sci. Comptes Rendus* **265**:729–732.

Boni, M. 1978. Considerazioni paleogeografiche su alcune mineralizzazioni a solfuri della Formazione di "Gonnesa." *Soc. Geol. Ital. Mem.* **19**:713–720.

Boni, M. 1980. Posterzynische Paläokarst—Erze in SW Sardinien. *Erzmetall.* **33**(3):169–172.

Boni, M. 1986. The Permo-Triassic vein and paleokarst ores in southwest Sardinia. Contribution of fluid inclusion studies to their genesis and paleoenvironment. *Mineralium Deposita* **21**:53–62.

Boni, M., and G. C. Amstutz. 1982. The Permo-Triassic paleokarst ores of south-west Sardinia (Iglesiente-Sulcis). An attempt at reconstruction of paleokarst conditions. In *Ore Genesis, the State of the Art,* eds. G. C. Amstutz et al., pp. 73–82. Berlin: Springer-Verlag.

Boni, M., A. Gandin, P. Marinaccci, and D. Stanzione. 1981. A paleoenvironmental, sedimentological and geochemical approach to the "Gonnesa Formation" (Lower Cambrian) (Buggerru area, SW Sardinia). *International Association of Sedimentology, 2nd European Meeting Bologna,* Abstract, pp. 14–16.

Brouwer, H. 1966. Ricerche tettoniche sul Paleozoico dell'Iglesiente e del Sulcis (Sardegna Sud-occidentale). *Acc. Naz. Lincei, VIII, Cl. Sc. Fis., Mat. and Nat.* **40**(6):1063–1074.

Brouwer, H. 1986. The Sardic tectonic phase in SW Sardinia: A concept rejected. Final Meeting IGCP, Project no. 5, Sardinia, Abstract.

Calvino, F. 1961. Mineralizzazioni di età alpina in Sardegna. Barite di Santoru e fluorite di Monte Cardiga. *Res. Ass. Min. Sarda* **65**(8):1-20.

Carannante, G., T. Cocozza, B. D'Argenio, and I. Salvadori. 1974. Caratteri deposizionali e diagenetici della "dolomia rigata" del Cambrico inf. della Sardegna. *Soc. Italiana Mineralogia i Petrologia Rend.* **30**(2):1159-1173.

Carannante, G., T. Cocozza, and B. D'Argenio. 1981. Tectono-sedimentary trends and paleogeography of Sardinia, during late Pre-Cambrian and Cambrian. *International Association of Sedimentology, 2nd European Meeting, Bologna,* pp. 227-230.

Carmignani, L., T. Cocozza, N. Minzoni, P. C. Pertusati, and C. A. Ricci. 1979. E' la Corsica il retropaese della catena ercinica in Sardegna? *Soc. Geol. Ital. Mem.* **20**:47-55.

Carmignani, L., T. Cocozza, C. Ghezzo, P. C. Pertusati, and C. A. Ricci. 1982. Lineamenti del basamento sardo. Guida alla Geol. del Paleozoico Sardo. *Guide Geol. Regionali,* Soc. Geol. Ital., pp. 11-23.

Cocozza, T. 1979. The Cambrian of Sardinia. *Soc. Geol. Ital. Mem.* **20**:163-183.

Cocozza, T., A. Jacobacci, R. Nardi, and I. Salvadori. 1974. Schema stratigrafico-strutturale del Massiccio Sardo-Corso e minerogenesi della Sardegna. *Soc. Geol. Ital. Mem.* **12**(2):85-186.

Debrenne, F. 1964. Archeocyate: contribution à l'étude des faunes cambriennes du Maroc, de Sardaigne et de France. *Notes et Mem. Serv. Geol. du Maroc 179,* 265p.

Del Bono, G. L. 1965. Relazione generale su una nuova possibile interpretazione della serie Cambro-Ordoviciana dell'Iglesiente. *Res. Ass. Min. Sarda* **8**:1-25.

Di Simplicio, P., G. Ferrara, C. Ghezzo, G. Guasparri, R. Pellizzer, C. A. Ricci, F. Rita, G. Sabatini. 1974. Il metamorfismo e il magmatismo paleozoico nella Sardegna. *Soc. Italiana Mineralogia i Petrologia Rend.* **30**:979-1068.

Fanni, S., A. Gandin, S. M. Grillo, F. Lippi, G. Marras, A. Salvadori, and S. Tocco. 1981. Piattaforma carbonatica cambriana della Sardegna sud-occidentale: sedimentazione e deposizioni metallifere. *Soc. Geol. Ital. Mem.* **22**:123-137.

Ferrer, A., X. Font, and M. Viladevall. 1985. Tipologia de las mineralizaciones de baritina de las Cadenas Costeras Catalanas (el ejemplo del sector Cangost, Barcelona). *Barcellona Rev. Investigationes Geologica* **40**:103-113.

Gandin, A., G. Padalino, S. Tocco, and M. Violo. 1973. Un esempio di deposizione stratiforme di barite nella "dolomia rigata" del Cambrico della Sardegna sud-occidentale. Tentativo di correlazione tra l'ambiente di sedimentazione e la precipitazione del solfato di bario (nota I). *Soc. Geol. Ital. Boll.* **92**:329-354.

Gandin, A., G. Padalino, and M. Violo. 1974. Correlation between sedimentation environment and ore-prospecting. Sedimentological and ore-genesis studies of Cambrian "arenarie" and "dolomia rigata" formations (Sardinia-Italy): Deposition and concentration of barite in an evaporitic environment. *Soc. Italiana Mineralogia i Petrologia* **30**:251-330.

Garbarino, C., S. H. Grillo, C. Marini, A. Mazzella, F. Melis, G. Padalino, S. Tocco, and M. Violo. 1984. The Paleozoic metallogenic epochs of the Sardinian microplate (western Mediterranean): An attempt at synthesis of geodynamic evolution and mineralizing processes. *Soc. Italiana Mineralogia i Petrologia Rend.* **39**(1):193-228.

Gasperi, G., and R. Gelmini. 1979. Ricerche sul Verrucano - 4: il Verrucano della Nurra (Sardegna Nord occidentale). *Soc. Geol. Ital. Mem.* **20**:215-231.

Ghezzo, C., I. Memmi, and C. A. Ricci. 1982. - Lineamenti geopetrologici del cristallino sardo. Le plutoniti e la loro successione negli eventi intrusivi. *Min. Petrogr. Acta* **18**:205-234.

Junker, B., and H. H. Schneider. 1979. L'infracambriano della Sarolegna Suboccidentale. *Soc. Geol. Ital. Mem.* **20**:461.

Lagny, Ph., and H. Rouvier. 1976. Les gisements Pb-Zn en roches carbonatées sous inconformité: gisements paléokarstiques ou gisements dans le paléokarst? *Mem. Serv. Soc. Geol. France* **7**:5-69.

Land, L. S., and G. K. Hoops. 1973. Sodium in carbonate sediments and rocks: A possible index to the salinity of diagenetic solution. *Jour. Sed. Petrology* **43:**614-617.

Leleu, M. 1966. Le karst et ses incidences métallogéniques. *Sci. Terre* **11:**(4):385-413.

McMoore, H. J. 1969. Influence of structures on the base metal deposits of southwest Sardinia, Italy. *Inst. Mining and Metallurgy Trans.* **78B:**135-147.

Marcello, A. 1969. Déposition supergénique de galène: Les exemples de la Sardaigne. *Proceedings of the Meeting on Remobilization of Ores and Minerals,* pp. 293-303. Cagliari: Assoc. Mineraria Sarda.

Marcello, A., S. Pretti, and I. Uras. 1983. La barite in Sardegna. *Ind. Mineraria* **5:** 1-32.

Mazzella, A., M. F. Melis, R. Porcu, A. Rivoldini, and M. Violo. 1970. La distribuzione geochimica del fluoro nella sequenza Ordoviciano-Siluriano della Sardegna. *Res. Ass. Min. Sarda* **2:**39-62.

Orgeval, J. 1979. Contribution à l'étude des minéralisations plombozincifères du socle carbonaté Cambrien de la mine des Malines (Gard, France). Ph. D. dissertation. Montpellier: Universite de Sciences et Technologie di Languedoc.

Padalino, G., S. Pretti, S. Tocco, and M. Violo. 1972. Some examples of lead-zinc-barite deposition in karstic environment. *2nd International Symposium on Mineral Deposits of the Alps, Ljubljana,* pp. 195-204.

Padalino, G., S. Pretti, D. Tamburrini, S. Tocco, I. Uras, M. Violo, and P. Zuffardi. 1973. Ore deposition in karst formation with examples from Sardinia. In *Ore in Sediments,* eds. G. C. Amstutz and A. J. Bernard, pp. 209-220. Berlin: Springer-Verlag.

Padalino, G., S. Tocco , and M. Violo. 1980. Different genetic environments may produce similar occurrences of stratiform barite: Examples from Sardinia (Italy). In *5th IAGOD Symposium Proceedings,* pp. 459-466. Stuttgart: E. Schweizerbart'she Verlagsbuchandlung.

Poll, J. J. K. 1966. The geology of the Rosas-Terraseo area (Sulcis South-Sardinia). *Leidse Geol. Med.* **35:**117-208.

Poll, J. J. K., and H. J. Zwart. 1964. On the tectonics of the Sulcis area, South-Sardinia. *Geol. Mijnbouw* **43:**144-146.

Pouit, G., C. Bouquet, and J. B. Bois. 1979. Les principaux niveaux mineralises (Zn-Pb-Cu-Ba) du Paleozoique des Pirénees centrales: elements de synthése. *[France] Bur. Recherches Géol. et Minières Bull.* **2:**23-34.

Pretti, S., F. Salvadori, and M. Violo. 1978. A new approach to the investigation on karst and ore-bodies in Cambrian carbonate rocks of Sardinia (Italy). *Res. Ass. Min. Sarda* **1:**67-85.

Spjeldnaes, N. 1961. Ordovician climatic zones. *Norks Geol. Tidsskrift* **41:**45-77.

Tamburrini, D. 1968. I giacimenti baritici sardi; caratteri geo-giaci-mentologici e minerari. *Symposium sulle bariti della Sardegna,* pp. 17-30. Cagliari: Assoc. Mineraria Sarda.

Tamburrini, D., and M. Violo. 1965. Il giacimento a barite di Mte Barega-Mte Arcau (Iglesiente, Sardegna). *Ric. Sci.* **35**(II-A):814-848.

Tamburrini, D., and P. Zuffardi. 1969. Field evidence of supergene remobilization of barium (and possibly of barite) in Sardinia. *Proceedings of the Meeting on Remobilization of ores and minerals,* pp. 305-314. Cagliari: Assoc. Mineraria Sarda.

Zamarreño, I. 1972. Les litofacies carbonatados del Cambrico de la zona Cantabrica (NW Espana) y su distribution paleogeografica. *Trabajos de geologia* **5:**3-118.

Zuffardi, P. 1969. La métallogénèse du plomb, du zinc et du baritine en Sardaigne: un exemple de permanence, de polygénétisme et de transformisme. *Annal. Soc. Geol. Belgique* 92:3-15.

Zuffardi, P. 1976. Karst and economic mineral deposits. In *Handbook of Stratabound and Stratiform Ore Deposits,* vol. 3, ed. K. H. Wolf, pp. 175-206. Amsterdam: Elsevier.

10

The Barite Ore Fields of Kuroko-type of Japan

Katsumi Marumo

The Kuroko-type deposits of Japan are situated in the specific geologic province (the Green Tuff region) characterized by thick volcanic and sedimentary piles, which resulted from subsidence and subsequent violent submarine volcanism of Miocene age (Fig. 10–1). The Kuroko-type deposits are defined as stratabound polymetallic sulfide-sulfate deposits comprised of stratiform massive orebodies and underlying stockwork orebodies. They are subdivided on the basis of the total sulfide to sulfate ratio and size of the stratiform orebodies relative to the stockwork ones into the following four subtypes: (1) the typical Kuroko deposit, (2) the Kuroko-type network deposit, (3) the Kuroko-type gypsum deposit, and (4) the Kuroko-type barite deposit. Among them, the typical Kuroko deposits of the Hokuroku district and the Kuroko-type barite deposits of the southwestern Hokkaido district are now major suppliers of barite in Japan.

The volume *Geology of Kuroko Deposits*, edited by S. Ishihara (1974) and published by the Society of Mining Geologists of Japan was the culmination of research on the Kuroko deposits carried out by Japanese geologists. Subsequent investigations on the typical Kuroko deposits of the Hokuroku district, carried out by the members of the United States–Japan–Canada Cooperative Research Project commenced in 1978. The results of collaborative efforts appeared in *The Kuroko and Related Volcanogenic Massive Sulfide Deposits*, Economic Geology Monograph 5, edited by H. Ohmoto and B. J. Skinner (1983).

The geologic, mineralogic, and geochemical aspects of the Hokuroku barite ore field described in this chapter owe much to the papers presented in these publications as well as to other papers. The descriptions of the barite ore field

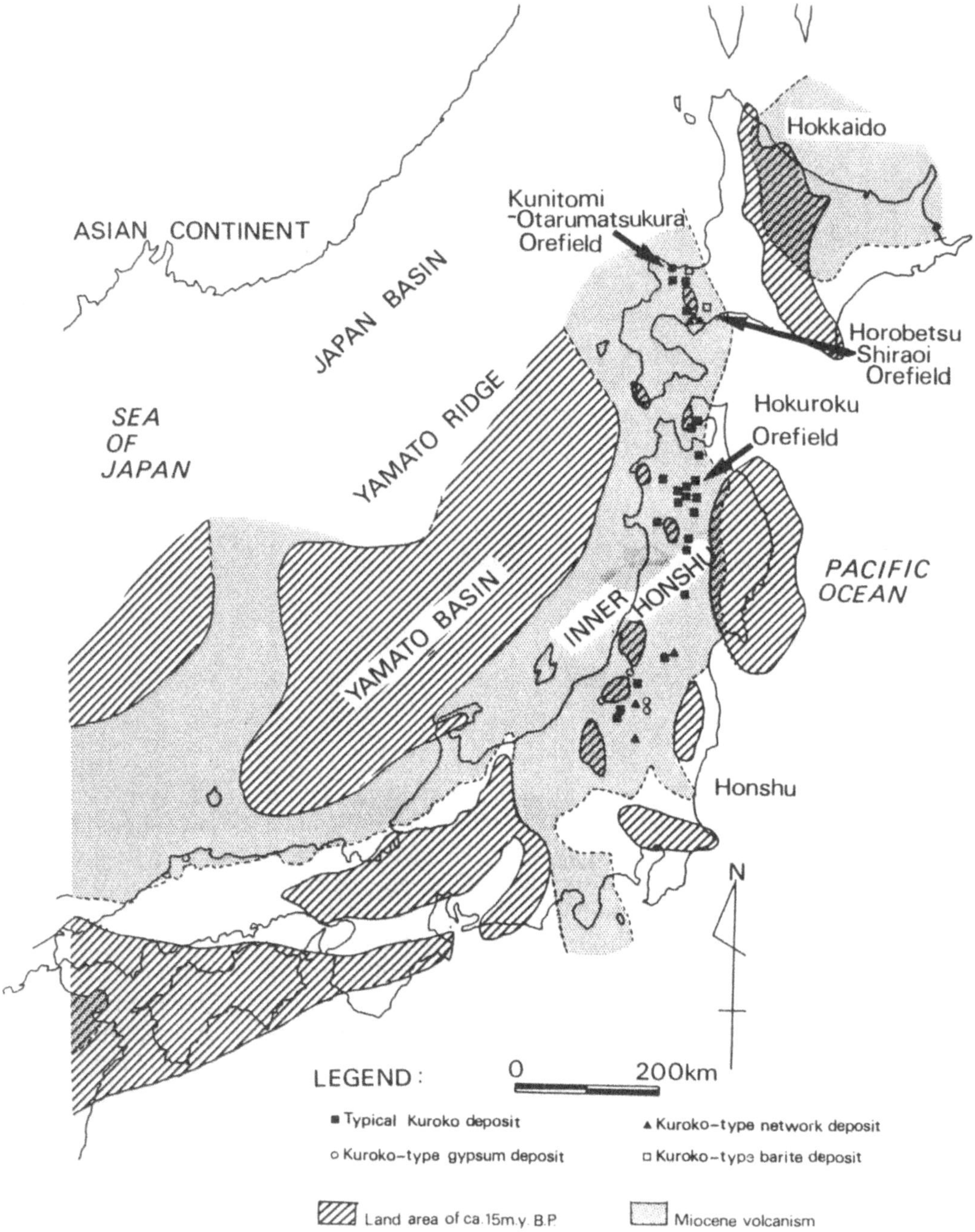

Figure 10–1. Location of major Kuroko-type deposits (after Sato, 1974) in Japan relative to those of landmasses, seas, and active volcanism ca. 15 m.y. ago (after Ohmoto, 1983, p. 11).

of the southwestern Hokkaido are based on the summaries of Igarashi et al. (1974), Marumo et al. (1985), Marumo and Sawai (1986), and other papers.

GENERAL GEOLOGY

STRUCTURE OF THE GREEN TUFF REGION

From a review of the geology, geophysics, and paleontology of the Sea of Japan and its surrounding area, Ohmoto (1983) concluded that the Green Tuff region was generated by tectonic and igneous activity (the Green Tuff activity) that accompanied formation of the Sea of Japan during the period from 65 m.y. ago to the present. The Green Tuff region is divided into four subprovinces that are aligned roughly parallel to the west coast of Honshu. From west to east, these subprovinces are arranged as follows: Japan basin, Yamato Ridge, Yamato basin, and Inner Honshu provinces (Figs. 10–1 and 10–2e).

The Japan basin is a deep sea, with a maximum depth of about 3.6 km. The area with a depth greater than 3 km is 250–500 km in width and approximately 1000 km in length. The continental crust is absent in most parts, and 2–2.5 km of sediments lie directly on the oceanic crust.

The Yamato Ridge province is shallower than 2–2.5 km in water depth and has an area of approximately 200 × 400 km. The ridge consists of more than a 15 km thick oceanic crust, a 4–7 km thick continental crust, and thin sediment (up to 1 km thick) in ascending order.

The Yamato basin is a deep sea, with a maximum depth of about 3 km. The area with depth greater than 2.5 km is 100–200 km in width and approximately 400 km in length. Sediments up to 1.5 km thick are underlain by 3–7 km thick continental crust and about 8 km thick oceanic crust.

The Inner Honshu province is between 100 km and 250 km in width and extends along the length of Honshu island. The elevation increases from about 1 km below sea level at the western edge to about 1 km above sea level at the eastern edge. The thickness of both the continental and the basaltic crusts also increases from west to east: from about 5 km to about 15 km for the continental crust and from about 8 km to about 20 km for the basaltic crust.

SEQUENCES OF GEOLOGIC EVENTS IN THE GREEN TUFF REGION

The sequences of geologic events in the Green Tuff region proposed by Ohmoto (1983) are summarized as follows:

1. Until about 65 m.y. ago, Japan was a part of the eastern coastal province of the Asian continent under which the Pacific plate subducted (Fig. 10–2a).
2. About 64 to 40 m.y. ago, the subduction of the Pacific plate changed from a north–northwest to west–northwest direction, accompanied by a steepening of the subduction angle. Thereby, back-arc spreading commenced in the area presently occupied by the Japan basin, accom-

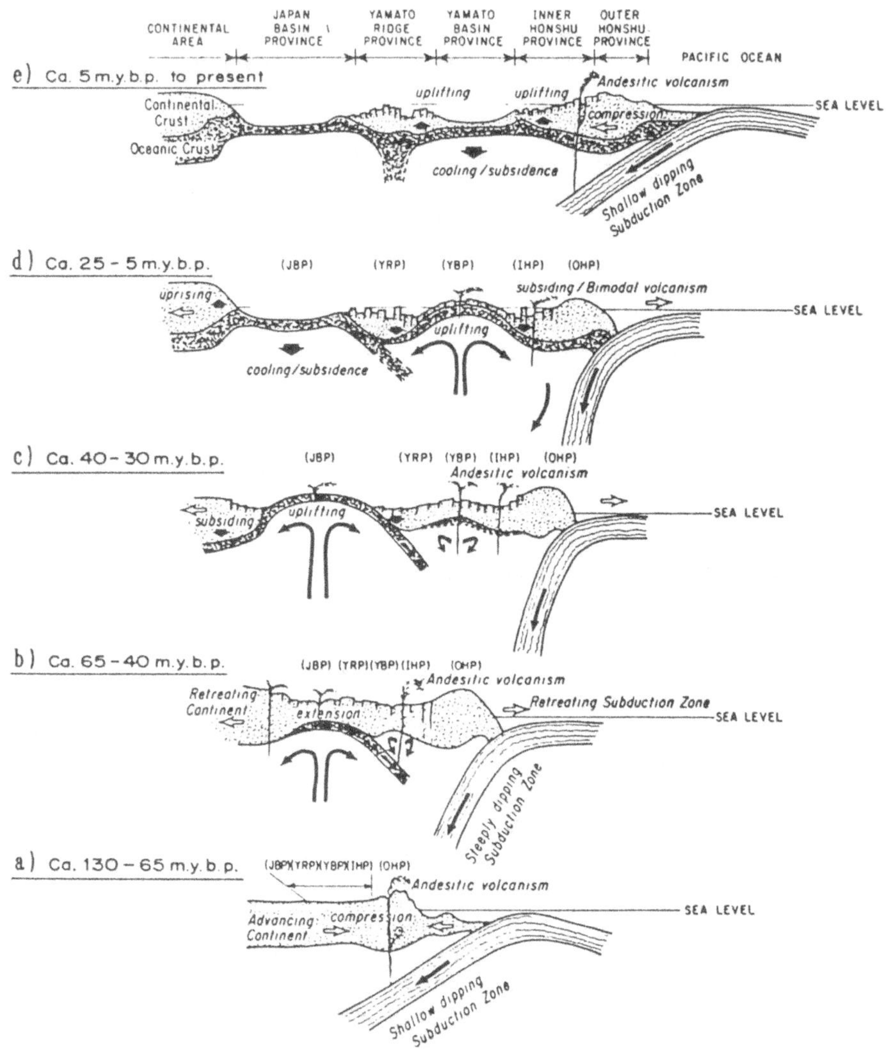

Figure 10–2. Proposed model for the plate tectonic history of the Sea of Japan during the last 130 m.y. (after Ohmoto, 1983, p. 17).

JBP: Japan basin province; YRP: Yamato Ridge province; YBP: Yamato basin province; IHP: Inner Honshu province; OHP: Outer Honshu province.

panying the bending of Japan and its migration away from mainland Asia (Fig. 10–2b). Back-arc spreading in the Japan basin province continued until about 30 m.y. ago (Fig. 10–2c).

3. Cooling of the asthenosphere under the Japan basin province began about 30 m.y. ago, causing gradual subsidence of this province. This, together with a eustatic rise of sea level, caused the formation of the first sea (the proto-Sea of Japan) in this province about 25 m.y. ago. Cessation of spreading in the Japan basin province resulted in intensification of back-arc spreading in the Yamato basin province and initi-

ation of bimodal volcanism in the inner zone of Japan about 25 m.y. ago (Fig. 10–2d).

4. Spreading in the Yamato basin province continued up to about 5 m.y. ago, causing this province to be uplifted and remain under subaerial conditions. On the flanks of the spreading center, that is, the Inner Honshu and Yamato Ridge provinces, sinking of cooler asthenosphere to counterbalance the ascent of hotter asthenosphere and the extensional movements of plates caused brittle failure of crustal blocks along preexisting basement fractures. This created a series of troughs and fault-bounded, small-scale (30 × 50 km) basins within the troughs in the Inner Honshu and Yamato Ridge provinces (Fig. 10–2d). The Hokuroku basin, which hosts major Kuroko deposits, is an example of such a basin.
5. The dip of the subducted Pacific slab shallowed again about 5 m.y. ago, resulting in cessation of back-arc spreading and subsidence of the Yamato basin province, uplifting of the Inner Honshu and the Yamato Ridge provinces, and a return to a compressional stress regime and andesitic volcanism (Fig. 10–2e).

GEOLOGY OF THE HOKUROKU BARITE ORE FIELD

The regional geology of the Hokuroku ore field has been reviewed in detail by Tanimura et al. (1983). The Hokuroku barite ore field is characterized by several fault-bounded basin structures. In the largest Middle Miocene sedimentary basin, the Hokuroku basin, typical Kuroko deposits (Hanaoka, Matsumine, Shakanai, Matsuki, Ezuri, Fukazawa, Uwamuki, Uchinotai, Ainai, Furutobe) occur in concentrated clusters, each cluster having an areal extent of about 1.5 × 3 km and being separated from each other by 5–10 km. The Tsugaruyunosawa Kuroko-type network deposit occurs in the southern margin of the Ikarigaseki Late Miocene sedimentary basin (Fig. 10–3). One of the important structural features in the Hokuroku basin is a series of steeply dipping normal faults that strike north–south, northwest–southeast, and northeast–southwest, such as the Tsutsumizawa, Ohshigenai, Hanawa, and Ashinazawa faults. Many of these faults appear to have been active from pre-Kuroko time to the present.

The Hokuroku ore field comprises the basement rocks (phyllites, cherts, and slates) of probable Paleozoic age, the Tertiary volcanics (the compositions range from basaltic to felsic) and marine detrital sediments, and the Quaternary tuffs and nonmarine detrital sediments. The Tertiary and Quaternary formations compiled by Tanimura et al. (1983) are presented in Figure 10–4.

The oldest Tertiary formation that unconformably overlies basement rocks is called the Menaichizawa formation in the western and central parts of the basin, and the Sasahata formation in the eastern part. The Menaichizawa formation is composed of brecciated andesitic lavas and andesitic pyroclastic rocks with interbedded conglomerates and sandstones. The Sasahata formation is comprised of conglomerates, sandstones, tuffs, and black mudstones.

The Hotakizawa formation, which in most areas lies conformably on the

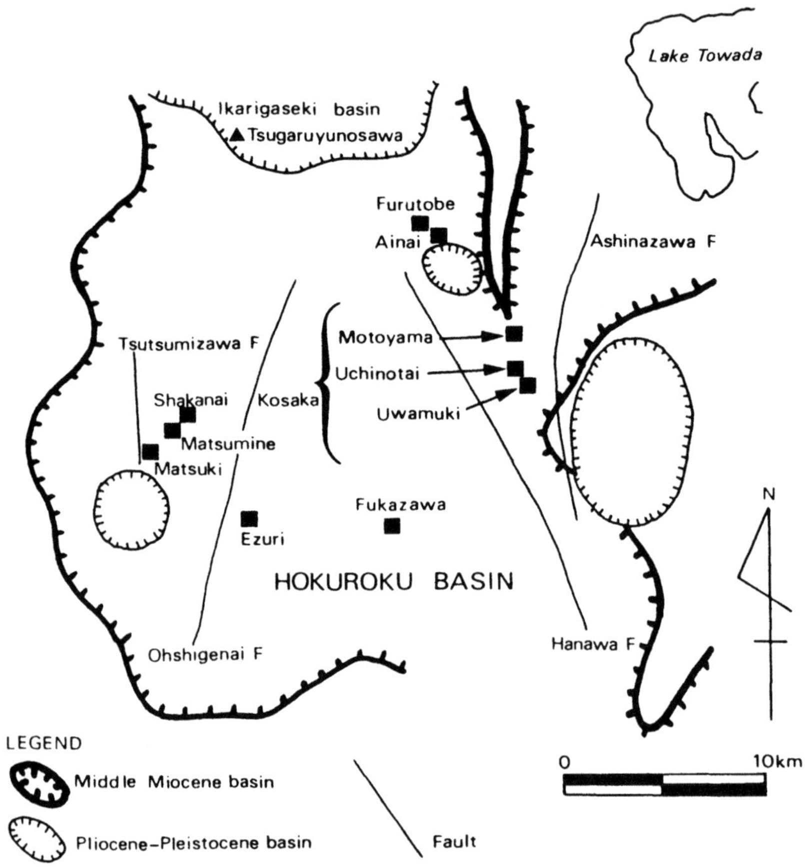

Figure 10-3. Miocene sedimentary basins and location of Kuroko-type deposits in the Hokuroku ore field (after Sato et al., 1974).

Menaichizawa formation, is characterized by a thick (up to 650 m) sequence of basaltic lavas and tuff breccias (B3) with minor intercalated mudstones (M3) and felsic tuffs (T5). The basalts erupted mainly along major fracture systems in the northern, western, and southern marginal zones of the Hokuroku basin.

The Hanaoka, Yukisawa, and Uwamuki formations are characterized by a thick (up to 600 m) sequence of felsic lavas (D4 and D3) and tuffs (T4 and T3) with minor intercalated mudstones (M2-b). The Hanaoka and Yukizawa formations conformably overlie the Hotakizawa formation. The Uwamuki formation lies unconformably on the Sasahata or the basement units. These formations can be broadly divided into the lower part, where dacite lava flows predominate over dacitic tuff breccias, and the upper part, which is composed mainly of tuff breccias and tuffs with minor dacite domes. The typical Kuroko deposits occur in the uppermost part of the upper layer.

The Tsutsumizawa, Kagoya, and Akamori formations overlie the Hanaoka, Yukisawa, and Uwamuki formations, respectively. These formations are characterized by a sequence of alternating tuffs (T2) and mudstone units (M2-a

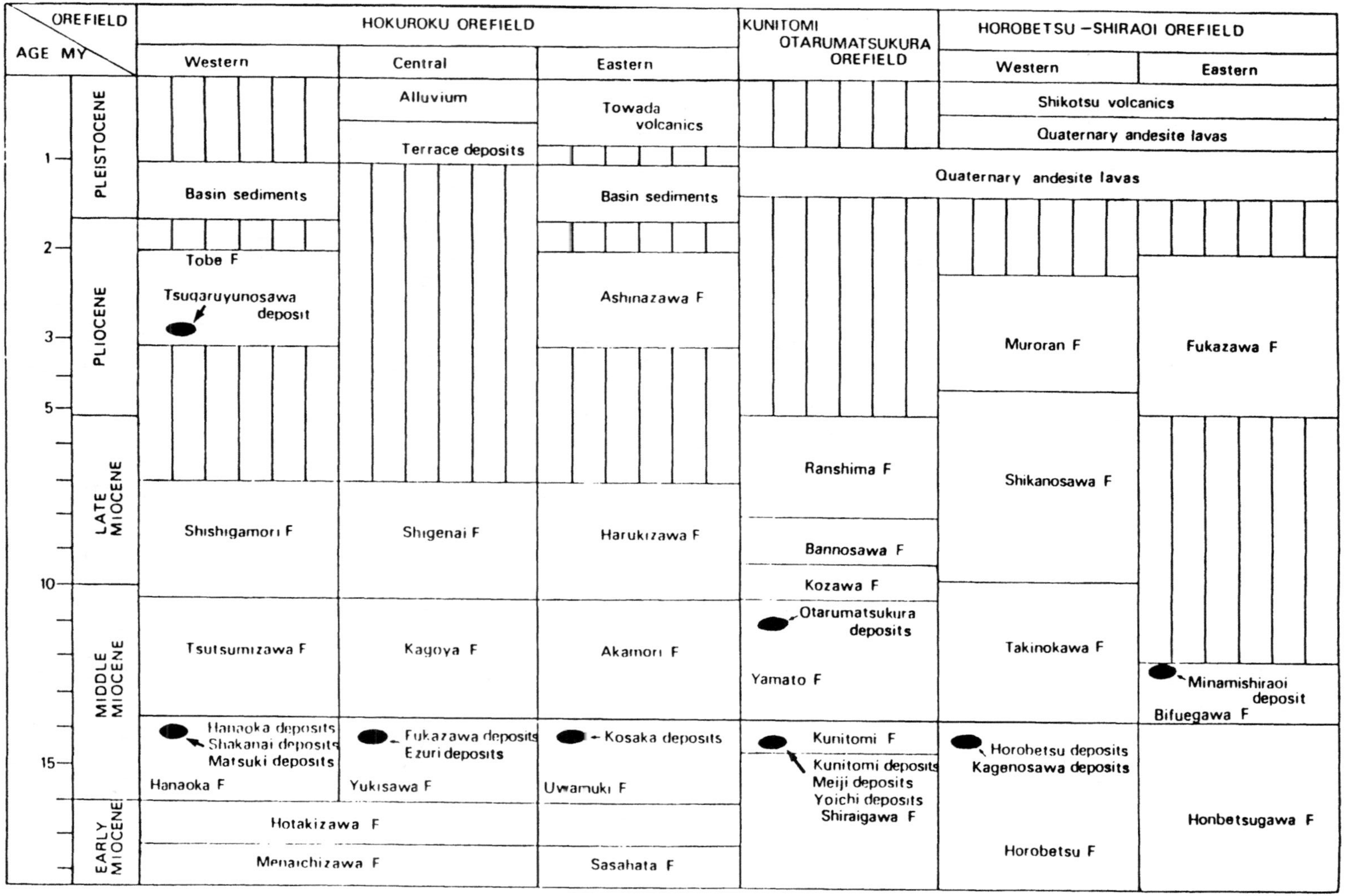

Figure 10–4. Comparison of stratigraphic succession for the Hokuroku ore field (after Tanimura et al., 1983), Kunitomi-Otarumatsukura ore field (after Hasegawa, 1986), and Horobetsu-Shiraoi ore field (after Yawata, 1986).

and M1-b), by the virtual absence of felsic lavas, and by the presence of basalt lavas (B2) in many areas.

The Shishigamori, Shigenai, and Harukizawa formations lie conformably on the Tsutsumizawa, Kagoya, and Akamori formations, respectively. They are comprised of interbedded felsic tuffs and mudstones.

The Tobe formation lies unconformably on the Shishigamori formation. The lower part of this formation is characterized by so-called abnormal sediments, which are comprised of very poorly sorted sediments containing rocks of all older formations. The upper part is comprised mostly of sandstones, dacitic tuffs, tuff breccias, lithic tuffs, and welded tuffs with occasional peats and lignite. The Tsugaruyunosawa Kuruko-type network deposit occurs in the lower part of this formation (Inoue and Mihashi, 1962).

GEOLOGY OF THE BARITE ORE FIELD IN SOUTHWESTERN HOKKAIDO

Southwestern Hokkaido is geologically the northern extension of the Inner Honshu province of the Green Tuff region. Thick volcanic piles and normal sediments of Miocene period occur above the basement rocks of Paleozoic to lower Mesozoic strata.

As shown in Fig. 10–5, southwestern Hokkaido is divided into two subprovinces on the basis of tectonic features and the total ratio of normal sediments to pyroclastics (Yajima, 1979). The first subprovince is upheaved areas (Shakotan Peninsula, Shizukari–Toyoura, and Otaru–Muroran) composed mainly of the Miocene pyroclastics. This subprovince is characterized by violent volcanism and tectonic movements of early to late Miocene ages. The second subprovince is characterized by sedimentary basins with thick piles of normal marine sediments of Miocene age. This subprovince is distributed from Kunitomi to Toya, and Sapporo to Shiraoi. The Kuroko-type deposits of southwestern Hokkaido are divided into the following three subtypes: (1) the typical Kuroko deposits (Kunitomi, Yoichi, Meiji, and Toya-takarada), (2) the Kuroko-type network deposits (Date, Horobetsu, Kagenosawa, and Morino), and (3) the Kuroko-type barite deposits (Otarumatsukura, Shiraoi, and Minamishiraoi). Most of the deposits were mined out by 1970; only the Minamishiraoi mine is still producing barite as a by-product of kaolin clay.

GEOLOGY OF THE KUNITOMI–OTARUMATSUKURA BARITE ORE FIELD

This ore field contains Kunitomi, Yoichi, Meiji, and Otarumatsukura Kuroko-type deposits. Among them, the Kunitomi, Yoichi, and Meiji typical Kuroko deposits occur in the marginal parts of a sedimentary basin surrounded by the Shakotan Peninsula and Otaru-Muroran upheaved areas. The Otarumatsukura Kuroko-type barite deposits are situated on the northwestern flank of the Otaru–Muroran upheaved area (Fig. 10–5).

The Kunitomi–Otarumatsukura Kuroko ore field comprises the basement rocks (phyllites, slates, and quartz diorite) of probable Paleozoic to early Mesozoic ages, the volcanics (the composition ranges from basaltic to felsic) and normal marine sediments of Tertiary age, and tuffs and nonmarine detrital sediments of Quaternary age (Hasegawa, 1986). The correlation of the Tertiary

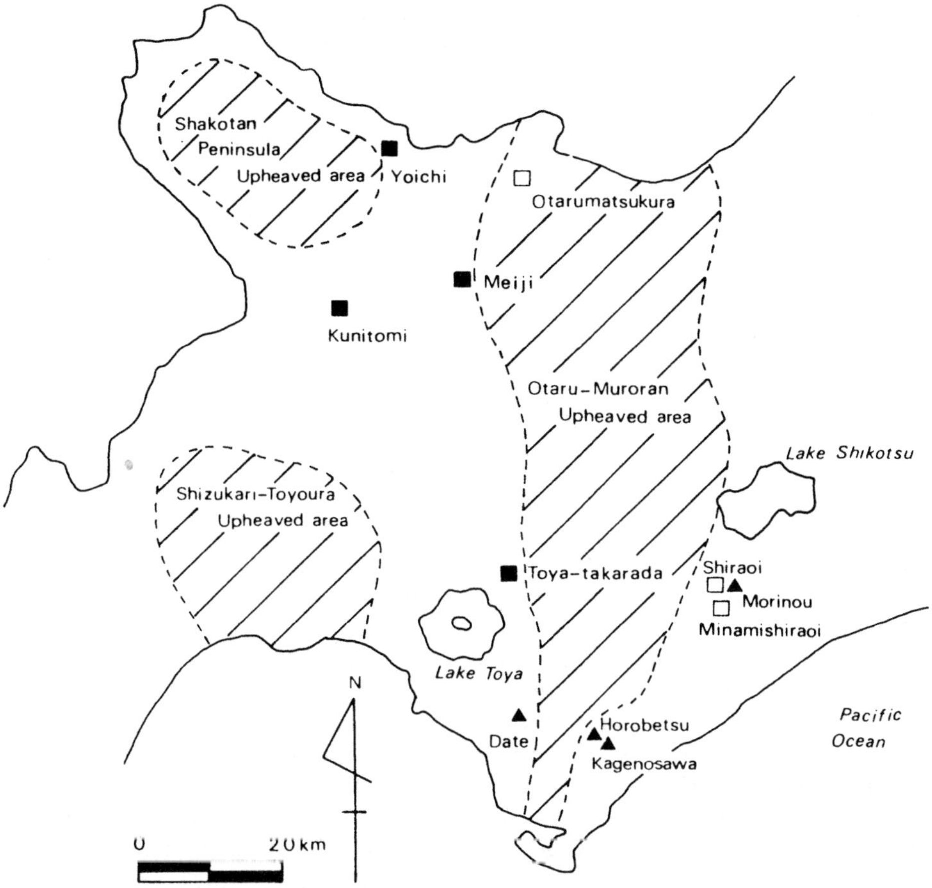

Figure 10–5. Geologic subprovince and location of Kuroko-type deposits in the southwestern Hokkaido (after Yajima, 1979).

formations of this ore field to the Hokuroku ore field is shown in Figure 10–4.

The oldest Tertiary group that unconformably overlies the basement is called the Furubira group and consists of the Shiraigawa formation, Kunitomi formation, Yamato formation, and Kozawa formation in ascending order.

The Shiraigawa formation is divided into the lower part, where conglomerates and mudstones are predominant over volcanics, and the upper part, which is composed of thick (up to 600 m) andesitic lavas and pyroclastics. The Kunitomi formation, which in most areas lies conformably on the Shiraigawa formation, is characterized by thick (up to 700 m) felsic lavas and pyroclastics with minor intercalated basalt lavas and mudstones. The Kunitomi, Yoichi, and Meiji typical Kuroko deposits occur in this formation. The Yamato formation, which conformably overlies the Kunitomi formation, is characterized by thick (up to 600 m) andesitic lavas and pyroclastics with minor intercalated basalts and mudstones. This formation hosts the Otarumatsukura Kuroko-type barite deposits. The Kozawa formation, which in many cases lies conformably on the Yamato formation, is composed of felsic lavas and pyroclastics with minor intercalated mudstones and sandstones.

The Kutchan group overlies the Furubira group in conformable or unconformable relation. The lower part of the group, the Bannosawa formation, consists mainly of mudstones and sandstones with intercalated andesitic pyroclastics. The upper part, the Ranshima formation, consists mainly of andesitic pyroclastics with minor intercalated mudstones and sandstones.

GEOLOGY OF THE HOROBETSU-SHIRAOI BARITE ORE FIELD

This ore field contains the Horobetsu, Kagenosawa, Date, and Moriro Kuroko-type network deposits, the Toya-takarada typical Kuroko deposit, and the Shiraoi and Minamishiraoi Kuroko-type barite deposits. These deposits occur in the marginal parts of a sedimentary basin surrounding the southern part of the Otaru–Muroran upheaved area (Fig. 10–5).

The Horobetsu–Shiraoi ore field comprises the basement rocks (phyllites, slates, and quartz diorite) of probable Paleozoic to early Mesozoic ages, the volcanics (the composition ranges from andesitic to felsic), tuffs and normal marine sediments of Tertiary age, and felsic tuffs, marine–nonmarine sediments, and andesitic lavas of Quarternary age (Yawata, 1986). The correlation of the Tertiary formations of this ore field to the Hokuroku ore field is shown in Figure 10–4.

The middle Miocene formation that unconformably overlies the basement is called the Horobetsu formation in the Horobetsu and Kagenosawa mining area, Osarugawa formation in the Toya-takarada and Date mining area, and Honbetsugawa formation in the Shiraoi–Minamishiraoi mining area. These formations are characterized by thick (up to 1700 m) sequences of andesitic to felsic lavas and pyroclastics with minor dacite domes and intercalated mudstones. The Horobetsu and Kagenosawa Kuroko-type network deposits occur in the Horobetsu formation. The Toya-takarada typical Kuroko deposit and Date Kuroko-type network deposit occur in the Osarugawa formation.

The Takinosawa–Shikanosawa, Soushunai, and Bifuegawa formations overlie the Horobetsu, Osarugawa, and Honbetsugawa formations, respectively. These formations comprise mostly interbedded andesitic to dacitic lavas and pyroclastics and mudstones. The Shiraoi and Minamishiraoi Kuroko-type barite deposits and the Morino Kuroko-type network deposit occur in the Bifuegawa formation.

The Pliocene Muroran, Kitayuzawa, and Fukazawa formations unconformably overlie the Takinosawa–Shikanosawa, Soushunai, and Bifuegawa formations, respectively. They are composed mainly of andesitic to felsic tuff breccia and tuffs with intercalated mudstones and sandstones.

ECONOMIC GEOLOGY

OUTLINE OF ORE DEPOSITS

THE KOSAKA TYPICAL KUROKO DEPOSITS

The geology of the Kosaka mining area consists of Paleozoic basement rocks, Tertiary formations (Sasahata formation, Uwamuki formation, which hosts the deposits, Akamori formation, Harukizawa formation, and Ashinazawa formation in ascending order), and a lot of intrusive rocks (Fig. 10–6).

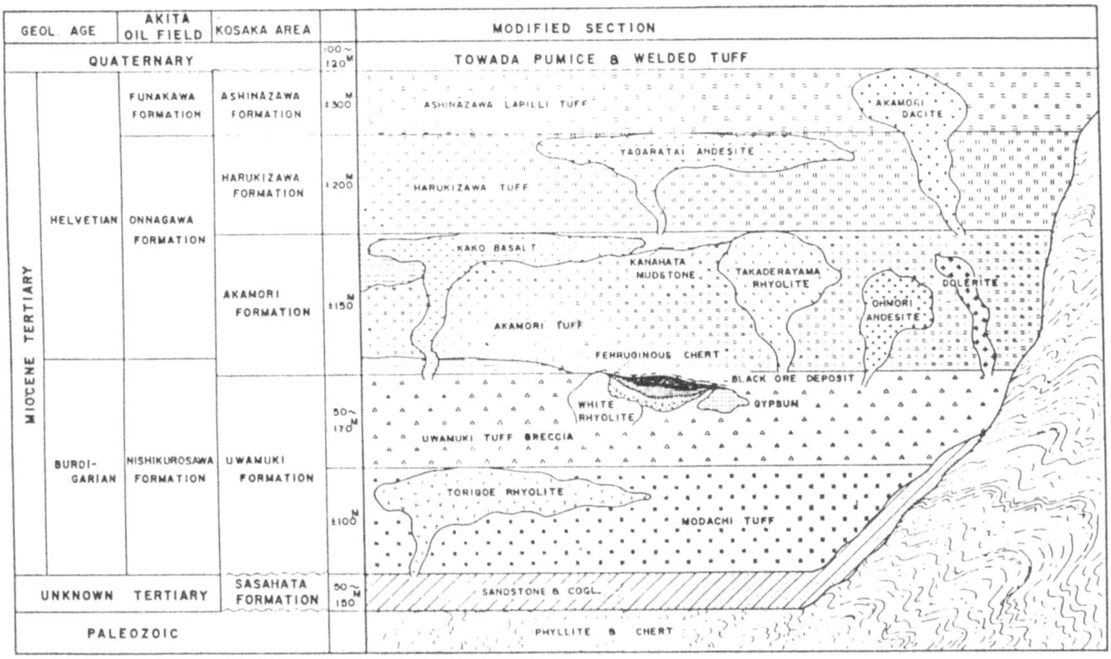

Figure 10–6. Generalized stratigraphic section through the Kosaka mining area (after Oshima et al., 1974).

The Kosaka mine comprises three groups of ore bodies, the Motoyama, Uchinotai, and Uwamuki deposits (Fig. 10–7). The Motoyama deposit was mined by the open-pit method from 1908 to 1920. The Uchinotai and Uwamuki deposits, which were discovered between 1959 and 1962, are now producing base metals, barite, and sericite. The present production is 52,000 metric tons (Au, 0.7 g/t; Ag, 79 g/t; Cu, 0.95 percent; Pb, 1.60 percent; Zn, 4.95 percent; $BaSO_4$, 10.9 percent) per month. The quantity of barite in the Kosaka deposits is approximately 3.0×10^6 metric tons (Table 10–1). The average ore grades of these deposits are also listed in Table 10–1.

Figure 10–8 shows a generalized profile of the ore deposits. Each deposit consists of black ore, yellow ore, siliceous ore, and gypsum ore. These ores show a distinct zoning, that is, black ore on top, with the yellow ore and siliceous ore lying successively in descending order. The gypsum ore is present underneath and peripheral to these sulfide ores. The black and yellow orebodies are generally in bedded form, whereas those of gypsum and siliceous ore are massive (Oshima et al., 1974).

The black ore consists mainly of barite, sphalerite, galena, chalcopyrite, pyrite, and tetrahedrite with small quantities of bornite, electrum, and silver minerals. The vertical variation of mineral assemblage within a black orebody is considerable and more or less regular. An example of such variation of mineral assemblage within black ore and overlying a barite orebody of the Uwamuki No. 2 deposit presented by Matsukuma and Horikoshi (1970) is shown in Figure 10–9. The barite contents in black ore increase upward, sometimes concentrated in compact seams 1 m thick. At the top of the black ore, sandy, loose beds of barite are also found. In the hanging wall of the black ore deposits, along the bedding of

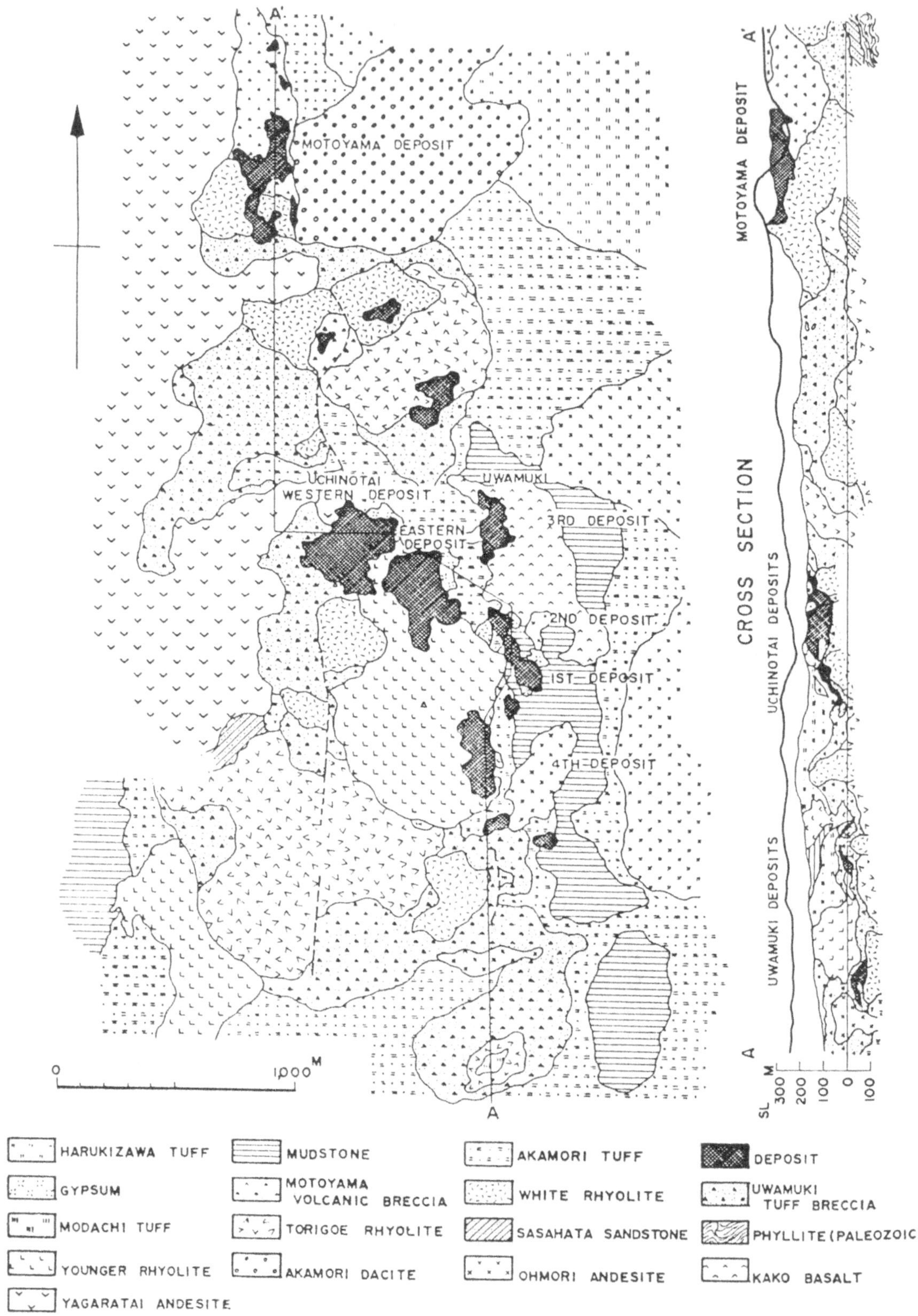

Figure 10–7. Geologic map of the Kosaka mining area (after Oshima et al., 1974).

Table 10–1. Barite Contents of the Kosaka and Fukazawa Typical Kuroko Deposits and the Minamishiraoi Kuroko-type Barite Deposit

Orebody	*Ore Type*	*Reserves (tons)*	*$BaSO_4$ (wt%)*	*$BaSO_4$ (metric tons × 10^5)*
Kosaka				
Uchinotai	Black ore	2.1×10^6	37	7.7
	Semiblack ore	1.5×10^6	31	4.6
	Yellow ore	1.7×10^6	5	0.8
				13.1
Uwamuki	Black ore	1.2×10^6	31	3.7
	Yellow ore	0.7×10^6	4	0.3
	Siliceous ore	1.7×10^6	10	1.7
				5.7
Motoyama	Black ore	1.8×10^6	37	6.7
	Semiblack ore	1.3×10^6	31	3.9
	Yellow ore	1.5×10^6	5	0.7
	Siliceous ore	2.5×10^6	2	0.5
				11.8
		Total $BaSO_4$ 3.1×10^6 metric tons		
Fukazawa				
Tsunokakezawa	Black ore	5.3×10^6	20	10.6
Manjaku	Black ore	1.3×10^6	20	2.6
Kanayamazawa	Black ore	1.0×10^6	15	1.5
				14.7
Minamishiraoi	Barite bed	1.8×10^5	90	1.6
	Siliceous breccia	3.5×10^5	14	0.5
				2.1

Sources: Data for Kosaka and Fukazawa typical Kuroko deposits from Farrell and Holland (1983); data for Minamishiraoi Kuroko-type barite deposit from Marumo and Fujinuki (1988).

the Akamori tuff, lens-shaped ferruginous chert, consisting mainly of quartz, hematite, and barite, is found.

The bottom part of the black ore changes to yellow ore through semiblack ore, which is defined as the ore of transitional composition between black and yellow ores. The yellow ore consists mainly of chalcopyrite, pyrite, and quartz. Barite is very rare in yellow ore, occurring as a small corralled island in quartz. The boundary between the yellow ore and the underlying siliceous ore is apparent because a thin clay layer is intercalated between them.

The siliceous ore contains disseminated idiomorphic pyrite, chalcopyrite, and quartz, with a small amount of sericite and chlorite. The ore occurs in the footwall of the black ore and yellow ore, and it grades into barren "white rhyolite" by increasing the amount of sericite and chlorite. The siliceous ore developed in white rhyolite is of irregular cylindrical shape (Fig. 10–8).

THE FUKAZAWA TYPICAL KUROKO DEPOSITS

The geology of the Fukazawa mining area consists of the Menaichizawa formation, Yukizawa formation, Kagoya formation, and Shigenai formation

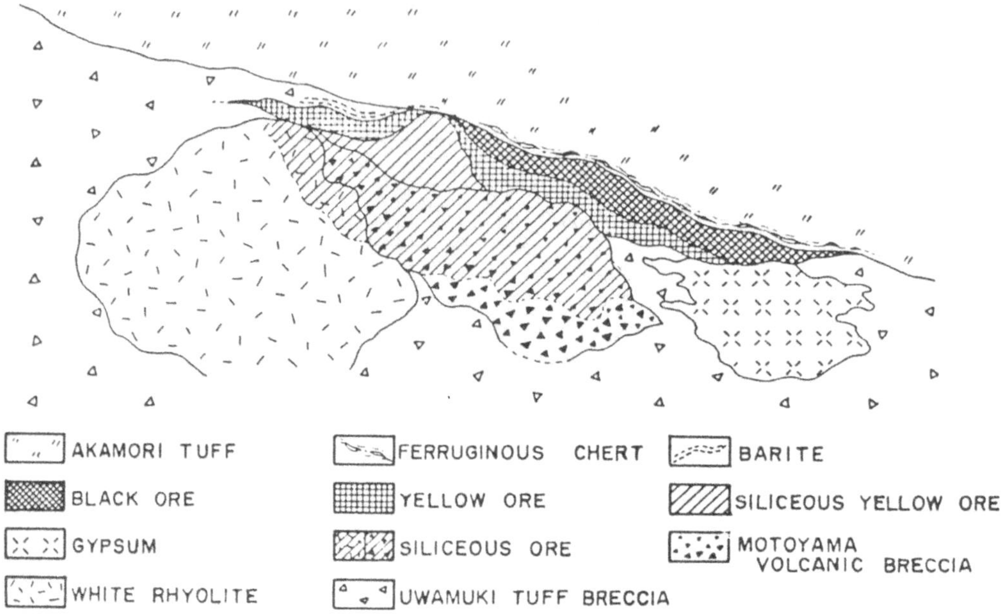

Figure 10–8. Generalized profile of the Kosaka deposits (after Oshima et al., 1974).

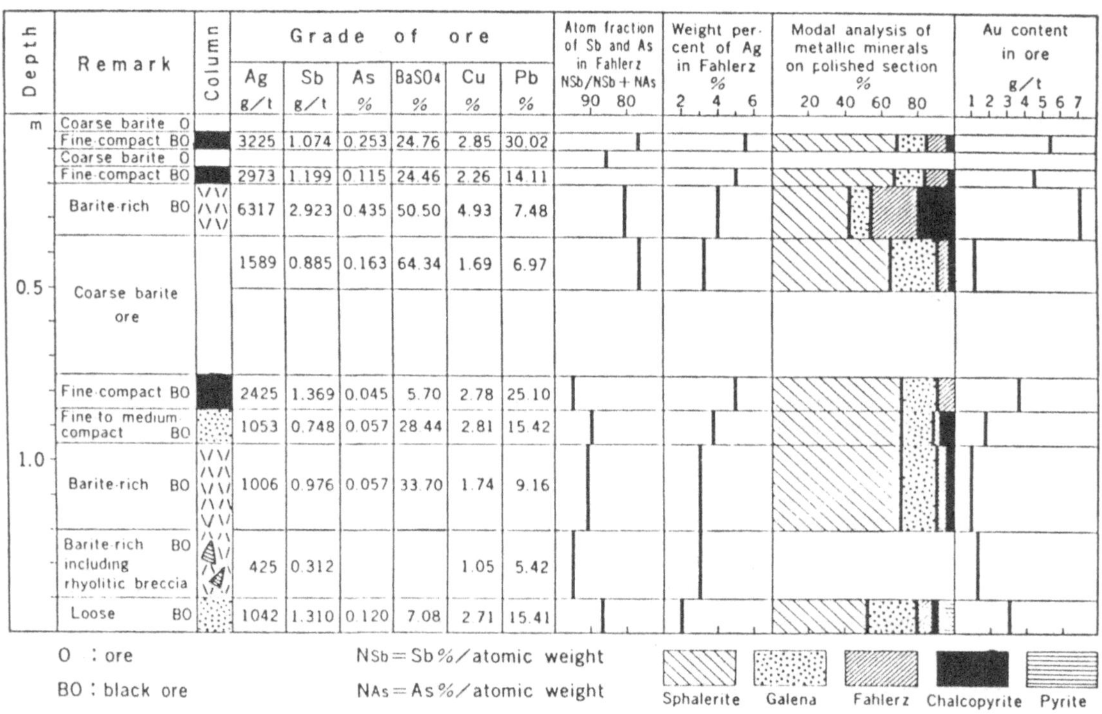

Figure 10–9. An example of vertical variation of mineral assemblage within a black orebody, Uwamuki No. 2 deposit, Kosaka mine (after Matsukuma and Horikoshi, 1970, p. 168).

in ascending order (Figs. 10–10 and 10–12). The deposits are localized above acidic pyroclastic rocks of the Yukizawa formation.

The Fukazawa mine comprises three main ore deposits (from northwest to southeast, Manjaku, Kanayamazawa and Tsunokakezawa) and a few unnamed smaller deposits (Figs. 10–11 and 10–12). The development of these deposits, with a production rate of 10,000 metric tons per month, began in 1973. The present production is 20,000 metric tons (Au, 0.8g/ton; Ag, 85g/ton; Cu, 1.10 percent; Pb, 2.80 percent, Zn, 10.80 percent, $BaSO_4$, 10.0 percent) per month. The quantity of barite in the Fukazawa deposits is approximately 1.5×10^6 metric tons (Table 10–1). The average ore grades of these deposits are also listed in Table 10–1.

Although mineral assemblages and mineral zoning of the Fukazawa deposits are similar to those of the Kosaka deposits, some differences are observed in the mode of occurrence and the characteristics of ore. Figure 10–13 shows an east–west section of the Tsunokakezawa ore deposit, which is one of the largest in the Fukazawa mine. The deposit shows a vertical zonal arrangement, with siliceous ore, yellow ore, black ore, barite bed, and ferruginous chert zones in ascending order (Ito and Ohmori, 1975). Among these ores, the siliceous ore and yellow ore are poorly developed compared to the black ore. The black ore portion of the deposit has maximum dimensions of 500 m (length) by 300 m (width) by 5 m (height).

The black orebody consists of compact black ore, brecciated black ore, and disseminated black ore. The compact ore is most abundant throughout these deposits. Grain size of the compact ore is finest in the uppermost parts and in general becomes coarser downward. Tetrahedrite and barite predominate in

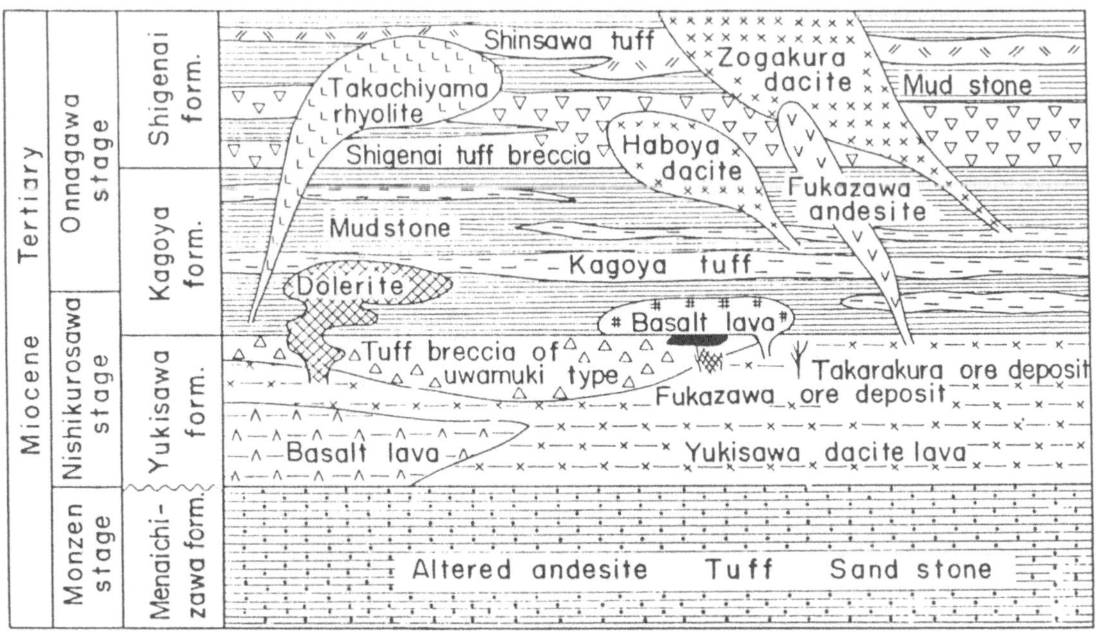

Figure 10–10. Generalized stratigraphic section through the Fakazawa mining area (after Tanimura et al., 1974).

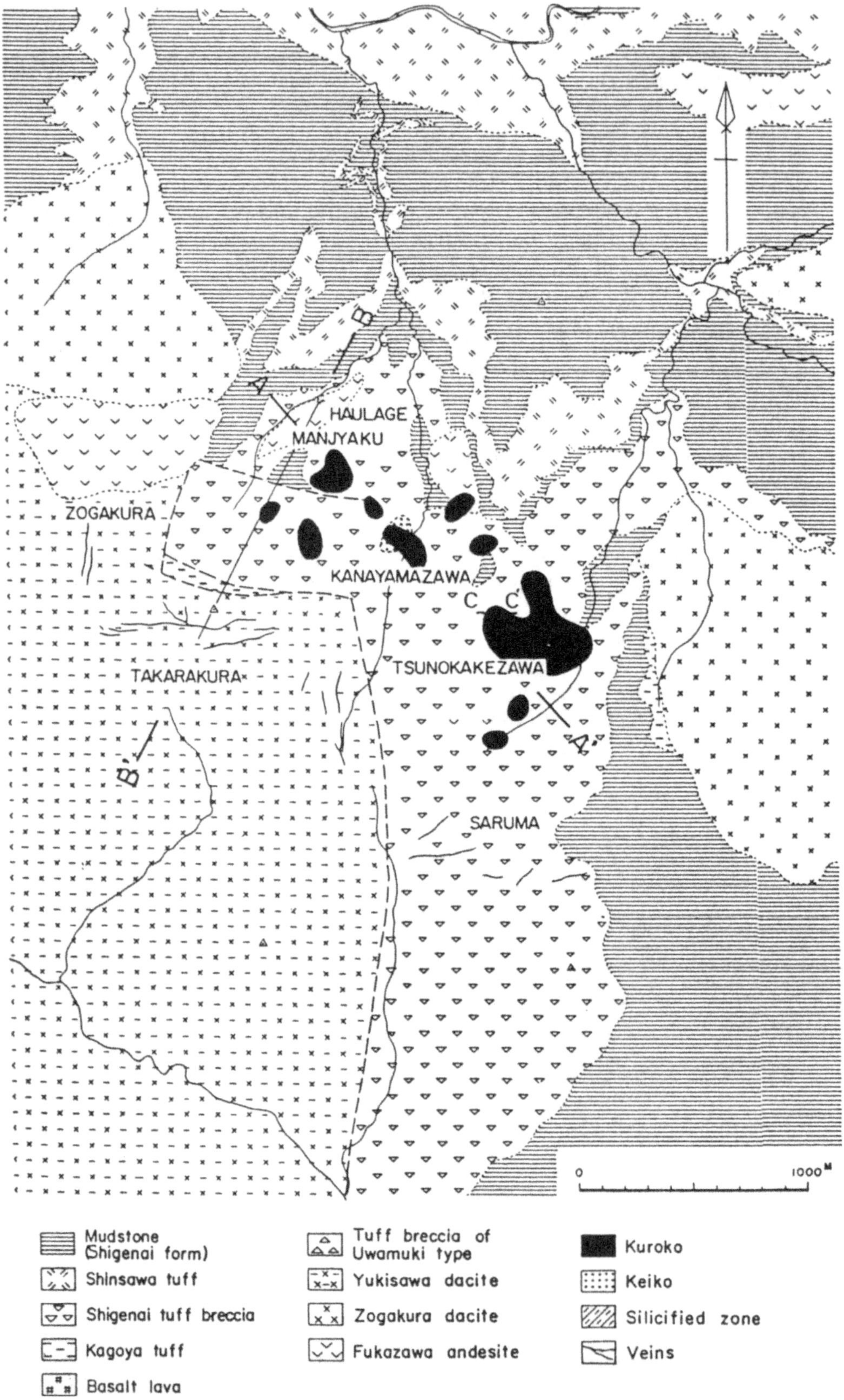
HAULAGE
MANJYAKU
ZOGAKURA
KANAYAMAZAWA
TAKARAKURA
TSUNOKAKEZAWA
SARUMA
A
A'
B
B'
C
C'
0
1000M
Mudstone (Shigenai form)
Shinsawa tuff
Shigenai tuff breccia
Kagoya tuff
Basalt lava
Tuff breccia of Uwamuki type
Yukisawa dacite
Zogakura dacite
Fukazawa andesite
Kuroko
Keiko
Silicified zone
Veins

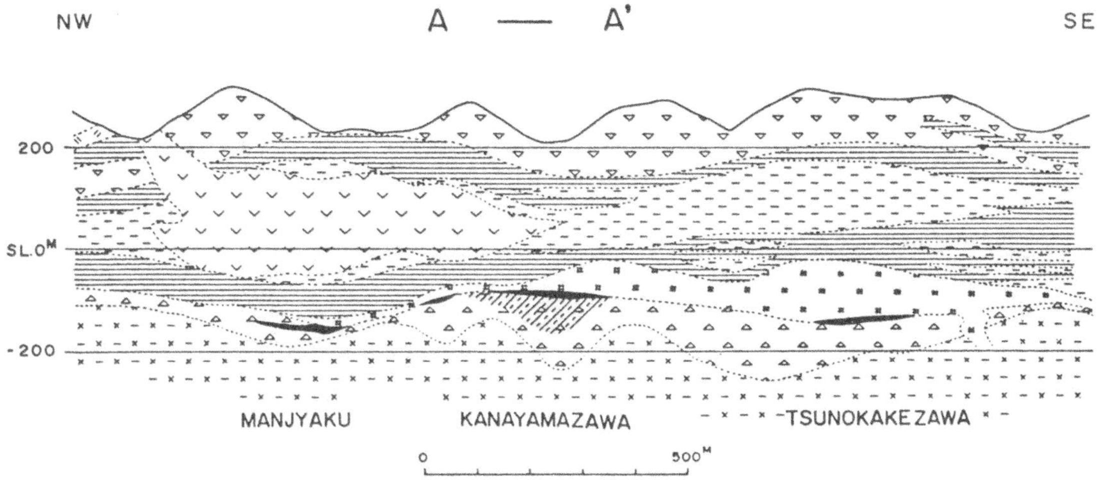

Fig. 4. The NW–SE section of the Fukazawa mine area.

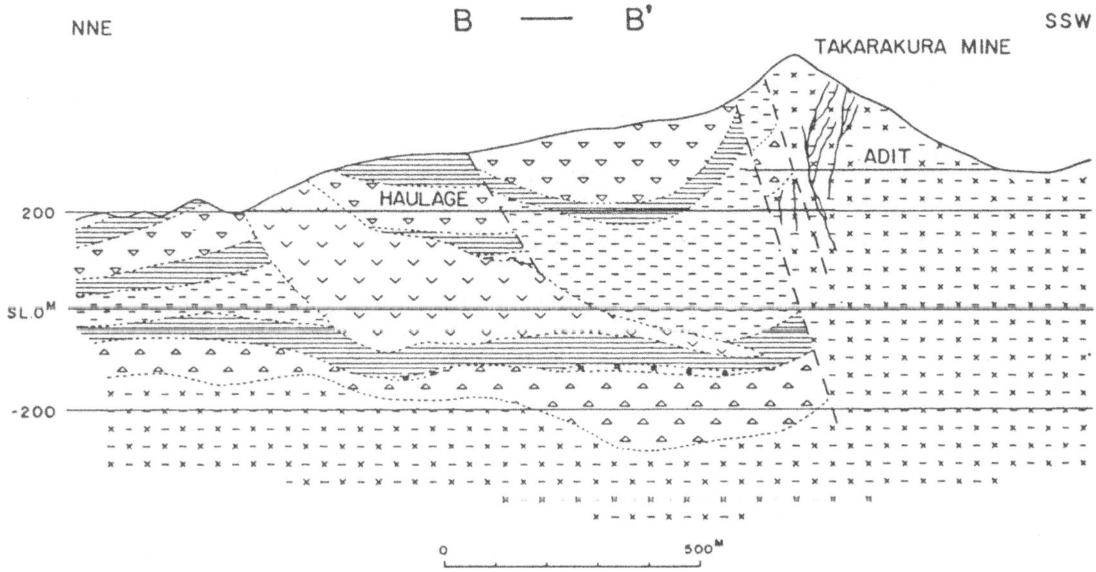

Figure 10–12. Northwest–southeast geologic section of the Fukazawa deposits (after Tanimura et al., 1974).

the uppermost part, and chalcopyrite is dominant in the lower parts. Brecciated black ore occurs only in limited portions of the southeastern and western parts of the Tsunokakezawa orebody. This ore consists of angular to subangular black ore fragments up to 10 cm across, and a matrix of black ore fragments of sand size and a small amount of clay minerals. Disseminated black ore, which consists of sphalerite and galena disseminating in argillized tuffs,

---

Figure 10–11 (at left). Geologic map of the Fukazawa mining area (after Tanimura et al., 1974).

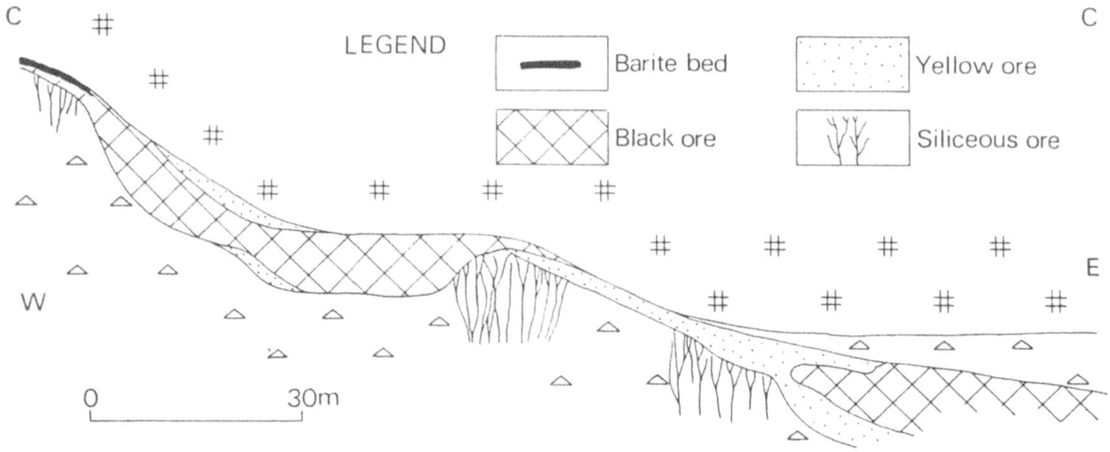

Figure 10–13. East–west section of the Tsunokakezawa ore deposit of the Fukazawa mine (after Ito and Ohmori, 1975).

are found in the eastern part of the Tsunokakezawa orebody. This ore occurs at a slightly lower horizon than that of the major Kuroko ores (Tanimura et al., 1974).

Yellow ore containing 2.5–3.5 percent Cu occurs as irregularly distributed pods between the black ore and siliceous ore. The upper yellow ore occurring within the hanging wall layered tuff and basalt is considered to be evidence of post-Kuroko mineralization (Sato et al., 1979). Siliceous ore, which comprises chalcopyrite, pyrite, and quartz with a minor amount of sericite, is predominant in the Kanayamazawa ore deposit. This ore is disseminated or stockwork in texture, funnel shaped, and present in lithic tuff and brecciated dacite.

THE OTARUMATSUKURA KUROKO-TYPE BARITE DEPOSITS

The geology of the Otarumatsukura mining area consists of the Yamato formation (andesitic tuff, volcanic breccia, and lava in ascending order), which overlies the Kunitomi formation (Kuroko hosting formation in the southwestern Hokkaido). Figure 10–14 shows a geologic cross section through the Otarumatsukura Kuroko-type barite deposits and Meiji typical Kuroko deposits.

As shown in Figure 10–15, the Otarumatsukura mine consists of seven ore deposits (Matsukura, Daigo, Daishi, Daiichi, Daini, Shoei, Ashahi, and Daisan from west to east). The mine was worked from 1932 until 1971. The monthly production of refined barite in 1963 was 1650 metric tons, which was more than half of the total barite production in Japan. Total mine production included 323,173 metric tons of barite refined ore and 2438 metric tons of zinc concentrates (Zn = 43.4 percent).

Each ore deposit consists of upper stratiform massive barite ore and lower barite-bearing silicified breccia. The stratiform massive barite ore occurs along the boundary between footwall andesitic volcanic breccia and hanging wall andesite lava. The massive barite ore, which contains a large quantity of barite

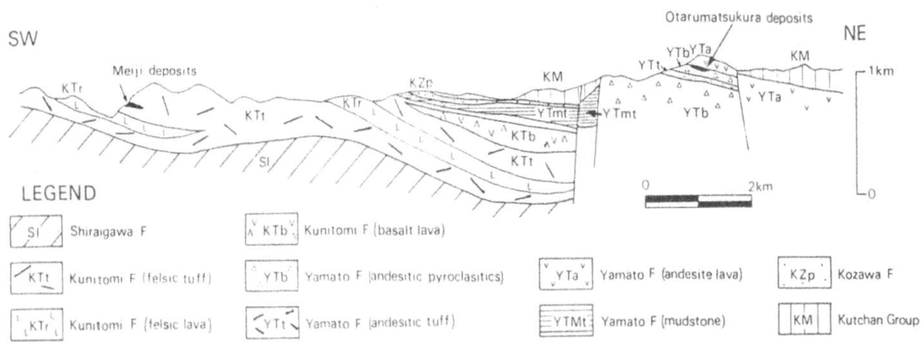

Figure 10-14. Geologic cross section through the Otarumatsukura Kuroko-type barite deposits and Meiji typical Kuroko deposits (after Hasegawa, 1986).

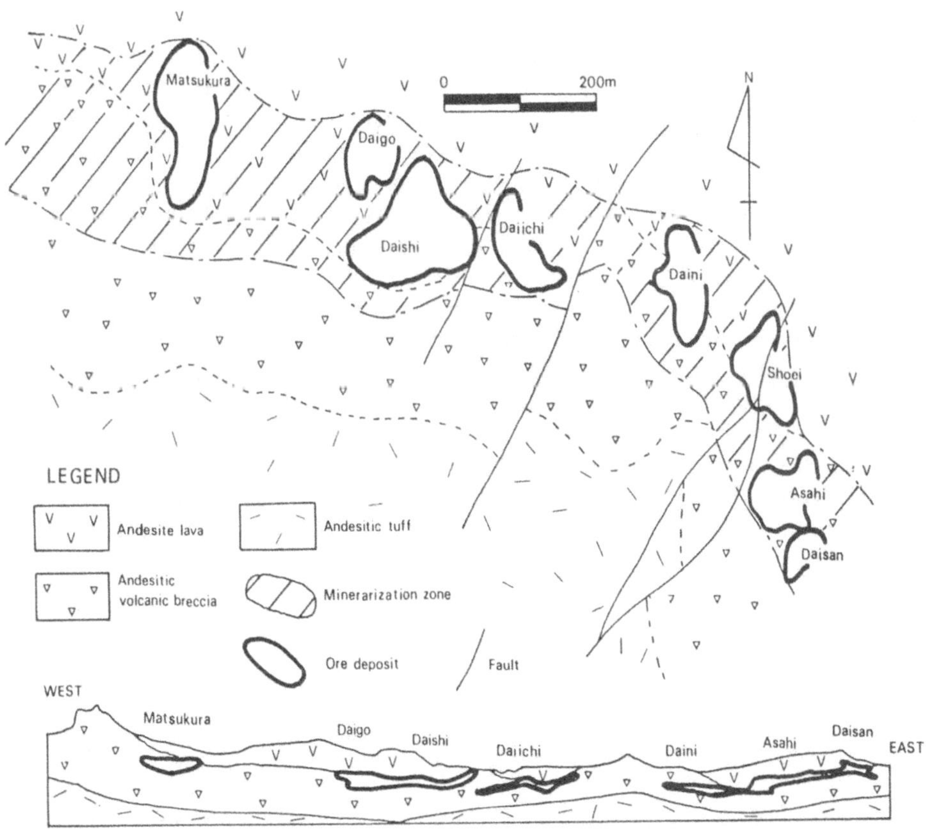

Figure 10-15. Geologic map and generalized cross section of the Otarumatsukura deposits (after Igarashi et al., 1974).

(more than 90 percent), is 5–15 m thick and has a lenticular form elongated in the north-south direction. The ore consists of idiomorphic to semi-ideomorphic barite crystals, less than 0.4 × 1.7 mm in size, with a small amount of quartz and sericite.

The barite-bearing silicified breccia occurs in the footwall andesitic volcanic breccia. It consists of networks, disseminations, and veins of barite (ideomorphic barite crystals 0.3 × 0.1 mm in size). Sphalerite showing network veins occurs. Accompanied by a small amount of pyrite and marcasite, galena is seen replacing barite (Igarashi et al., 1974).

SHIRAOI–MINAMISHIRAOI KUROKO-TYPE BARITE DEPOSITS

The geology of the Shiraoi–Minamishiraoi mining area is characterized by volcanism of Miocene to Plio-Pleistocene ages. The mine comprises three deposits: the Shiraoi barite deposit, the Minamishiraoi barite–kaolin deposit, and the Morino Kuroko-type network deposit. Among them, the Minamishiraoi deposit is mined by the open-pit method with a monthly production rate of 500 metric tons of barite and 1000 metric tons of kaolin. The quantity of barite in the Minamishiraoi deposit is approximately 2.1 × 10^5 metric tons (Table 10–1).

As shown in Figure 10–16, the Minamishiraoi deposit occurs in and on Miocene andesite lavas and andesitic pyroclastics (Bifuegawa formation). The deposit consists of an upper stratiform massive barite orebody, a lower barite-bearing silicified breccia orebody, and a footwall clay zone (Fig. 10–17).

The stratiform massive barite orebody occupies an area of 100 × 100 m and averages 5 m in thickness. The ore is compact and fragile with more than 90 percent $BaSO_4$ content; it consists mainly of barite (ideomorphic barite crystals 0.04 × 0.2 mm in size) with small quantities of quartz, pyrite, marcasite, sphalerite, and galena. These sulfide contents increase downward and grade into the lower barite-bearing silicified breccia orebody.

The barite-bearing silicified breccia orebody occurs in andesite pyroclastics. It consists of networks, disseminations, and veins of barite (ideomorphic barite crystals 0.05 × 0.5 mm in size), pyrite, marcasite, and chalcopyrite. The ore is silicified but contains a small quantity of clay minerals (dickite, nacrite, and mica/montmorillonite mixed-layer minerals).

The footwall clay zone occurs in andesite lava and its pyroclastics. The clay zone consists of kaolinite, nacrite, dickite, montmorillonite, mica/montmoril-

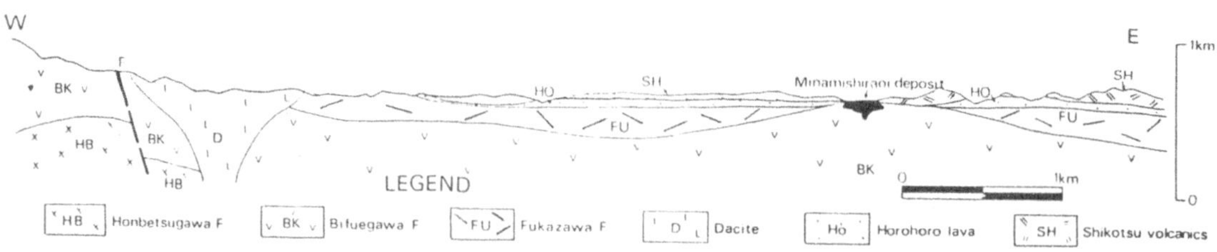

Figure 10–16. Geologic cross section of Shiraoi mining area (after Metal Mining Agency of Japan, 1978).

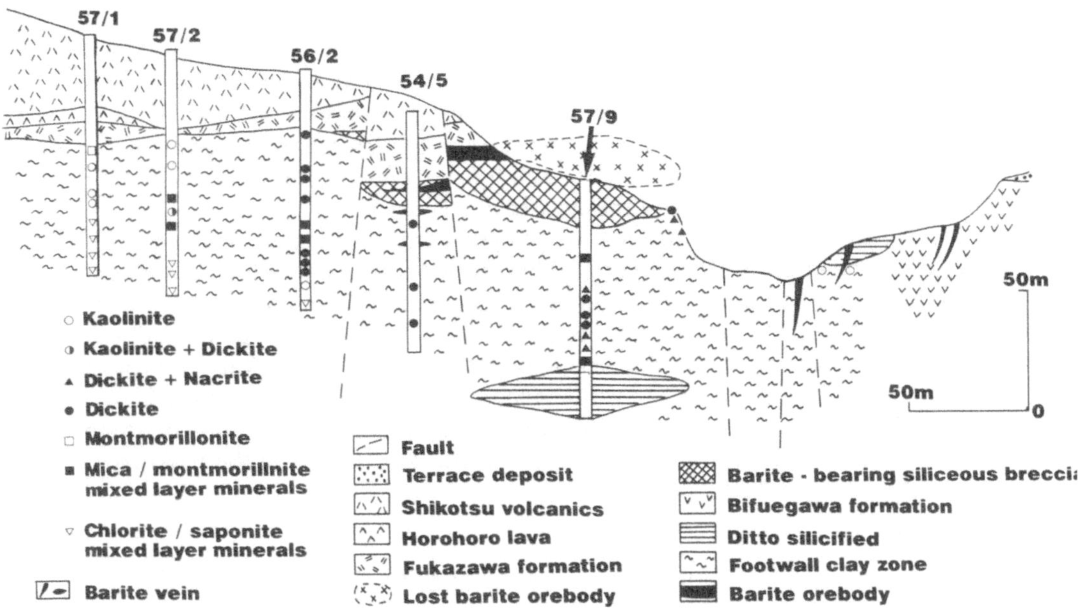

Figure 10–17. Geologic cross section of the Minamishiraoi deposit (after Marumo et al., 1985).

lonite mixed-layer minerals, and chlorite/saponite mixed-layer minerals together with pyrite, marcasite, chalcopyrite, sphalerite, tetrahedrite, and galena. The clay mineral–sulfide associations are subdivided into two zones in spatial distribution. That is, nacrite, dickite, mica/montmorillonite mixed-layer minerals, which are associated with pyrite and chalcopyrite, are distributed in the central zone below the barite-bearing silicified breccia, whereas kaolinite and montmorillonite, which are associated with marcasite and sphalerite, occur in the peripheral zone. The pipelike aggregates of pyrite and barite occur near the barite-bearing silicified breccia orebody. The size of the pipes is mostly 3 cm across and 1–2 m long (Marumo et al., 1985).

MINERALIZATION AGES OF DEPOSITS

Although interpretations that the Kuroko deposits occur in a single stratigraphic horizon (aged about 13 m.y.) have been made often in the literature on Kuroko deposits, Tanimura et al. (1983) showed that these deposits are not contemporaneous on the basis of radiometric ages of igneous rocks and occurrences of age-definitive fossils in the Hokuroku ore field. Deposits in other districts may also have formed at times much different from 13 m.y.

For example, the age of the Tobe formation is estimated to be about 3 m.y. on the basis of a fission-track age value of 2.5 $\pm$ 0.7 m.y. and a K-Ar age value of 3.8 $\pm$ 0.7 m.y. for tuff samples (Tanimura et al., 1983). Thus, the mineralization age of the Tsugaruyunosawa Kuroko-type network deposit occurring in the Tobe formation would be 3 m.y. or much younger (see Fig. 10–3).

An age of around 12 m.y. for the peak of hydrothermal activity at the Fukazawa deposit is suggested from the oldest K-Ar age value of 12.9 ± 0.6 m.y. for a D3 dacite sample and two K-Ar age values of 11.7 ± 0.6 m.y. and 11.6 ± 2.3 m.y. for sericites in the T3 tuff units. The suggestion of ca. 11 m.y. for the peak of igneous activities in the Kosaka mining area is based on K-Ar age values of 11.3 ± 0.6 m.y., 11.0 ± 0.6 m.y., and 10.7 ± 0.5 m.y. for T3 tuff and D3 dacite samples (Tanimura et al., 1983).

Although the Minamishiraoi Kuroko-type barite deposit occurs in the Bifuegawa formation overlying the Honbetsugawa formation, which correlated with the Kuroko-bearing formation in the Horobetsu–Shiraoi district, sericites from the Minamishiraoi deposit, together with those of the Horobetsu and Kagenosawa Kuroko-type network deposits and the Toya-takarada typical Kuroko deposit, give K-Ar age values generally in the range of 12.3–14.2 m.y. These K-Ar age values indicate that the age of the mineralizations of these deposits were similar to those of the typical Kuroko deposits in the Hokuroku basin (Marumo and Sawai, 1986).

Mineral Textures and Parageneses of Barites

On the basis of microscopic observations of major ore minerals (sphalerite, pyrite, galena, tetrahedrite, and chalcopyrite) and gangue minerals (barite and quartz) from typical Kuroko deposits, Eldridge et al. (1983) divided these minerals into three categories (Fig. 10–18). These groups are termed facies 1 (fine-

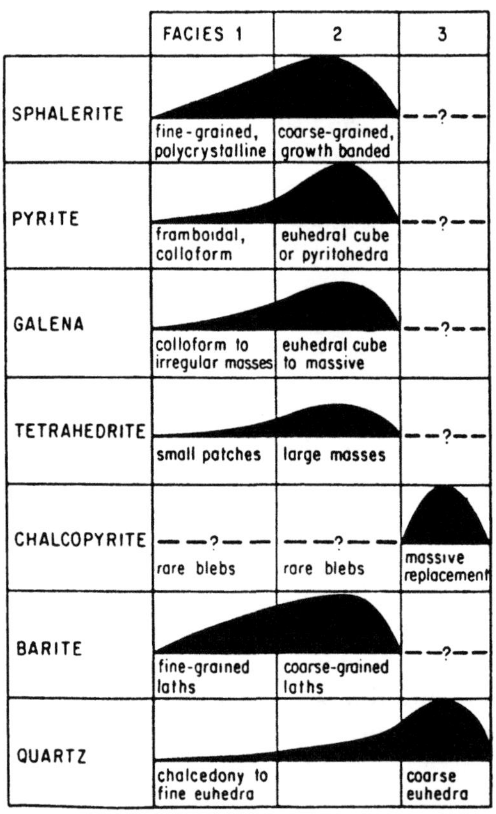

Figure 10–18. Diagram showing the dominant minerals in each facies (after Eldridge et at., 1983, p. 247).

grained; usually <50 μm) and facies 2 (relatively coarse-grained; commonly up to a few millimeters and rarely as much as 20 mm). Chalcopyrite, having a replacement relationship to the facies 2 sulfides, has been placed in facies 3 (<1 μm blebs to masses of several millimeters).

Facies 1 barite appears as either lath-shaped or acicular crystals in black ore. Commonly, several crystals are intergrown and some clusters of radiating crystals are found. Facies 2 barite crystals typically contain copious sulfide fragments and have numerous large (up to 100 μm in diameter) fluid inclusions. In general, facies 1 barite formed contemporaneously with facies 1 sulfides, and facies 2 barite with facies 2 sulfides. Clasts of closely intergrown facies 1 sulfides and barite are typically overgrown by facies 2 sulfides and barite. Most of the barites formed prior to, but some after, chalcopyrite mineralization as indicated by some facies 1 and facies 2 barite crystals encrusting clasts containing facies 3 mineralization. The ore solutions that produced facies 3 ore affected facies 1 and 2 barites as well as the sulfides. Barite does not show rounded outlines against chalcopyrite but sometimes does against quartz. Some barite laths have been pseudomorphically partially replaced by quartz. Downward through the black ores, the ratio of facies 1 barite to facies 2 barite decrease and replacement by quartz increases (Eldridge et al., 1983).

ISOTOPE GEOCHEMISTRY OF BARITE

STRONTIUM ISOTOPE

Strontium isotopic compositions in barites for the Motoyama, Uchinotai-nishi, Uchinotai higashi, Uwamuki 2, and Uwamuki 4 orebodies of the Kosaka deposits are plotted according to their relative stratigraphic positions in Figure 10-19. There is no significant change in the isotopic compositions over a stratigraphic interval of 200 m (Farrell and Holland, 1983).

Strontium isotopic compositions in barites for the Tsunokakezawa and Koezuruko orebodies of the Fukazawa deposits are illustrated in Figure 10-20. The $^{87}Sr/^{86}Sr$ values of the barites from the black ores and hanging-wall barite

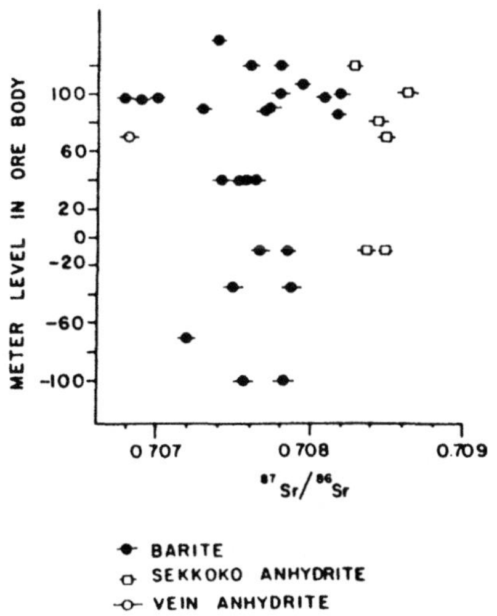

Figure 10-19. Strontium isotope data for the Motoyama, Uwamuki, Uchinotai-nishi, and Uchinotai-higashi of the Kosaka deposits. Data are plotted according to the relative stratigraphic positions of the samples in the district (after Farrell and Holland, 1983, p. 313).

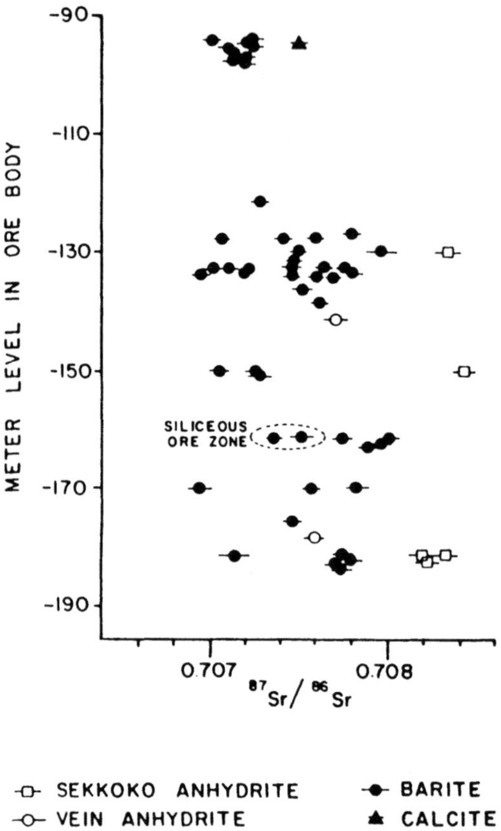

Figure 10–20. Strontium isotope data for the Fukazawa deposits. Barite and calcite analyses above the -110 mL level are from the Koezuruko 1 ore deposit; all other analyses are from the Tsunokakezawa ore deposit (after Farrell and Holland, 1983, p. 311).

sands are slightly lower than those of the anhydrite and gypsum from the gypsum ore and cover a somewhat wider range, 0.7070 to 0.7080. Barites in the siliceous ore zones have $^{87}Sr/^{86}Sr$ values that are indistinguishable from those of barites in the black ores or hanging-wall barite sands (Farrell and Holland, 1983).

The $^{87}Sr/^{86}Sr$ values of barites from the Fukazawa and Kosaka deposits are higher than those of Miocene volcanics (0.7040–0.7060) and lower than sea water (0.7088) and the Paleozoic basement (0.7088–0.7300).

Oxygen and Sulfur Isotope

Whereas the $\delta^{34}S$ values in barites are controlled by the source of the sulfur and the chemistry of the hydrothermal fluids transporting it (Kajiwara, 1971; Ohmoto, 1972), and barite $\delta^{18}O$ values are more indicative of the source of the water in the fluids (Sakai et al., 1970) and the precipitation temperature (Kusakabe and Robinson, 1977).

Oxygen and sulfur isotopic compositions in Kuroko barites from black ores, yellow ores, and siliceous ores of Kosaka and Ainai deposits are analyzed by Watanabe and Sakai (1983). Their isotopic data fall in a narrow range: 18.5 to 24.5 per mil for sulfur and 6.0 to 10.0 per mil for oxygen (Fig. 10–21). These compositions indicate that the hydrothermal fluids for barites

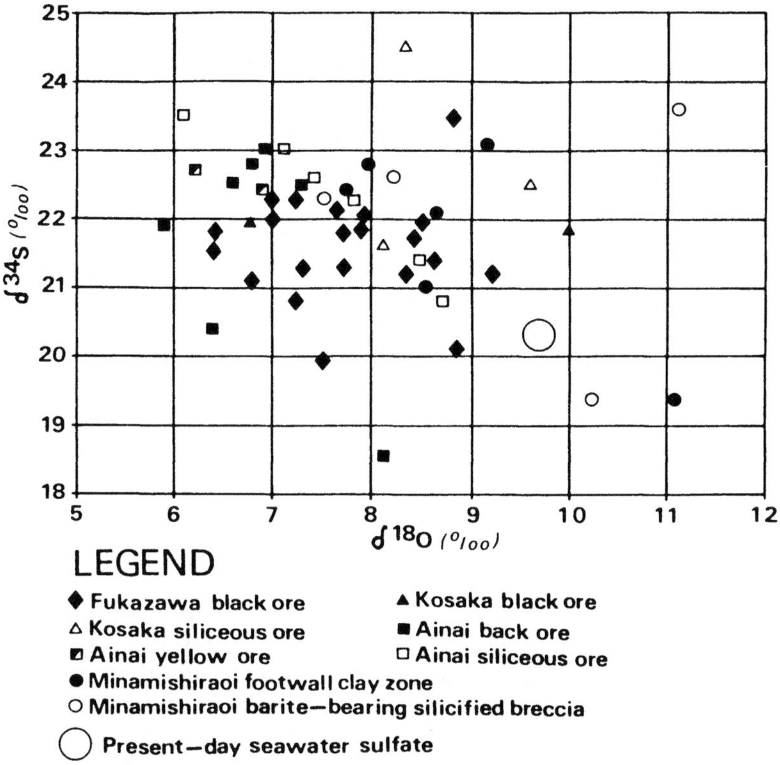

Figure 10–21. The $\delta^{18}O$ versus $\delta^{34}S$ plot for barites from Kuroko-type deposits.

were predominantly Miocene sea water. Kusakabe and Chiba (1983) showed that there is no systematic variation in the oxygen and sulfur isotope ratio of barite occurring in three different modes: veins and infillings of a matrix of black ore sulfide; clastic and nodular in black ore; and bedded barite overlying the black ore unit, at the Fukazawa deposits.

A barite from the Tsugaruyunosawa Kuroko-type network deposit is characterized by high sulfur isotope (38.5 per mil) and oxygen isotope (19.9 per mil) values, indicating that this barite sample was precipitated from hydrothermal fluids generated by mixing of meteoric water with isotopically heavy waters related to the volcanogenic sediments (Watanabe and Sakai, 1983). The participation of meteoric water in the mineralization is supported from the occurrences of abnormal sediments, peats, and lignite in the Tobe formation, which hosts the deposit.

The oxygen and sulfur isotope compositions in barites from the Minamishiraoi Kuroko-type barite deposit are superimposed on Figure 10–21. Although there is some overlap, coarse-grained barite occurring as veins in the barite-bearing silicified breccia and footwall clay zone appears to be significantly richer in $\delta^{18}O$ than barites from other Kuroko deposits (Marumo et al., unpublished data).

MODELS FOR THE FORMATION OF THE DEPOSITS

GEOCHEMICAL MODEL FOR KUROKO-BARITE GENESIS

Based on the sulfur and oxygen isotopic analyses in barite associated with Kuroko-type deposits, Watanabe and Sakai (1983) and Kusakabe and Chiba (1983) concluded that SO_4^{2-} in the Kuroko-forming fluids was largely of sea water origin.

Although sea water contains enough SO_4^{2-} to produce barite in Kuroko-type deposits, it does not contain enough barium ion to precipitate barite for deposits. As pointed out by Farrell and Holland (1983), the amount of barium contained in an average white rhyolite dome 200 m in diameter and 150 m thick and 500–2000 ppm barium concentration is 7×10^3 to 27×10^3 metric tons. This sum is less than the amount of barium present in some single Kuroko orebodies (Table 10–1). Thus, the Kuroko barites are not generated by individual white rhyolite domes.

One of the promising suppliers of barium for barite is Miocene volcanics and basement rocks. From a lead isotopic study of ores and country rocks in the Hokuroku ore field, Fehn et al. (1983) concluded that the lead in the Kuroko ore fluids was derived from the Miocene volcanics and underlying basement rocks, whereas Ohmoto et al. (1983) arrived at the same conclusion from a comparison of heavy metal contents in the central sericite–chlorite alteration zone with those in the peripheral lower grade alteration zones. Ohmoto et al. (1983) calculated that an average cluster of Kuroko deposits containing about 1 million tons of barite could be produced by the barium leached from 20 km^3 sized country rock.

Similarly, 2.1×10^5 metric tons of barite of the Minamishiraoi Kuroko-type barite deposit could be produced by leaching of barium from 3 km^3 sized country rock. The strong Na_2O- and CaO- depletion of the footwall clay zone suggests that barium was probably leached from this clay zone and its peripheral altered rock (Marumo and Fujinuki, in press).

In the Kosaka district, the Paleozoic cherts, slates, and phyllites lie within 300–500 m of the Kuroko horizon. A drill core from the Paleozoic basement in this district has been bleached and veined by pyrite and quartz. The accompanying strong strontium depletion suggests that barium was probably leached from this rock unit as well as from the Miocene volcanics lying between the basement and ore horizon (Farrell and Holland, 1983).

A hypothetical geologic cross section through Kuroko hydrothermal convection cells presented by Farrell and Holland (1983) is shown in Figure 10–22. On the basis of the strontium isotopic ratios of barite in the deposit, Farrell and Holland proposed a convection system that leached out rock barium to precipitate barite at the mixing sites of hydrothermal fluids with sea water. When sea water is convected downward through the Miocene volcanics into the basement to change into hydrothermal fluids, it will probably leach barium and strontium from the wall rocks. The strontium isotopic composition of sea water will approach those values of the wall rock units through which it passes. Thus, during reaction with Miocene volcanics having lower

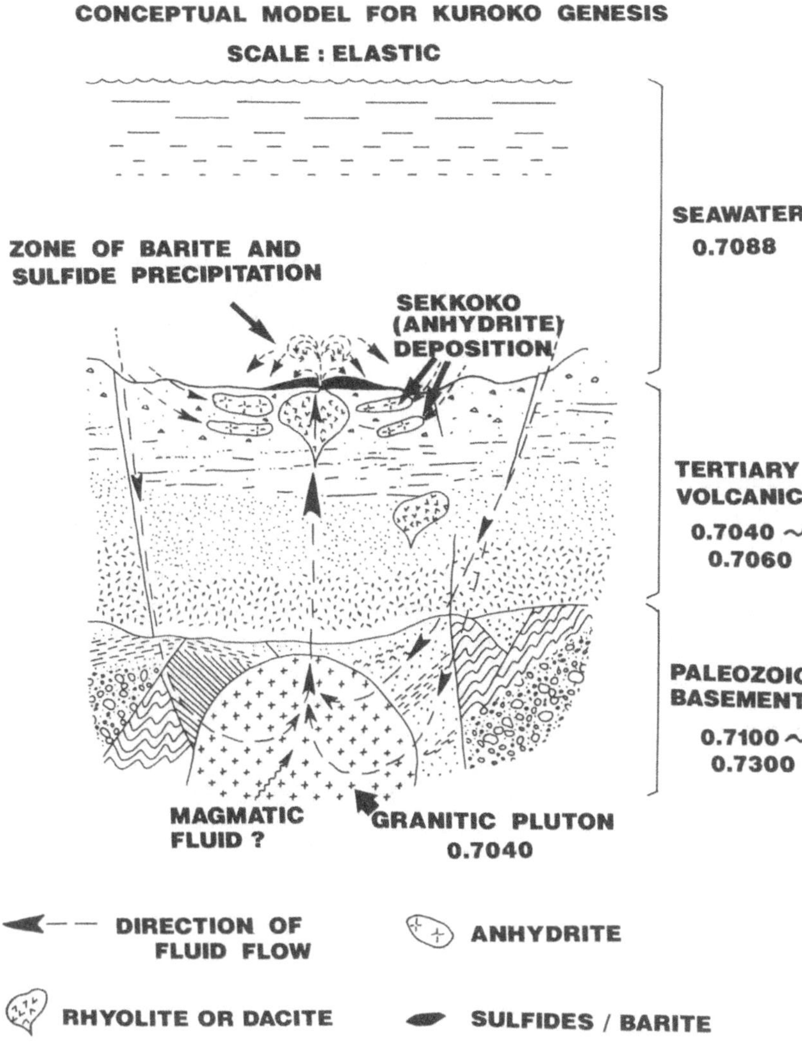

Figure 10–22. Hypothetical geologic cross section through the Kuroko hydrothermal convection cell (after Farrell and Holland, 1983, p. 318).

$^{87}Sr/^{86}Sr$ values than sea water, the $^{87}Sr/^{86}Sr$ value of sea water decreased. During reaction with basement rocks, whose $^{87}Sr/^{86}Sr$ value is so much higher than that of sea water, the $^{87}Sr/^{86}Sr$ value of the ^{87}Sr-depleted sea water probably increased and may well have exceeded that of original sea water. The solution passed through the Miocene volcanics a second time before barite precipitation, and may also have passed through and interacted with one of the deep granitic plutons that are apparently required to supply the heat necessary to drive the Kuroko convection systems. The combined effect of these interactions, as well as the possible addition of magmatic fluids associated with granitic plutons, probably account for the absence of $^{87}Sr/^{86}Sr$ values in barite that are higher than the $^{87}Sr/^{86}Sr$ ratio of sea water.

FACTORS CONTROLLING SIZES AND SULFIDE TO BARITE RATIOS OF THE KUROKO-TYPE DEPOSITS

The Kosaka, Fukazawa, Otarumatsukura, and Minamishiraoi deposits were formed in areas where active volcanism occurred before ore deposition. The sea bottom, therefore, consisted mainly of relatively hard, compact, and well-fissured volcanic rock masses. As pointed out by Kuroda (1983), such well-fissured rock masses allow rising of hydrothermal solution and mixing with sea water to precipitate barite and sulfides on and near the sea floor. The similarity of the barite occurrences of these deposits, that is, barite-rich ore occurs at the upper part of the deposit, would come mainly from these analogous sea-bottom conditions.

Although mineral assemblages and mineral zoning in black-type ores are similar among these deposits, the sizes, total sulfide to barite ratios, and black ore to yellow ore ratios of these deposits are different deposit by deposit. Ohmoto et al. (1983) proposed that both the type and size of an ore deposit within a mining area are related to the relative distance from a heat source; with increasing distance from the heat source, deposits become smaller and change from composite to simple in type (Fig. 10–23). These changes may simply reflect the fact that both the temperature of the discharging fluid at any given stage of hydrothermal activity and the total amount of fluid discharged over the life of the hydrothermal activity decreased generally with increasing distance away from the heat source.

This background indicates the lower temperature of formation for the Mi-

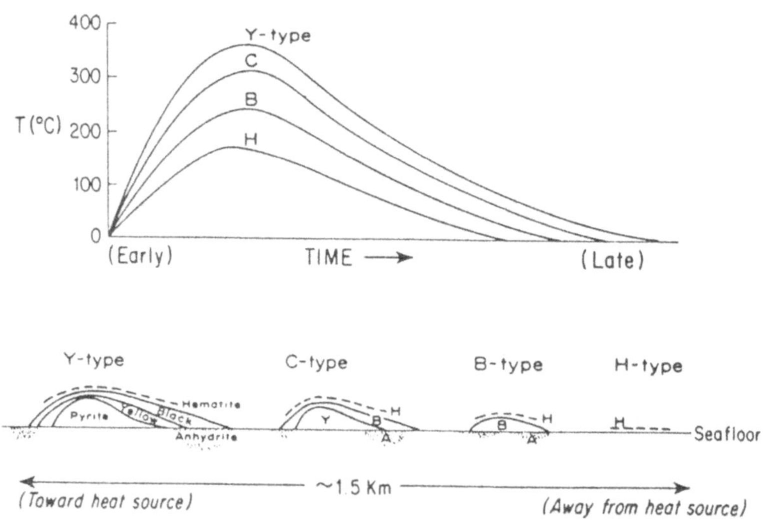

Figure 10–23. A model relating the distribution of ore types to the thermal history and to the relative distance from heat source (after Ohmoto et al., 1983, p. 597).

Y = yellow ore; C = composite ore; B = black ore; H = hematite ore.

namishiraoi barite deposit, which is smaller and simple in type with respect to other typical Kuroko deposits. The first evidence for the lower formational temperature of this deposit comes from the oxygen isotopic compositions in barites. As described previously, some of the Minamishiraoi barites show significantly higher oxygen isotopic composition than do those of black ore of the other typical Kuroko deposits. The temperature dependence of oxygen isotope fractionation factors for the system presented by Kusakabe and Robinson (1977) shows that barite tends to have a higher oxygen isotopic composition with decreasing temperature when it precipitates from sea water. Thus, the trend of oxygen isotopic compositions of barite will be explained by the lower temperature of precipitation.

The second evidence is the occurrence of kaolin minerals, montmorillonite, mica/montmorillonite mixed-layer minerals, and marcasite in the footwall alteration zone of the Minamishiraoi deposit instead of sericite, chlorite, and pyrite, which prevail in the hydrothermal alteration zones enveloping typical Kuroko deposits (Urabe et al., 1983). Since progressive transformation of montmorillonite to sericite through mixed-layer minerals with increasing temperature has been observed in the active geothermal fields (Steiner, 1968; Muffler and White, 1969), the occurrences of montmorillonite and mixed-layer minerals will indicate the lower formational temperature of the Minamishiraoi alteration zone than those of ore-enveloping alteration zones of typical Kuroko deposits.

The presence of kaolin minerals also implies the lower formational temperature of the footwall alteration zone based on the following two reasons: (1) using a diagram for the stability fields of sericite, chlorite, and kaolinite in terms of the activities of Mg^{+2} and K^{+} to that of H^{+}, Riverin and Hodgson (1980) predicted that kaolinite can be precipitated instead of sericite and chlorite in the alteration zone of the volcanogenic massive sulfide deposits in the case where temperatures of hydrothermal fluids are less than 175°C and the flux of fluid is so great; and (2) judging from the mode of distribution of kaolin minerals and their hydrogen isotopic compositions, Marumo (in prep.) concluded that these kaolin minerals were precipitated when ore fluid originated from sea water moved upward and altered wall rock. Since the flux of the hydrothermal solution must be great enough when it passes well-fissured volcanic rock masses, the prediction of Riverin and Hodgson, together with the kaolin occurrences and their hydrogen isotopic data, strongly suggest the lower temperature of the hydrothermal solution.

REFERENCES

Eldridge, C. S., P. B. Barton, and H. Ohmoto. 1983. Mineral textures and their bearing on formation of the Kuroko orebodies. In *The Kuroko and Related Volcanogenic Massive Sulfide Deposits*, ed. H. Ohmoto and B. J. Skinner, pp. 241–281. New Haven, Conn.: Economic Geology Publishing Company.

Farrell, C. W., and H. D. Holland, 1983. Strontium isotope geochemistry of the Kuroko deposits. In *The Kuroko and Related Volcanogenic Massive Sulfide Deposits*, ed. H. Ohmoto and B. J. Skinner, pp. 302–319. New Haven, Conn.: Economic Geology Publishing Company.

Fehn, U., B. R. Doe, and M. H. Delevaux. 1983. The distribution of lead isotopes and the origin of Kuroko ore deposits in the Hokuroku district, Japan. In *The Kuroko and Related Volcanogenic Massive Sulfide Deposits*, ed. H. Ohmoto and B. J. Skinner, pp. 488-506. New Haven, Conn.: Economic Geology Publishing Company.

Hasegawa, K. 1986. Geology of Kutchan-Otaru district: Geology of the Miocene Tohoku-Honshu arc, *Houbun-do, Sendai* **1**(2):1-6 (in Japanese).

Igarashi, T., K. Okabe, and J. Yajima, 1974. Massive barite deposits in west Hokkaido. In *Geology of Kuroko Deposits*, ed. S. Ishihara, pp. 39-44. Society of Mining Geologists of Japan, Mining Geology Special Issue No. 6.

Inoue, T. and U. Mihashi, 1962. On the geological survey in the Ikarigaseki-mura and the eastern part of Owani-mati, Minami-Tugarugun, Aomori prefecture, in the northeast of Japan. In *Report of the Research Institute of Underground Resources*, pp. 60-73. Mining College, Akita University, Japan (in Japanese with English abs.).

Ito, T. and Y. Ohmori, 1975. Recent exploration of the Fukazawa Kuroko deposits. *Min. Geology* (Japan) **25**:187-194 (in Japanese with English abs.).

Kajiwara, Y. 1971. Sulfur isotope study of the Kuroko ores of the Shakanai No. 1 deposits, Akita Prefecture, Japan. *Geochem. Jour.* (Japan) **4**:157-181.

Kuroda, H. 1983. Geologic characteristics and formation environments of the Furutobe and Matsuki Kuroko deposits, Akita prefecture, northeast Japan. In *The Kuroko and Related Volcanogenic Massive Sulfide Deposits*, ed. H. Ohmoto and B. J. Skinner, pp. 149-166. New Haven, Conn.: Economic Geology Publishing Company.

Kusakabe, M., and H. Chiba. 1983. Oxygen and sulfur isotope composition of barite and anhydrite from the Fukazawa deposit, Japan. In *The Kuroko and Related Volcanogenic Massive Sulfide Deposits* ed. H. Ohmoto and B. J. Skinner, pp. 292-301. New Haven, Conn.: Economic Geology Publishing Company.

Kusakabe, M., and B. W. Robinson. 1977. Oxygen and sulfur isotope equilibria in the $BaSO_4$-HSO_4^--H_2O) system from 110 to 350°C and applications. *Geochim. Cosmochim. Acta* **41**:1033-1040.

Marumo, K., and T. Fujinuki. 1988. Barium and base metal mineralization of the Minamishiraoi barite deposit, Hokkaido. In *Research on Evaluation of Deep Seated Mineral Resources*, ed. T. Nakajima. Report 269, Geological Survey of Japan, in press.

Marumo, K., and O. Sawai. 1986. K-Ar ages of some vein-type and Kuroko-type deposits in the southwestern Hokkaido, Japan. *Min. Geology* (Japan) **36**:21-26 (in Japanese with English abs.).

Marumo, K., T. Hasaka, J. Miyazaki, and K. Ikeda. 1985. Mineralogical characteristics of the footwall clay zone of the Minamishiraoi barite deposit, Hokkaido. *Min. Geology* (Japan) **35**:227-237 (in Japanese with English abs.).

Matsukuma, T., and E. Horikoshi. 1970. Kuroko deposits in Japan, a review. In *Volcanism and Ore Genesis*, ed. T. Tatsumi, pp. 153-179. Hongo, Tokyo: University of Tokyo Press.

Metal Mining Agency of Japan. 1978. *Report on Detailed Survey in the Chitose District During the Fiscal Year Showa 53,* Tokyo, 88p (in Japanese).

Muffler, P. L. J., and D. E. White. 1969. Active metamorphism of Upper Cenzoic sediments in the Salton Sea geothermal field and the Salton Trough, southeastern California. *Geol. Soc. America Bull.* **80**:157-182.

Ohmoto, H. 1972. Systematics of sulfur and carbon isotopes in hydrothermal ore deposits. *Econ. Geology* **67**:551-578.

Ohmoto, H. 1983. Geological setting of the Kuroko deposits, Japan: Part 1, Geologic History of the Green Tuff region. In *The Kuroko and Related Volcanogenic Massive Sulfide Deposits*, ed. H. Ohmoto and B. J. Skinner, pp. 9-24. New Haven, Conn.: Economic Geology Publishing Company.

Ohmoto, H., M. Mizukami, S. E. Drummond, and C. S. Eldridge. 1983. Chemical processes of Kuroko formation. In *The Kuroko and Related Volcanogenic Massive Sulfide Deposits*, ed. H. Ohmoto and B. J. Skinner, pp. 570-604. New Haven, Conn.: Economic Geology Publishing Company.

Oshima, T., T. Hashimoto, H. Kamono, and S. Kawabe. 1974. Geology of the Kosaka mine, Akita prefecture. In *Geology of Kuroko Deposits*, ed. S. Ishihara, pp. 89-100. The Society of Mining Geologists of Japan, Mining Geology Special Issue No. 6.

Riverin, G. and C. J. Hodgson. 1980. Wall-rock alteration at the Millenbach Cu-Zn mine, Noranda, Quebec. *Econ. Geology* **75**:424-444.

Sakai, H., S. Osaki, and M. Tsukagishi. 1970. Sulfur and oxygen isotopic geochemistry of sulfate in the black ore deposits of Japan. *Geochem. Jour.* (Japan) **4**:27-39.

Sato, T., S. Tanimura, and T. Ohtagaki. 1974. Geology and ore deposits of the Hokuroku district, Akita prefecture. In *Geology of Kuroko Deposits*, ed. S. Ishihara, pp. 11-18. The Society of Mining Geologists of Japan, Mining Geology Special Issue No. 6.

Sato, Y., Y. Ohmori, and H. Miyamoto. 1979. Recent exploration of the Fukazawa Kuroko deposits and some interesting modes of occurrence of the ores. *Min. Geology* (Japan) **29**:175-185 (in Japanese with English abs.).

Steiner, A. 1968. Clay minerals in hydrothermally altered rocks at Wairakei, New Zealand. *Clay and Clay Min.* **16**:193-213.

Tanimura, S., T. Shimoda, and T. Sawaguchi. 1974. On the Fukazawa ore bodies, Akita prefecture. In *Geology of Kuroko Deposits*, ed. S. Ishihara, pp. 147-155. The Society of Mining Geologists of Japan, Mining Geology Special Issue No. 6.

Tanimura, S., J. Date, T. Takahashi, and H. Ohmoto. 1983. Geologic setting of the Kuroko deposits, Japan: Part 2, Stratigraphy and structure of the Hokuroku district. In *The Kuroko and Related Volcanogenic Massive Sulfide Deposits*, ed. H. Ohmoto and B. J. Skinner, pp. 24-39. New Haven, Conn.: Economic Geology Publishing Company.

Urabe, T., S. D. Scott, and K. Hattori. 1983. A comparison of footwall-rock alteration and geothermal systems beneath some Japanese and Canadian volcanogenic massive sulfide deposits. In *The Kuroko and Related Volcanogenic Massive Sulfide Deposits*, ed. H. Ohmoto and B. J. Skinner, pp. 345-364. New Haven, Conn.: Economic Geology Publishing Company.

Watanabe, M. and H. Sakai. 1983. Stable isotope geochemistry of sulfates from the Neogene ore deposits in the Green Tuff region, Japan. In *The Kuroko and Related Volcanogenic Massive Sulfide Deposits*, ed. H. Ohmoto and B. J. Skinner, pp. 282-291. New Haven, Conn.: Economic Geology Publishing Company.

Yajima, J. 1979. Neogene mineralization of the Teine-Chitose district, West Hokkaido, Japan. *Geol. Surv. Japan Bull.* **30**:645-674.

Yawata, M. 1986. Geology of Funkawan-shouwashinzan-shikotsuko district: Geology of the Miocene Tohoku-Honshu arc, *Houbun-do, Sendai* **1**(3):1-6 (in Japanese).

11

Barite Nodules and Concretions

Milka K. de Brodtkorb

Different nodular forms and concretions are known in continental and marine sediments. The continental sediments are interesting only from a genetic point of view, whereas those formed in marine environments may be of economic interest. Two ore fields of the marine type are described here: Canota, Argentina, and Zamora, Spain.

General Characteristics

Sedimentary rocks that contain disseminated barite as oolites, pisolites, nodules, rosettes, "dollars," anhedral to subhedral crystals, and irregular masses are common. The concretions and rosettes from the Lower Permian of Oklahoma are very well known as are the rosettes and dollars from the Cretaceous of Egypt and the concretions from the Triassic of Bidston Hill, England (Revelle and Emery, 1951). Barite nodules and concretions were also found in some clay sediments and deposits. Near Ovid, Colorado, Mitchell (1960) found nodules in poorly consolidated sediments composed of sand, silt, and clay of Tertiary age. The barite nodules occur in definite horizons associated with clay layers. They are 1.5–3 cm in diameter and when broken some reveal a subradial structure and others show cavities up to 5 mm across that have clear gemmy crystals. Mitchell (1960) mentioned that bluish tabular barite crystals also occur in concretions in clays in Banner County, Nebraska.

In Rudabánya, Hungary, Gyula (1973) observed pisolitic, spheric, and fungiform concretions, 1–7 cm in diameter, in a friable sandy and clayey aleurite of Upper Miocene age. In some places 50 to 60 barite concretions were extracted from 1 m^3 of material. These were impure concretions, commonly with concentric textures, and contained kaolin, calcite, spherosiderite, and vegetable fragments. In Gyula's opinion, the Ba comes from hydrothermal veins in the area and was transported in solution and deposited in swampy environments.

In Wroclaw, Poland, Kowalski et al. (1974) found barite concretions together with carbonate concretions in Pliocene clay layers. These layers consist mainly of beidellite, illite, kaolinite, quartz, and calcite. The concretions have an average size of 3–15 cm and occur every 1.5–2 m in a definite clay horizon. They present spheroidal, elongated shapes with either smooth or rough surfaces. The fine-grained barite mass of the concretions presents voids partially filled with barite crystals up to 4 mm long. Under the microscope it can be observed that they are composed of 96 percent barite, 3 percent quartz, small glauconitic grains, and iron oxides. The chemical analyses yielded the following results: 95 percent $BaSO_4$ and 0.22 percent $SrSO_4$. The clays, on the other hand, have values of 720 ppm Sr and 370 ppm Ba. According to Kowalski et al. (1974), the concretions were formed by a selective diagenetic differentiation of elements within the clay series.

Barite nodules found in the province of Neuquén, Argentina, are 5–10 cm in diameter and occur in clay deposits that formed as a product of tuff alteration. Radial-fibrous barite and aragonite nodules, 3–4 cm in diameter were also found sporadically in clayey sediments of the Cacheuta formation, province of Mendoza, Argentina.

The barite concretions found 1235 m deep, off Colombo, Sri Lanka, by the British surveying ship *Investigator*, are well known as an example in marine environments. Jones (1887) described them as irregularly rounded, varying in shape from almost spherical to roughly cylindrical, ranging in size from 2 cm to 10 cm. Barite is present as small aggregates together with glauconite and remains of foraminifera and radiolaria. Jones estimated a $BaSO_4$ content of 75 percent in the nodules whose specific gravity was determined as 3.77.

On the Dutch Siboga expedition to the East Indies, in 1899 and 1900, barite concretions were dredged from a depth of 304 m, near the Kai Islands. These concretions were found in a blue mud and have a maximum diameter of 6 cm. They are flat or irregular in shape, with smooth surfaces; some are compact, whereas others are loose and porous. The texture is fine grained, and clay minerals, opal, iron oxides, and foraminiferal tests accompany the barite. Some are geodelike with small barite crystals inside (Bøggild, 1916).

In 1938 the Geological Society of America started a geological research program to study the ocean floor off the coast of California (Revelle and Emery, 1951). A series of dredgings were performed along a probable fault scarp, northeast of the island of San Clemente. In one of these dredgings, about 100 kg of mud containing approximately 250 concretions were extracted 650–800 m deep. These are poorly sorted, tubular, flat, subspherically and irregularly shaped concretions, with lengths ranging from 1 cm to 25 cm and weights from 2 g to 1300 g, with an average weight of 60 g. Some have rough surfaces. Others are relatively smooth. Many concretions have a hard and compact outer shell, concentrically zoned, surrounding a softer core. They were unbedded in marine mud and contain up to 77 percent barium sulfate.

Under the microscope it can be observed that the concretions contain small radial aggregates of barite of closely interlocking crystals, often best developed inside the test of foraminifera, as well as clays, detrital minerals, siliceous sponge spicules, foraminiferal tests, and calcareous echinoid spines. Analysis of the enclosing mud revealed 11 percent sand, 36 percent silt, 22 percent clay, and 31 percent colloids. The barium content varies from 62 percent to 77 percent, the highest amount being in the outer layer of a zoned concretion. The porosity ranges from 13.8 percent to 33.6 percent. With respect to the foraminifera, it was observed that 68 percent of the species do not occur on the present sea floor at the same depths as barite outcrops, and they probably represent an early Pliocene fauna. Revelle and Emery (1951) suggest that the tubular shape and the concentric zoning of many of the nodules might have originated through cementation, by barite, of sediments surrounding a percolating channel. According to Revelle and Emery, the nodules formed as a consequence of a reaction between ascending hot solutions and interstitial sea water present in the sediments. The calculated chlorinity of the interstitial water in the concretions is 33 percent, considerably higher than the normal chlorinity of bottom sea water. The occurrence of the barite nodules in California is close to a major fault, which would allow hot solutions from a deeper source to flow into sea water.

The Shinguri bed belongs to a series of beds with continental morphology (borderland), which occur in the Sea of Japan (Kim et al., 1970). They are Upper Tertiary–Quaternary in age. In 1968 barite nodules with andesite fragments and some other rocks were dredged from depths of 150–160 m. The nodules are 1–20 cm in diameter, with spheroidal shapes and bluish-greenish gray and brown colors. The nodule cores are formed by radial barite crystals up to 0.1 mm long and anhedral calcite and goethite grains.

In the outer rims, the nodules contain, besides the minerals present in the core, diatoms, minor amounts of radiolaria, fragmented quartz, feldspar grains, clays, and silts. Calcite veinlets grow denser in the cores and branch out toward the outer parts. Kim et al. (1970) suggest that the nodules originated after the diatomite muds were replaced by barite. They mention that barite nodules are known in other beds, also associated with Kuroko deposits and that in the explanation of the origin of recent nodules could be the genesis for older ones.

Lipkina and Tsoy (1980) found barite concretions on the southeastern part of the Yamato Ridge, Sea of Japan, where more than 40 were dredged from depths of 1200–1300 m. Spheric nodules, 3–7 cm in diameter, were found together with many rolls and dumbbell-shaped concretions, ranging in length from 3 cm to 32 cm and from 1.5 cm to 7 cm in cross section. The concretions are covered by black manganese hydroxide films. They most probably were recently exposed on the sea floor by erosion of the host rock. The $BaSO_4$ content of the nodule cores reaches 98 percent. On the other hand, the concentration decreases to 74 percent in the shells due to the presence of about 3 percent SiO_2, 1 percent Fe_2O_3, 0.1 percent FeO, 2 percent Al_2O_3, 1 percent $K_2O + Na_2O$, and not more than 0.01 percent TiO_2 and 0.1 percent P_2O_5. The most common density ranges from 3.56 to 3.65. Lipkina and Tsoy (1980) conclude that the barite concretions formed in a diatomaceous sediment, inheriting its bedding and retaining its fossils. A submarine fumarole field probably existed in the late Cretaceous, where barium- and sulfur-saturated hydrothermal solutions passed through the diatomaceous sediments and produced intricately shaped barite concretions. The concretions show evidence of solution in sea water, and also that they were repeatedly rewashed and moved along the sea floor.

There are also numerous references to barite nodules in ancient marine sediments, some of which will be mentioned in the following paragraphs.

Barite nodules occur at numerous locations in the Appalachians, but not in economic amounts. Carpenter and Fagan (1969) describe nodules from the base of the Athens Shale (Middle Ordovician) found in seven localities over a distance of 100 km in northeastern Tennessee and southern Virginia. They are spheroidal to ellipsoidal in shape, and their size ranges from less than 1.5 cm to more than 30 cm in diameter. The nodules contain barite crystals up to 1 mm in size, disseminated in a black shale matrix, together with some pyrite. Carpenter and Fagan propose a diagenetic origin for the nodules through the interaction of ascending barium chloride solutions with sulfate-bearing interstitial sea water, present in the sediments.

Smith (1974) mentions that several tons of fetid barite nodules were found scattered in a field in Berks County, Pennsylvania. The nodules occur as millimeter-sized, tabular barite crystals or coarse crystals radiating from a common center. They have a very strong odor when broken. Smith proposes that the shale chips included in some nodules indicate that they formed within the shales that contain them.

Clark (1986) mentions that barite nodules in the Appalachians occur in Middle Ordovician shales in Tennessee and southern Virginia (Athens Shale), in Lower to Upper Devonian shales in Virginia and Pennsylvania (Need-more, Marcellus, Millboro, and Harrel Shales) and in Upper Devonian shales in New York (Java and Perrysburg formations).

The nodules Clark studied are characterized by alternating concentric calcareous and barite layers and varying grain sizes, that suggest to her a diagenetic origin with variations in growth rates and chemical conditions. The nodules occur in shales and mudstones that were deposited in mostly stagnant, oxygen-deficient marine environments; these shales contain slightly higher amounts of barium, manganese, cobalt, nickel, and zinc than do the shales that contain no nodules. Although barium is present in normal sea water in small amounts only (approximately 0.06 ppm), anoxic sea water of low to normal pH can contain substantial dissolved barium. Clark (1986) presents a model for precipitation of barite where barium sulfate from oxygenated waters reach deeper anoxic waters, possibly through the marine biological cycle.

In north central Ohio, Holden and Carlson (1979) also found septarian concretions over a distance of 8 km through discontinuous exposures along the Vermilion River, in the Cleveland Shale of Upper Devonian age. They are flattened in the bedding planes, averaging 6 cm in length. The concretions are composed of black, finely crystalline barite, surrounded by a fine-grained pyrite-rich shell; the septa are mainly composed of coarsely crystalline barite.

Some concretionary barites have been found in the Walton mine, Nova Scotia, Canada's largest producer of barite (Dawson, 1966). The main orebody occurs as a pipe-shaped barite mass that plunges to the east at 35°. At depth it is underlain by a sulfide orebody that contains lead, zinc, copper, and silver minerals. The concretionary barite forms long sausage-shaped nodules (Dawson, 1966) of black or dark brown color, due to the presence of hydrocarbons, and are up to 12 cm long and 4 cm across. They lie in a white, coarse barite matrix. The junction of the black nodules with the surrounding white barite is sharp, each concretion having a rim of coarse white crystals. Under the microscope the nodules show a concentric banding. The cores are made up of white barite up to 0.25 mm long that grade into a dark annular zone with 0.02–0.03 mm long crystals that bear a dark brown material along their boundaries. Following this banding are five or six successive annular bands, 0.5–4 mm wide, each band comprising a coarser, white inner zone and a finer, dark outer zone.

Also in Canada, in the Twitya River area, conspicuous spherical to subspherical barite nodules occur in carbonaceous shales and argillaceous dolomitic limestones of Upper Devonian–Mississippian age (Laznicka, 1976). They are made up of gray radiating barite crystals. Microscopic study of the barite and of the sedimentary fabrics in its host rock has demonstrated that the crystallization of nodules was a late diagenetic event. According to Laznicka (1976), the Paleozoic, black shale facies of the Eastern Cordillera contains numerous occurrences of barite and several large deposits of bedded barite occur, in presumably the same stratigraphic unit, in a 100 km radius around the Twitya River locality of nodular barite.

Suprychov (1967) studied the barite diagenetic concretions in the Albian deposits of the Mountain Crimea. In the Upper Albian clays, which are relatively deep-sea facies, the concretions were found in the different areas, around barite concretions, 2–30 cm in diameter. The textural and paragenetic peculiarities of these concretions testify to their formation as early diagenetic in which Ba was previously scattered in the sediment. Syneresis fractures in the concretions are filled with newly formed barite crystals. The concretions contain hydromica and carbonates (5–15 percent) as impurities, as well as some grains of iron sulfides, Sr (0.1–0.3 percent) and traces of Mn, Pb, and Tl.

CANOTA ORE FIELD

The Canota ore field is located approximately 37 km north–northwest of the city of Mendoza (province of Mendoza), Argentina. (Brodtkorb et al., 1987)

It comprises the La Victoria mine, which is rudimentarily worked at present, the Pirucha mine, which was mined in the past for approximately 30,000 tons of barite, and the Ramoncito and Don Manuel properties (Fig. 11–1).

GEOLOGIC SETTING

The geology of the area is composed of a series of flysch and distal platform deposits of the Empozada formation (Ordovician) and Villavicencio formation (Ordovician–Devonian).

In the study area, the Empozada formation is made up of gray and black shales, rich in organic matter, that locally are intensely bleached by weathering. An abundant fauna of graptolites, mainly of the genus *Climacograptus* and *Glyptograptus* (Alfaro and Fernandez, 1985), confirm a Caradocian age of the shales. The Villavicencio formation is the dominant unit in the area

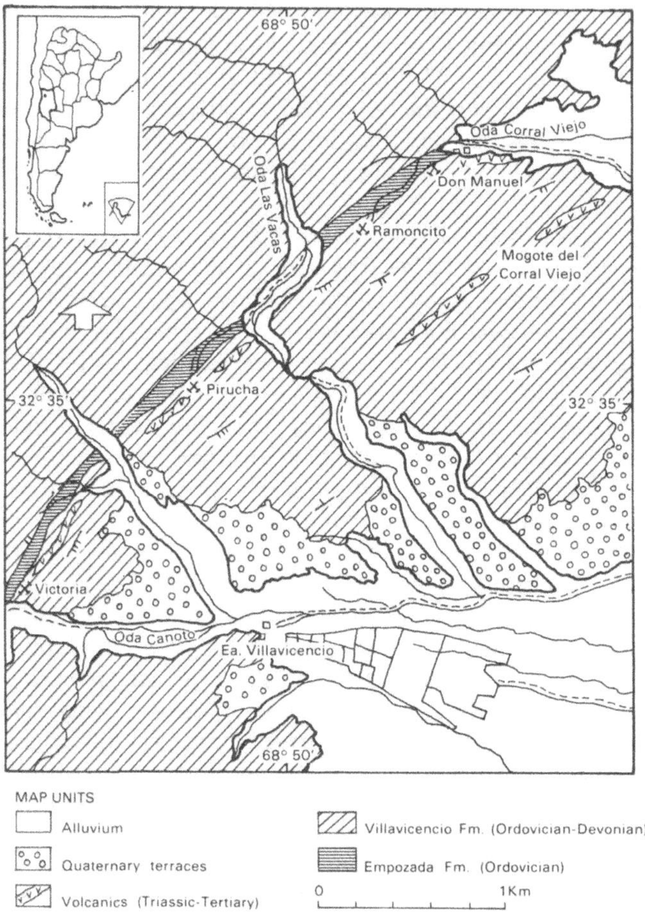

Figure 11-1. Geologic map of the Canota ore field.

and is composed of a rhythmic flysch sequence of quartzitic sandstones and greenish-gray shales.

The Empozada shales, the unit bearing the barite deposits, form a 4 km long belt, with a north-northeast trend, which extends from the downstream side of the Corral Viejo creek in the northeast, to the Canota creek in the southwest where the shales are covered by modern fluviatile sediments (Fig. 11-1). The belt has a maximum width of 100 m in Corral Viejo and a minimum of 20 m in the La Victoria mine. In the Corral Viejo area, the shales strike at north 70–80° east and dip at 30–55° south, whereas in Canota they strike at north 20–30° east and dip in a north-northwest direction. Thus, the barite-bearing beds lie almost subhorizontally in the Don Manuel and Ramoncito deposits and almost vertically in the Pirucha and La Victoria mines.

On the northern side of Corral Viejo an almost vertical fault puts the Empozada formation shales in contact with the Villavicencio formation, whereas on the southern side, the flysch deposits conformably overlie the black shales. There are some acidic dikes and sills of probable Triassic–Tertiary age in the area. Evidence of synsedimentary volcanism is being studied at present.

The geologic setting indicates a depositional environment for the Caradocian deposits characterized by a low energy platform with restricted circulation, which is typical of graptolite environments with distal euxinic conditions, sheltered from wave action. These deposits were laid close to a basinal slope, facing an oceanic environment to the west (Fig. 11-2). The Villavicencio facies represents a turbiditic sequence, which further west is closely related to mafic rocks that were interpreted as initial rift oceanic magmatism based on their geochemical properties (Kay et al., 1984).

MODES OF BARITE OCCURRENCE

In this belt barite is present predominantly as nodules, small lenses, and massive beds, as well as small crystals disseminated through certain shale beds. The barite-bearing levels are a half meter to several meters thick and are scattered throughout the sequence alternating with barren levels. Following the direction of strike, the barite-bearing levels tend to pinch out locally and appear again in the area that extends from the Don Manuel mine to the La Victoria mine.

In the area between the Don Manuel and Ramoncito deposits, the nodular levels are the most common form of barite accompanied by small barite lenses and beds bearing barite crystals. Conversely, massive barite beds are more important than beds bearing barite nodules or those scattered throughout the Ordovician shales and quartzitic sandstones.

Other characteristics of the Canota district barites are their dark gray and black colors and their fetid odor when broken.

BARITE NODULES

The nodules occur as a string aligned along the two planar directions and are conformable with the black shales' fissility. This set is repeatedly found throughout the shale sequence. The shale planar structure takes the shape of

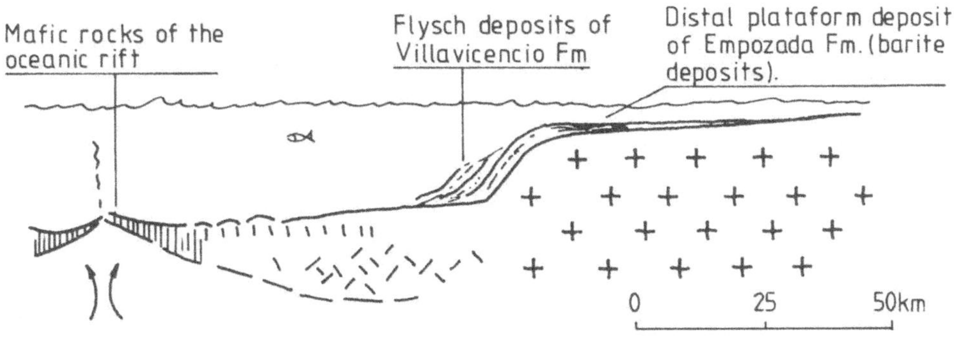

Figure 11-2. Schematic diagram showing the relationship of the passive margin of southwestern Gondwanaland with the proposed oceanic rift during Late Ordovician times. The barite deposits are located in the vicinity of the platform edge.

the lower and upper surface of the nodules (Figs. 11-3 and 11-4). Within the shale, there are some levels in which the amount of barite nodules varies in the horizon (Fig. 11-5).

The nodules present ellipsoidal to subspheroidal shapes with a length to width to thickness ratio ranging from 1.25:1:0.3 to 1.5:1:0.8; the most frequent length axis being 15-20 cm. Shorter (between 3 cm and 15 cm) or larger than average nodules are seldom found. Two nodules, 30-35 cm long, when broken showed concretional structures of septarian type (Fig. 11-6).

The nodules present two different forms of internal crystallization. The first (Fig. 11-7) shows a homogeneous crystallization in the whole nodule with a grain size of 0.3-0.4 mm, whereas in the second, the crystal size increases up to 5 mm from the center in marginal zones (Fig. 11-4).

Nodules with such varying growth present a rough external surface or crust in contact with the shale. This is followed by an outermost layer with small barite crystals in a shaly matrix that forms some sort of halo (Fig. 11-4, zone H). The crust has an average thickness of 1 cm, whereas the halos can reach a thickness of 2-5 cm. The external surface or crust of the nodules is frequently covered with a white layer made up by illite, jarosite, gypsum, and alunite. In some barite levels, the amount of nodules is large enough to have two or three nodules coalesced to form undulating pseudolenses.

Relicts of primary sedimentation occasionally can be observed in nodules when cut with a saw and polished.

Under the microscope the nodules show a quite homogeneous composition. In sections cut through the nodule core, the barite appears as a granular mosaic with an average crystal size of 0.35 mm. Conversely, in nodules from the second type, whose crystal size increases from the core to the margins, the barite develops a prismatic habit with crystals 0.4-0.7 mm long in a very tight texture without any preferential orientation. In both cases, clay materials as well as jarosite, limonite, and organic matter are present in the intergranular

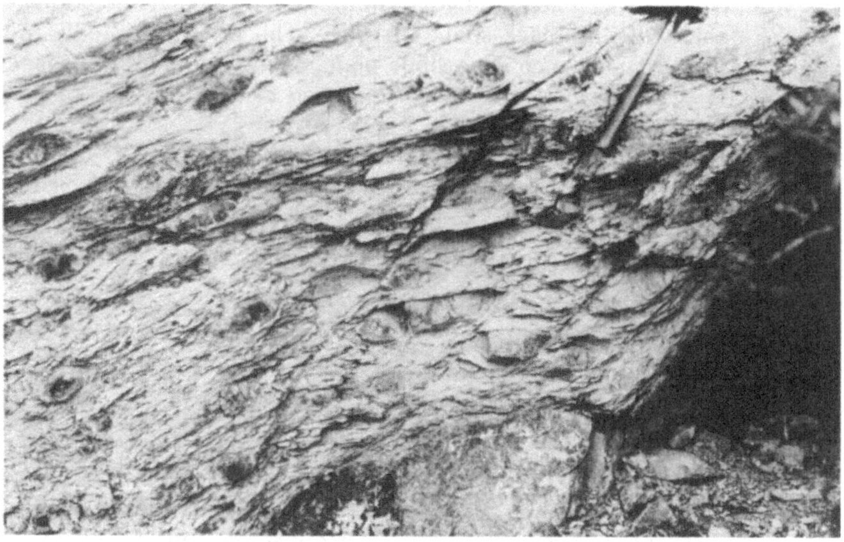

Figure 11-3. Front of mining activity at Don Manuel. Nodule crystal size increases up to 5 mm from the center (second type).

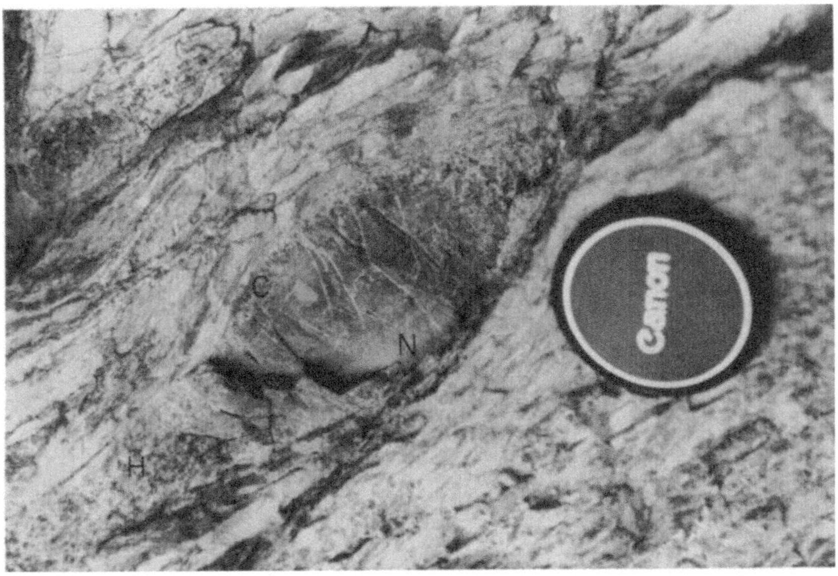

Figure 11-4. Detail of one nodule of Figure 11-3 from Don Manuel. N = core; C = crust; H = halo.

Figure 11-5. Front of mining activity at Ramoncito shows concentrations of nodules in shale.

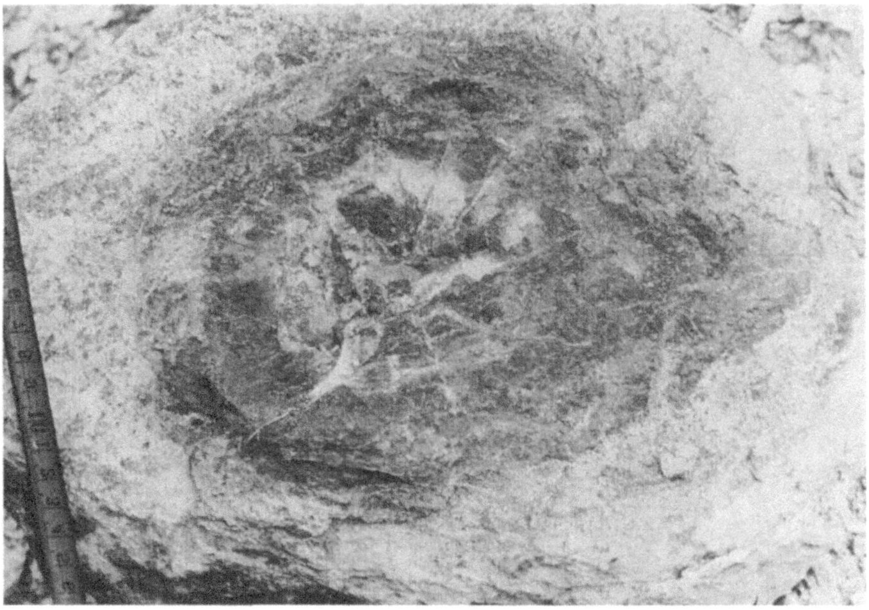

Figure 11-6. Septarian-type nodule from Don Manuel.

Figure 11-7. Don Manuel nodule, 18 cm in diameter, showing homogeneous crystallization in the whole nodule (first type).

spaces. In some cases, the barite grains show a poikilitic aspect given by the inclusions of quartz and feldspar grains that were enclosed by the barite during its crystallization. Fine veinlets (20–50 μm thick) crosscutting the barite mosaic are common. They are made up of clear barite and occasionally of quartz or both.

The specific gravity of the nodules varies from 3.95 to 4.15.

SMALL LENSES AND MASSIVE BEDS

We call "small lenses" those geometric forms with small thickness (5–30 cm) in relation to their length (1–8 m). Small lenses are found in the Don Manuel and Ramoncito properties, and are far less abundant than the nodules. Massive beds, on the other hand, are lenticular bodies 1–5 m thick and up to 15–30 m long. In the Pirucha and La Victoria mines, massive beds occur conformable to the stratification and associated with nodule-bearing levels. Here, the almost vertical layering makes the thickness of the massive lenticular beds the only observable dimension, because mining did not always uncover the other two. Mining exposed three massive beds in the La Victoria mine and seven in the Pirucha mine. It is worth mentioning that nodular barite levels occur at the bottom and top of massive beds in a similar way to those of the Vide de Alba deposit, Spain (Moro Benito and Arribas Moreno, 1981).

BEDS WITH BARITE CRYSTALS

Barite crystals occur in two different forms: (1) in crusts and halos, surrounding barite nodules (Fig. 11-4), where the barite crystals are scattered through a shaly matrix (as seen earlier), and (2) disseminated through layers of shale 5-10 cm thick, which alternate with beds of barren shales (Fig. 11-8). In these, the barite occurs as crystals and small lenticular aggregates. The individual crystals are, in the first case, subidiomorphic to idiomorphic and up to 0.5 cm in size. They grew without following the planar structure of the host rock. Also different degrees of aggregation of the barite grains can be observed under the microscope. Thus, two to three small grains constituting an incipient nucleus of aggregation can be found (as well) in small lenses with lengths ranging from 0.1 mm to 4–5 mm. They are made up of a granular association of barite crystals that integrate to the host rock as part of it being conformable to an incipient fissility.

RARE EARTH AND TRACE ELEMENT ANALYSES

Rare earth and trace elements have been determined by neutron activation (Table 11-1). The procedure used for the investigation and normalization here was that of Kramar and Puchelt (1982).

The chondrite-normalized REE patterns from the Canota district (Fig. 11-9) are similar to that of post-archean shales, when compared to data published by Taylor and Mc Lennan (1985). However the REE content of barite is con-

Figure 11-8. Layers with crystals (c) from Don Manuel.

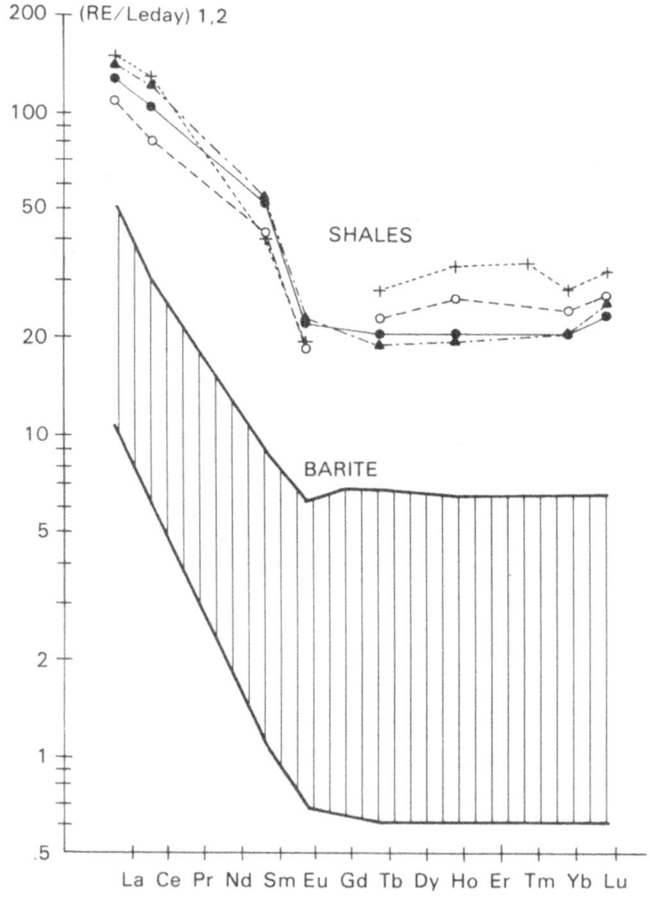

Figure 11-9. Chondrite-normalized REE patterns of barites and shales of the Canota ore field (see Table 11-1 for data).

Table 11-1. Trace Elements in Canota Barites and Shales

	Barite Nodule Don Manuel					Massive Barite Victoria	Shales			
							Don Manuel		Near Ramocito	Victoria
	2a	2b	2c	2d	2e	63	20	25	27	67
As	5.35 ± 0.36	3.18 ± 0.29	2.48 ± 0.29	4.11 ± 0.37	0.97 ± 0.25	5.02 ± 0.30	15.96 ± 0.97	1.72 ± 0.86	3.44 ± 0.76	45.60 ± 1.45
Ba			main element				20000 ± 2000	20000 ± 2000	16000 ± 100	38600 ± 400
Co	0.60 ± 0.05	0.31 ± 0.04	0.36 ± 0.05	0.34 ± 0.05	0.50 ± 0.05	0.20 ± 0.03	1.39 ± 0.11	1.44 ± 0.13	5.90 ± 0.19	0.35 ± 0.09
Cr	13.59 ± 3.52	5.23 ± 3.53	11.48 ± 3.49	7.23 ± 3.88	8.79 ± 3.72	6.60 ± 3.35	90.98 ± 3.70	91.75 ± 3.97	59.14 ± 2.47	41.41 ± 3.25
Cs	1.41 ± 0.07	0.88 ± 0.06	0.96 ± 0.06	0.87 ± 0.07	1.09 ± 0.07	0.63 ± 0.05	9.26 ± 0.19	9.69 ± 0.21	6.53 ± 0.14	5.58 ± 0.16
Hf	4.94 ± 0.30	19.80 ± 0.45	45.74 ± 0.77			0.095	5.41 ± 0.16	5.94 ± 0.19	2.59 ± 0.10	4.10 ± 0.18
Mo	<10	<10	5.0 ± 4.0	10.4 ± 8	<11	n.d.	32.16 ± 12.8	9.71 ± 9.05	<9	230.8 ± 14.2
Ni	10.37 ± 3.27	6.40 ± 2.94	4.65 ± 2.86	4.69 ± 3.25	6.45 ± 3.15	2.51 ± 2.23	9.53 ± 7.01	10.00 ± 7.40	71.54 ± 6.45	16.50 ± 6.76
Rb	15.0 ± 2.7	6.5 ± 2.0	7.54 ± 2.0	6.8 ± 2.4	8.12 ± 2.2	2.37 ± 1.57	148.5 ± 7.6	144.9 ± 8.1	86.0 ± 5.2	52.2 ± 6.7
Sb	0.20 ± 0.05	0.14 ± 0.05	90.49 ± 43.24	0.17 ± 0.06	96.29 ± 44.43	0.36 ± 0.05	1.33 ± 0.12	1.59 ± 0.14	1.20 ± 0.10	6.28 ± 0.23
Sc	3.25 ± 0.05	1.90 ± 0.03	2.10 ± 0.04	1.98 ± 0.04	2.48 ± 0.04	0.62 ± 0.02	17.22 ± 0.21	17.29 ± 0.22	10.31 ± 0.13	n.d.
Sr	645 ± 38	781 ± 40	901 ± 42	780 ± 45	933 ± 45	1700 ± 100	244 ± 29	197 ± 31	124 ± 19	194 ± 33
Ta	0.17 ± 0.02	79.96 ± 15.20	79.05 ± 16.00	0.11 ± 0.02	0.12 ± 0.02	0.055	1.19 ± 0.07	1.28 ± 0.07	0.63 ± 0.04	1.04 ± 0.07
Th	0.63 ± 0.23	0.21 ± 0.21	<0.2	<0.23	0.32 ± 0.24	<0.2	11.78 ± 0.25	11.89 ± 0.27	8.65 ± 0.19	9.56 ± 0.28
Zn	1.92 ± 1.18	1.64 ± 1.14	3.39 ± 1.88	1.75 ± 1.65	1.22 ± 1.07	1.81 ± 1.39	13.17 ± 3.7	21.90 ± 5.04	248.2 ± 14.3	5.77 ± 4.0
U	6.29 ± 0.47	3.25 ± 0.45	3.60 ± 0.44	2.52 ± 0.49	3.45 ± 0.51	2.24 ± 0.46	14.49 ± 0.65	16.84 ± 0.75	8.43 ± 0.47	21.38 ± 0.74
La	16.06 ± 0.51	8.31 ± 0.40	8.39 ± 0.40	5.98 ± 0.39	15.56 ± 0.53	3.20 ± 0.30	38.22 ± 1.08	42.71 ± 1.25	33.65 ± 1.01	44.94 ± 1.19
Ce	21.40 ± 1.32	12.94 ± 1.26	10.72 ± 1.20	8.13 ± 1.31	13.57 ± 1.38	4.32 ± 1.06	82.84 ± 2.06	94.27 ± 2.26	63.80 ± 1.49	100.1 ± 2.3
Pr	n.d.	n.d.	n.d.	n.d.	n.d.	n.d.	n.d.	n.d.	n.d.	n.d.
Nd	n.d.	n.d.	n.d.	n.d.	n.d.	n.d.	49.75 ± 2.74	53.24 ± 3.55	38.73 ± 2.04	37.62 ± 3.42
Sm	1.57 ± 0.06	1.37 ± 0.06	1.19 ± 0.06	1.18 ± 0.06	1.42 ± 0.06	0.18 ± 0.05	9.16 ± 0.13	9.41 ± 0.15	7.76 ± 0.12	7.71 ± 0.14
Eu	0.38 ± 0.03	0.42 ± 0.03	0.36 ± 0.03	0.34 ± 0.03	0.40 ± 0.03	0.047 ± 0.010	1.60 ± 0.08	1.67 ± 0.09	1.28 ± 0.06	1.32 ± 0.08
Gd	n.d.	n.d.	n.d.	n.d.	n.d.	n.d.	n.d.	n.d.	n.d.	n.d.
Tb	0.20 ± 0.03	0.32 ± 0.03	0.24 ± 0.03	0.32 ± 0.03	0.21 ± 0.03	0.022 ± 0.018	0.94 ± 0.09	0.88 ± 0.09	1.09 ± 0.07	1.32 ± 0.10
Dy	n.d.	n.d.	n.d.	n.d.	n.d.	n.d.	n.d.	n.d.	n.d.	n.d.
Ho	0.31 ± 0.18	n.d.	n.d.	0.66 ± 0.32	0.42 ± 0.31	0.27 ± 0.25	1.39 ± 0.30	1.35 ± 0.33	1.88 ± 0.31	2.38 ± 0.29
Er	n.d.	n.d.	n.d.	n.d.	n.d.	n.d.	n.d.	n.d.	n.d.	n.d.
Tm	0.57 ± 0.14	n.d.	n.d.	0.56 ± 0.13	0.1	0.23 ± 0.09	0.47 ± 0.09	0.49 ± 0.08	0.56 ± 0.06	1.03 ± 0.09
Yb	n.d.	n.d.	n.d.	1.08 ± 0.24	0.85 ± 0.24	0.54 ± 0.18	3.78 ± 0.17	4.01 ± 0.17	4.47 ± 0.09	5.11 ± 0.15
Lu	0.059	0.22 ± 0.20	n.d.	0.049 ± 0.040	n.d.	n.d.	0.70 ± 0.04	0.78 ± 0.05	0.83 ± 0.03	0.92 ± 0.06

Note: Analyses made by Prof. H. Puchelt, Institut für Petrographie und Geochemie, Karlsruhe University, German Federal Republic.

siderably below that of the coexisting shales. Nevertheless, the barite data parallel the shale patterns. The chondrite-normalized REE patterns show a strong concentration of LREE, more or less distinct negative Eu anomalies, and comparatively low HREE values.

According to investigations by Hauck (1984), barites from hydrothermal veins are characterized by a distinct positive Eu anomaly, whereas marine barites have either no Eu anomaly or a negative one.

The neutron activation data from shales also indicate a certain Zn accumulation (up to 248 ppm) in this environment, which can be related to some submarine or volcanic activity. Also the high Sb value may indicate such activity (Marchig et al., 1985).

Sulphur Isotopes

Various barite nodules from the Canota ore field were analyzed along their diameters and compared with those of Japan (Sakai, 1971). The results of two of them are shown in Figure 11–10. In comparison: present ocean sulfate = 20.0–20.3; Kuroko deposits = 22–24; and Ordovician evaporites = 26–34.

According to Sakai (1971), the barite concretions from the banks of the Sea of Japan off northeastern Honshu, Japan, of Middle Miocene age, are very enriched in heavy sulfur isotopes, suggesting the action of sulfate-reducing

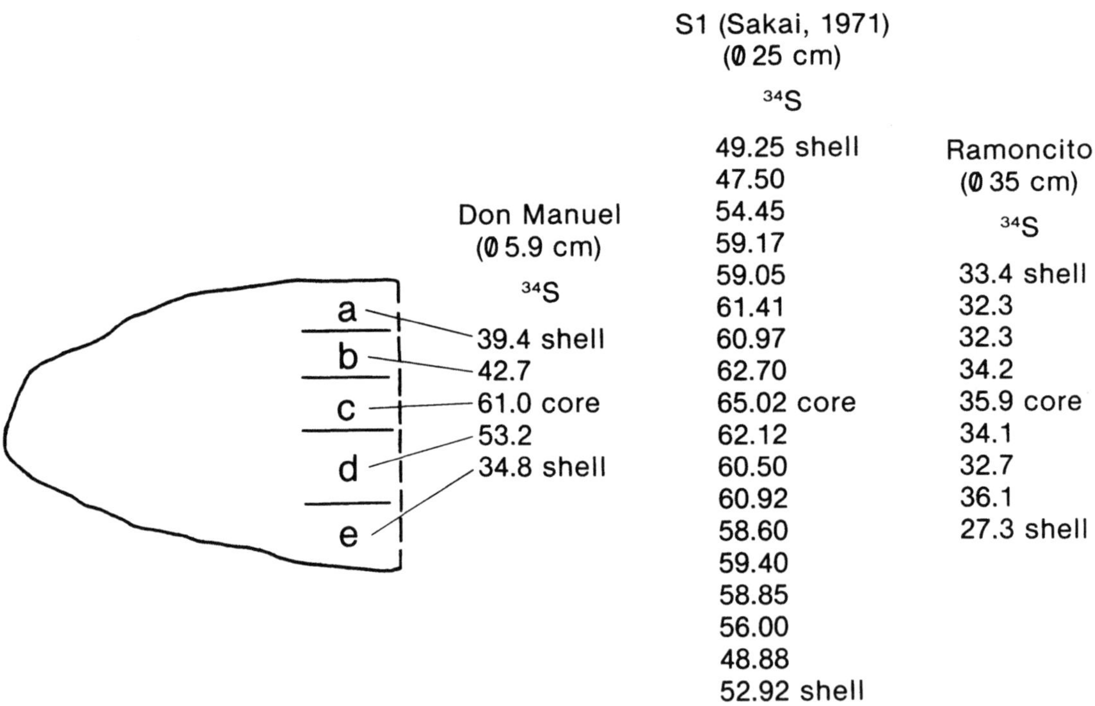

Figure 11–10. Sulphur isotope analyses of Don Manuel and Ramoncito made by Prof. Puchelt (Institut für Petrographie und Geochemie, Karsruhe University, Federal Republic of Germany).

bacteria in a close environment. Barite nodules from the California coast (Recent) and New Zealand (Cretaceous) are as heavy as many of the Japanese ones, showing a certain similarity in their depositional environments (Sakai, 1971). The nodules from Don Manuel can also be added to these data.

The Ramoncito nodule, however, is not so enriched in heavy sulfur isotopes. This type of nodule was predominantly formed by sea water-derived sulfates.

STRONTIUM ISOTOPES

If the $^{87}Sr/^{86}Sr$ isotopic ratios of barite nodules from the Canota district are plotted on a Burke et al. diagram (1982), they fall into a field that would indicate a marine origin for the Sr contained in the barite and more precisely that it derived from Upper Ordovician sea water.

	$^{87}Sr/^{86}Sr$
Barite nodule: Don Manuel	.70886 ±3
Barite nodule: Don Manuel	.70884 ±3
Barite nodule: Pirucha	.70885 ±2
Barite crystals from the layers: Don Manuel	.71125 ±2

Two samples of shales were analyzed and yielded the following results:

	Rb ppm	*Sr ppm*	$^{87}Sr/^{86}Sr$ *present*	$^{87}Sr/^{86}Sr$ *initial*	$^{87}Sr/^{86}Sr$
Shale: Don Manuel	15	220	.70999 ±3	.70873	.20
Shale: Pirucha	10	150	.70999 ±5	.70877	.19

It can be seen that the isotopic values of barite crystals from the layers of Don Manuel are significantly higher than the values of nodules and shales. If the crystals formed at the expense of the preceding barite, the isotopic ratio would be the same as that for nodular barite. Alternatively, it is possible that the crystals formed at the expense of Ba in shales that also contained rubidium and could have given way to solutions capable of depositing a radiogenic ^{87}Sr-rich barite (c.f., respective data in Table 11–1).

The isotope analyses were made by Prof. M. Barbieri (University of Rome, Italy).

FINAL REMARKS

The origin of the Canota barite raises various questions: the source of the barium, the origin of the sulfate, and the formation of the barite nodules, layers, and idiomorphic barite crystals. The high barium concentration in this facies cannot be derived from sea water alone. Barium must have been supplied to the ocean by hot springs much like the present hydrothermal activity in the oceanic rift systems. Recent studies on the chemical composition of ocean floor smoker chimneys have shown barium concentrations of 14.2 percent and 15 percent in the Juan de Fuca area (Chase et al., 1985) and recent sulfide ores from the Tyrrhenian Sea gave Ba concentrations up to 48 percent (Puchelt, 1986).

The oceanic environment must have been a marginally restricted basin close to the continental slope that was related to a late Ordovician oceanic rift. The shales of this environment have a high organic productivity, thus creating euxinic conditions in which part of the original marine sulfate was bacterially reduced. The sulfate was enriched in ^{34}S and formed the cores of some barite nodules. The other barium was either absorbed to the clay minerals or organic substances or stayed in the interstitial waters. During a later stage of this system, new sea water came into the basin and precipitated the rest of the barium onto the previously formed nodules now giving sulfur isotope values corresponding to Ordovician sea water sulfate. In some cases, nodules were predominantly formed by sea water-derived sulfates. Idiomorphic barite crystals grew in clay layers, where no nodules had formed previously. This process occurred diagenetically after a longer time lapse. Since some of the shale layers show increased rubidium values (c.f., Table 11–1), radiogenic strontium 87 could have formed simultaneously, which caused the high $^{87}Sr/^{86}Sr$ values observed in barite crystals. The occurrence of Sr isotopically zoned barite nodules is a frequent phenomenon in the Don Manuel site where it was observed in several horizons.

Since the barite is stratabound, the barite-bearing beds define a real sedimentary facies formed in a basin with heavy organic matter. The deposition of the barite layers and nodules were syn- to early diagenetic, but the growth of the barite crystals was late diagenetic.

Zamora Ore Field

Barite ores occur in the Zamora ore field (Spain) and are related to the Alcañices-Carbajales de Alba synclinorium (Moro Benito and Arribas Moreno, 1981). The deposits were exploited by open-pit and cut-and-fill methods. The petrographic, mineralogic, and metallogenic characteristics of the Vide de Alba and San Blas deposits are analyzed here.

The geology of the area is formed by a gneissic basement, probably of Cambrian–Precambrian age, overlain by Early Paleozoic sediments. The various formations had been affected by the Hercynian orogeny and crossed by different granitic intrusions. The barite ores are associated with a lower series, which is constituted of green schists, bituminous schists (ampelites), cherts (lidites), volcanic rocks, limestones, and quartzites of a Silurian–Devonian complex. Overlying these is an upper series formed by alternation of schists and graywackes of possible turbiditic origin.

Vide de Alba Deposit

The mineralization is lenticular, subvertical, and dips to the northeast. It is approximately 100–150 m long, more than 50 m deep, and its maximum width is 20–25 m. It has been mined by means of an open pit and some galleries, which follow the shape of the mineralization.

The underlying layer in the Vide de Alba deposit (also known as Ambiciosa mine) is a feldspar quartzite of Middle to Upper Silurian age, whereas the overlying layer is composed of graywackes, sandstones, and schists. The stratigraphic unit that contains the barite ores is composed of alternating pyroclastic rocks, lidites, ampelites, intraformational quartzitic breccias, and alternate layers of lidites with volcanic ashes (Fig. 11–11).

The massive barite in this deposit is a wide bed about 20 m thick, with a known depth of 50 m, in which thin argillaceous layers with iron oxides are observed. A laminar barite bed is present at the base of the sequence. The massive barite presents a compact, saccharoidal aspect, of gray color, and microscopically it has an equigranular texture with a grain size of about 0.15 mm. Accessory minerals are carbonates and pyrite. The nodular–lenticular barite layers have a thickness of about 2 m and present geopetal textures. The sediment where they lie is essentially constituted of microcrystalline quartz, barite microlites, idiomorphic and subidiomorphic pyrite crystals, clays, and carbonates.

The nodules are spherical and subspherical with fibroradial or granular textures and present accessory minerals such as grains of quartz, carbonates, and pyrite (Figs. 11–12 and 11–13). Also in these layers are frequent lenslike bodies and different types of crystals. All these kinds of barite show partial silicification or dolomitization.

Pyrite is an abundant mineral, not only frequently present with different kinds of barite, but also in the rocks that constitute the underlying and overlying series. Other sulfides of Cu, Pb, and Zn are observed.

The available information seems to indicate that the physicochemical conditions under which the barite and other associated minerals were deposited do not correspond to a unique period but to several alternating periods. This would also explain the rhythmic layout observed in the mineralization at the orebody, hand-specimen, and microscopic scales.

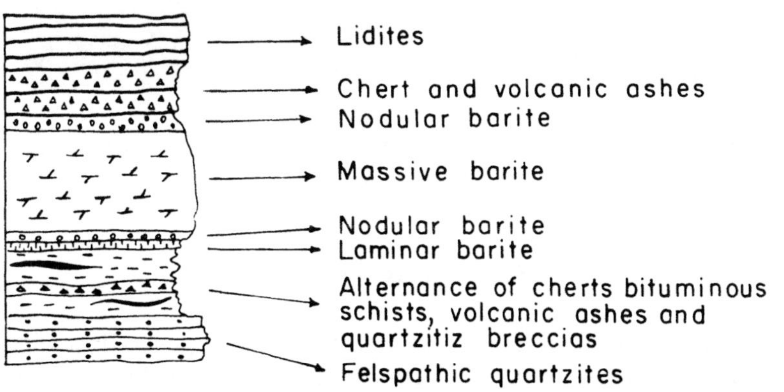

Figure 11–11. Generalized stratigraphic section for the Middle-Late Silurian of the Mina Ambiciosa area, bearing the mineralized horizons (after Moro Benito and Arribas Moreno, 1981).

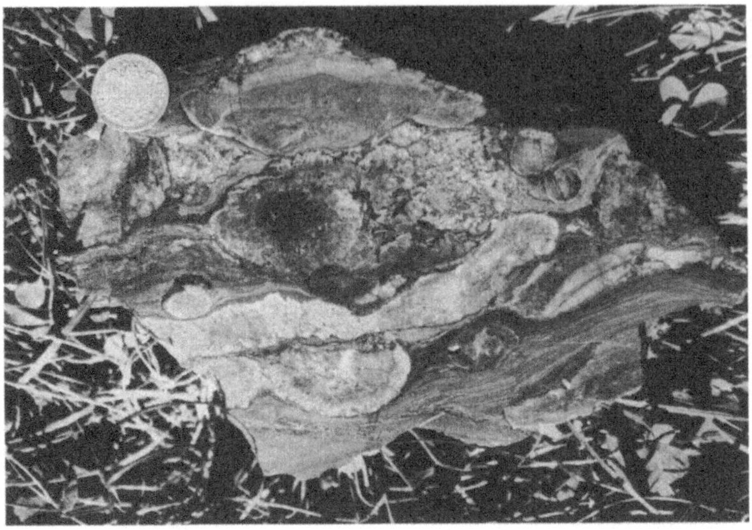

Figure 11-12. Spherical barite nodules from Vide de Alba of the Zamora ore field (courtesy of Moro Benito and Arribas Moreno).

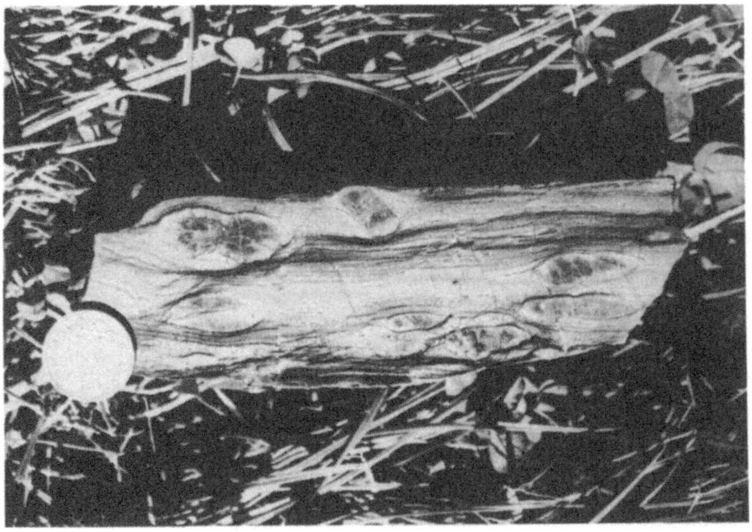

Figure 11-13. Barite nodules from Vide de Alba of the Zamora ore field (courtesy of Moro Benito and Arribas Moreno).

SAN BLAS DEPOSIT

This deposit (also known as Mary Carmen mine) is generally similar to the previously described deposit. The massive barite has a thickness of only 0.10–1 m and it alternates with layers of chert, limestone, and schist that contain variable quantities of $BaSO_4$ (Fig. 11-14).

The levels of dolomitic limestones with barite are abundant within the min-

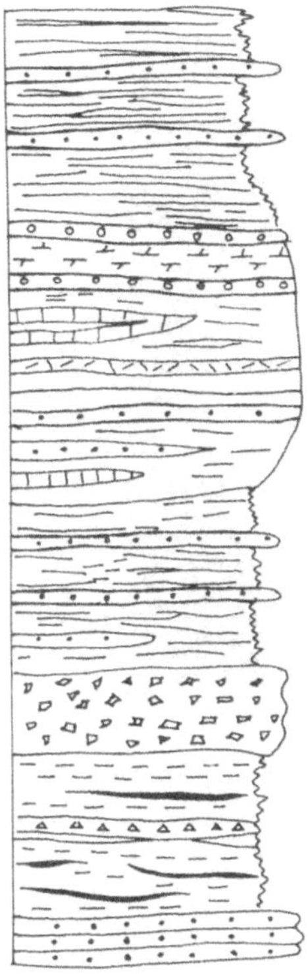

Figure 11-14. Location of the barite layers in the stratigraphic section of the Silurian-Devonian at the Mary Carmen area (after Moro Benito and Arribas Moreno, 1981).

eralized series and alternate with chert and massive barite layers. Barite crystals are tubular, idiomorphic, and occur in a sediment of micritic nature, where idiomorphic crystals of pyrite also are observed. Moro Benito and Arribas Moreno (1981) define them as baritic dolomicroesparite, presuming their origin from a mud. A recrystallization with the formation of barite and limestones was produced during the diagenesis. The bituminous schist, which is thinly bedded, has a banded structure and is constituted of alternating layers of sulfides, organic matter, cherts, and barite.

In regard to the barite nodule levels (Fig. 11-15) and the lenses, it is necessary to emphasize that similar to the Vide de Alba ore field, these are formed in the footwall and hanging wall of the massive barite beds. Within the barite nodules, spherical bodies are found that resemble possible radiolarian relicts (Fig. 11-16).

The sericitic quartzites with barite are constituted of granular barite, where the intergranular spaces are occupied by quartz, sericite, and pyrite.

The breccias that occur in the series are formed by angular or subangular fragments of chert, sericitic quartzite, sericitic schist, and barite-bearing dolomitic limestones, associated with barite nodules. The matrix of the breccia is made up of a silica-baritic matter with abundant pyrite.

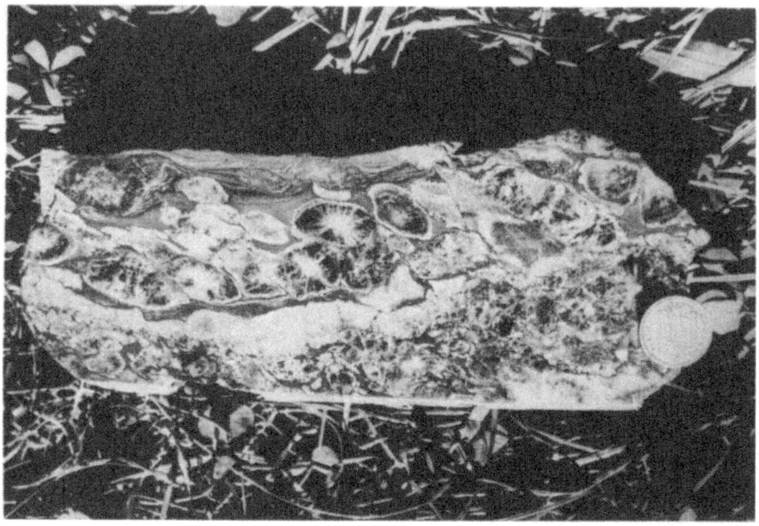

Figure 11–15. Fibroradial nodules from the Mary Carmen mine in the Zamora ore field (courtesy of Moro Benito and Arribas Moreno).

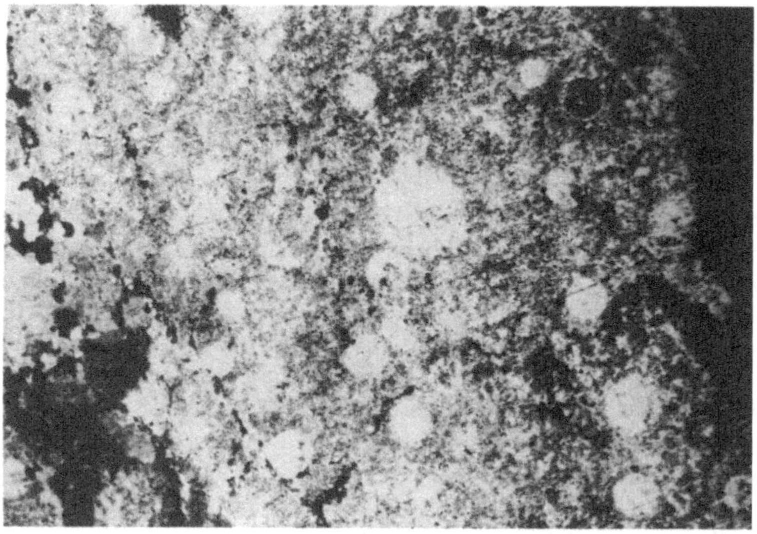

Figure 11–16. Spherical bodies, probably from radiolarians, preserved in a barite nodule with fibroradial texture from the Zamora ore field. Thin section × 25, crossed Nichols (courtesy of Moro Benito and Arribas Moreno).

GENETIC REMARKS

Moro Benito and Arribas Moreno (1981) proposed a syngenetic volcano-sedimentary origin for these deposits, whose original characteristics have been modified by diagenetic and metamorphic processes.

A striking congruence exists between the barite nodules and levels and the pelitic, carbonatic, and siliceous layers.

The chemical composition of barite ores and outcrops of Zamora province (Moro Benito and Arribas Moreno, 1981) is fairly uniform, which means that the geologic and physicochemical environmental characteristics were similar. They are also similar to deposits of other parts of the world, which, according to Moro Benito and Arribas Moreno, could indicate conditions of analogous precipitation. These characteristics of stratabound deposits are also common in Precambrian, Hercynian, Mesozoic, and Cenozoic deposits.

REFERENCES

Alfaro, M., and R. Fernandez. 1985. Una graptofauna del Ordovícico superior (Caradociano) de Estancia Canota (prov. de Mendoza). Ameghiniana. *Rev. Asoc. Paleon. Argentina* **22**(1–2):63–67.

Bøggild, O. B. 1916. Meeresgrundproben der Siboga Expedition. Siboga Expeditie LXV. (See in Revelle and Emery, 1951).

Brodtkorb, M. K. de, I. B. A. Schalamuk, M. Barbieri, H. Puchelt, S. Ametrano, R. Fernandez, and R. Etcheverry. 1987. Nodular barite deposits of Canota, Argentina. VII IAGOD Congress 1986. Lůlea. In Press.

Burke, W. H., R. E. Denison, E. A. Hetherington, P. B. Koepnick, H. F. Nelson, and J. B. Otto. 1982. Variation of $^{87}Sr/^{86}Sr$ through Phanerozoic time. *Geology* **10**:516–519.

Carpenter, R. H., and J. M. Fagan. 1969. Barite nodules in the Athens Shale in Northeastern Tennessee and Southwest Virginia. *Southeastern Geology* **10**(1):17–29.

Chase, R. L., J. R. Delaney, J. L. Karsten, H. P. Johnson, S. K. Junipe, J. E. Lupton, S. D. Scott, V. Tunnicliffe, S. R. Hammond, and R. E. Duff. 1985. Canadian-American Seamount Expedition, 1985: Hydrothermal vents on an axis seamount of the Juan de Fuca Ridge. *Nature* **313**:1–3.

Clark, S. H. B. 1986. The origin of some shale-hosted barite nodules of the Appalachian Basin. Abstract. 7th IAGOD Symposium. *Terra Cognita* **6**(3):514.

Dawson, J. B. 1966. Concretionary barytes from Walton mine, Nova Scotia. *Inst. Min. and Metal. Trans.* **75B**:162–164.

Gyula, R. 1973. Baritkonkréciók a rudabányai pannonból. *Hungarian Geol. Soc. Bull.* Földtain Közlöny **103**:136–140.

Hauck, M. 1984. Die Barytlagerstätten der Inselgruppe Milos/Agäis (Griechenland). Ph.D. dissertation. Univ. Karlsruhe.

Holden, W. F., and E. H. Carlson. 1979. Barite concretions from the Cleveland Shale in north-central Ohio. *Ohio Jour. Sci.* **79**(5):227–232.

Jones, E. J. 1887. On some nodular stones obtained by trawling off Colombo in 675 fathoms of water. *Jour. Asiatic Soc. Bengal* **56**:209–212.

Kay, S. M., V. A. Ramos, R. W. Kay. 1984. Elementos mayoritarios y trazas de las vulcanitas ordovícicas de la Precordillera Occidental: Basaltos de rift oceánico temprano(?) próximos al margen continental. *9th Geological Congress of Argentina*, vol. 2, pp. 48–66.

Kim, Y., M. Hoshino, H. Aoki, and H. Honma. 1970. Barite nodules collected from the Shiguri-Bank. *Chikyu Kagaku* **24**(2):49–52.

Kowalski, W., K. Szpila, and R. Wyrwicki. 1974. Konkrecje barytowe w ilach serii poznanskiej ze Stablowic pod Wroclawuiem. *Kwartalwik Geologiczny* **18**(2):370–376.

Kramar, U., and H. Puchelt. 1982. Reproducibility test for INAA determination with AGV 1, BCR 1 and GSP 1 and new data for 17 geochemical reference materials. *Geostandard Newsletter* **6**:221–227.

Laznicka, P. 1976. Barite nodules of possibly late diagenetic origin from Twitya River area, Mackenzie Mountains, Northwest Territories. *Canadian Jour. Earth Sci.* **13**:144–145.

Lipkina, M. I., and I. B. Tsoy. 1980. Barite concretions from the Yamato Ridge, Sea of Japan. *Doklady Akad. Nauk SSSR.* **254**:273–275.

Marchig, V., P. Möller, H. Backer, and P. Dulski. 1985. Foraminiferal ooze from the Galapagos rift area—hydrothermal impact and diagenetic mobilization of elements. *Marine Geology* **62**:85–104.

Mitchell, R. S. 1960. Small barite nodules from Ovid, Colorado. *Rocks and Minerals* **35**(1–2):9–11.

Moro Benito, M. C., and A. Arribas Moreno. 1981. Los yacimientos españoles de barita estratiforme y su significado metalogénico en el contexto mundial. *Rev. Española Geol. Min.* **7**(42):18–45.

Puchelt, H. 1986. Sulfid und Oxiderze in der Tyrrhenis in SONNE 41 Cruise Report. *Institut für Petrographie und Geochemie,* Karlsruhe, Germany, pp. 190–203.

Revelle, R., and K. O. Emery. 1951. Barite concretions from the ocean floor. *Geol. Soc. America Bull.* **62**:707–724.

Sakai, H. 1971. Sulphur and oxygen isotopic study of barite concretions from banks in the Japan Sea off Northeast Honshu, Japan. *Geochem. Jour.* **5**:79–93.

Smith, R. C. 1974. Fetid barite from Berks County, Pennsylvania. *Pennsylvania Geology* **6**:4–7.

Suprychov, V. A. 1967. On the barite diagenetic concretions in the Albian deposits of the Mountain Crimea. *Akad. Nauk. UKR.S.S.R.* **4B**:323–326.

Taylor, S. R., and S. M. McLennan. 1985. *The Continental Crust: Its Composition and Evolution.* Oxford: Blackwell, 312p.

12

Magnesite

Wolfgang Wetzenstein

Uses, Consumption, and Production of Magnesite

A wide range of industries use magnesia (MgO) as a raw material. It is prepared from either the natural mineral magnesite ($MgCO_3$) or magnesium hydroxide ($Mg(OH)_2$), which is extracted from sea water or brines. Two main commercial grades of magnesia are derived from crude magnesite: dead-burned magnesite (refractory magnesia) and caustic-calcined magnesite (light-burned magnesite).

Dead-Burned Magnesite or Refractory Magnesia

Dead-burned magnesite refers to a granular product calcined at temperatures of 1,450°C to 2,000°C to yield a sintered product by firing magnesite, magnesium hydroxide, or other material reducible to magnesia. About 75 percent of all refractory magnesia is used in one form or another by the steel industry. Magnesite bricks are used in Bessemer and basic open-hearth furnace linings; their use is much more extensive in the newer furnaces. For instance, Japan, the United States, and West Germany use magnesia-enriched dolomite blocks for the main lining of furnaces (oxygen converter processes such as LD). In recent years there has been a tendency to use higher amounts of magnesite. Although the initial cost of magnesite is higher, magnesite furnaces can burn longer without changing the lining. Depending on the process of steelmaking, the consumed amount of refractory magnesia varies between 4 lb and 12 lb per ton of steel (Wicken and Duncan, 1983).

Apart from the steel industry, magnesite refractories are also used by the cement, lime, glass, and nonferrous metals industries. The cement industry especially has become a large and growing consumer of magnesite refractories.

CAUSTIC-CALCINED MAGNESITE OR LIGHT-BURNED MAGNESITE

This product is obtained by calcination temperatures between 700°C and 1,000°C. Less than 10 percent ignition loss remains and the product displays absorption capacity or activity. The main uses are in the production of fertilizers, animal feed, electrical and thermal insulators, cements, pulp and paper, rayon, and chemical industries.

Until 1930 Austria was the main producer of crude magnesite. This panorama rapidly changed in favor of the U.S.S.R. World production of crude magnesite grew from about 3.7 million tons in 1951 to over 15 million tons in 1972. The current output has fallen to approximately 11 million tons per year. The same trends can be observed in world steel production: Between 1950 and 1963 world production doubled. Until 1969 there was an increase of another 50 percent. Since 1970 world steel production has remained on a more or less constant level. Table 12–1 lists the main producers of crude magnesite.

South Africa has produced magnesite for more than 30 years, but due to the irregular mineralization in pockets, the output has not been constant:

Table 12–1. World Production of Natural Magnesite (tpa)

	1972	*1981*
U.S.S.R.	3,500,000[a]	2,100,000
China	1,000,000[a]	2,000,000
Korea (North)	1,500,000	1,900,000
Austria	1,429,000	1,158,852
Greece	931,458	930,000
Turkey	332,000	772,966
Czechoslovakia	600,000[a]	664,000
Spain	268,000	500,000
India	273,000	453,410
Brazil	200,000	350,000
Yugoslavia	493,000	300,000
U.S.A.	[b]	150,000
Canada	180,000	[c]

Sources: Data from Anonymous (1974); Wicken and Duncan (1983); Harben and Bates (1984).

[a]Estimated.

[b]No production figures available but 569,000 tons of dead-burned magnesia from both natural and sea water/brine sources were shipped by producers during 1971. Capacity of Gabbs Mine: 160,000 tpa dead-burned, 80,000 tpa caustic-calcined.

[c]No production figures available.

Table 12-2. World Production of Magnesia from Sea Water and Brines (tpa/1981)

U.S.A.	650,000[a]	
	460,000[b]	
Japan	715,000[b]	
U.K.	250,000[b]	
Italy	180,000[b]	known production capacity: 220,000
Eire	175,000[b]	
Norway	120,000[b]	
U.S.S.R.	100,000[b]	
Netherlands	100,000[a]	
Mexico	50,000[b]	
	35,000[a]	
Canada	60,000[b]	
Israel	50,000[a]	
China	10,000[b]	

Sources: Data from Anonymous (1974); Wicken and Duncan (1983); Harben and Bates (1984).
[a]Brines
[b]Sea water

100,000 short tons in 1966, 69,000 short tons in 1972, and 40,000 short tons in 1979. The natural magnesite production of Zimbabwe averages about 50,000 tpa (tons per annum). The production of crude magnesite in Australia fluctuated between 40,000 tpa and 60,000 tpa in the early 1950s; only 20,000 tons were mined in 1972 (Anonymous, 1974). There are many potential magnesite producers all over the world, such as Nepal, Central America, Bolivia, Colombia, Ecuador, and Saudi Arabia.

To meet rising demand, production of sea water- and brine-won magnesia has increased remarkably since the early 1950s. Today approximately 3 million tons of magnesia are produced synthetically per annum. The main producers are listed in Table 12-2. Furthermore, a sea-water magnesia project based on dolomite was scheduled in Greece (Anonymous, 1975) with an initial target capacity of 100,000 tons/year and eventual expansion to 200,000 tons/year, a project that was never realized.

FORMATION AND DISTRIBUTION OF MAGNESITE DEPOSITS

Generally magnesite occurs in two physical forms: (1) cryptocrystalline, amorphous or microcrystalline magnesite with low iron content, and (2) crystalline, chiefly macrocrystalline or sparry magnesite with higher iron content. Environmental geologic conditions that lead to the formation of magnesite are very different. Redlich (1909) differentiated formation of magnesite from a saline environment; veins of magnesite in serpentinites, with opal, chalcedony, and quartz; formation in the aureole of serpentinites with the paragenesis magnesite, talc, and chlorite; metasomatic magnesites in limestone and dolomite;

and the formation of hydromagnesite (Ashcroft, British Columbia). Four types of magnesite deposits were described by Bain (1924): magnesite as a replacement mineral in carbonate rocks; a vein-filling material; sedimentary magnesite; and an alteration product of ultramafic rocks.

Formation of magnesite took place from the Proterozoic to Recent. From the physicochemical point of view, the genesis of magnesite can be explained by metasomatism. Two groups have to be distinguished: magnesites formed by CO_2 metasomatism and those formed by Mg metasomatism. Recent investigations of sediments from a hypersaline environment with or without brines, carbonate deposits in a lacustrine environment, and playa and meta-playa sediments have led to new understanding of the formation of magnesite deposits.

A practical nomenclature of magnesite types could be the following:

1. Cryptocrystalline magnesite in or affiliated with ultramafics,
2. Crystalline magnesites in sedimentary or metamorphic rocks,
3. Saline to hypersaline and/or evaporitic magnesites,
4. Magnesites and related Mg-bearing minerals in intramontain freshwater basins.

CRYPTOCRYSTALLINE MAGNESITE IN OR AFFILIATED WITH ULTRAMAFICS

The major magnesite deposits of economic interest occur in belts of ultramafic rocks (dunites, peridotites, pyroxenites) and result from the alteration of serpentine. The formation can be explained by CO_2 metasomatism (Johannes, 1970). Two genetic possibilities exist: ascendent hydrothermal activity or weathering products of serpentinitic rocks under appropriate climatic and oxidizing conditions. Serpentinization is considered to be the reactions:

$$\text{2 forsterite} + 3\ H_2O \rightleftharpoons \text{1 serpentine} + \text{1 brucite}$$
$$\text{1 serpentine} + 3\ CO_2 \rightleftharpoons \text{2 quartz} + \text{3 magnesite}$$

Under special surface and subsurface conditions, magnesite can be considered as a weathering product. The decomposition process of serpentinites takes place in different steps forming intermediate stages such as $Mg(OH)_2 \rightarrow MgCO_3 \times H_2O$, and finally $MgCO_3$. Products are typical vein fillings and nodules, characterized by high purity and extremely small grain size ($1\mu m$ to $10\mu m$) (Petrascheck, 1972) as well as by their relatively high and varying porosity (5 percent to 25 percent), which can be explained easily by their formation (Lesko, 1972). Most of the magnesite deposits of this type are found near the surface and have limited extent in depth.

Concretionary, cryptocrystalline magnesite nodules (diameters range from millimeters to decimeters) have been formed frequently under weathering surface conditions. These formations are known from the surroundings of the serpentinized ultrabasic rocks where these magnesites occur in younger sediments, for example, magnesite and sepiolite nodules in Tertiary sediments of Eskisehir, Turkey (Petrascheck, 1963) and Sepetci, Turkey (Petrascheck, 1972); magnesite nodules from Euboea and Vavdos, Greece (Petrascheck, 1972; Franz et al., 1982); Lydenburg, South Africa (Herzberg, 1976). Palaeonodules were described from Snarum, Norway and from Triassic sediments of the Alps (Petrascheck, 1972; Eisbacher, 1969). In his genetic classification, Ilič (1968) mentions magnesite deposits resulting from weathering ("infiltration type"). Ilič (1968, p. 159) explains the formation as:

Crusts of weathering of ultrabasic rocks are formed in arid and wet climates under the action of air, surface and ground waters, and living organisms. In these crusts the following three zones can commonly be distinguished from the surface down:

1. Zone of ochers;
2. Nontronite zone (often nickel-bearing);
3. Zone of serpentine with magnesite.

Under conditions of lateritization of ultrabasic rocks, descending solutions containing carbonic acid leach magnesium from the upper zones of the crust of weathering, transport it in the form of bicarbonate and deposit in fissures and cavities of the lowermost zone, thus forming magnesite stockworks and small nests. Deposits of this type have a rather small vertical extent (10 m–15 m).

The Halilovskoe deposit on the Urals (U.S.S.R.) can be taken as the prototype of infiltration magnesite deposits. On the Balkan Peninsula occurrences of infiltration magnesite have only been observed at two localities: at Goleš (Serbia) and at Gro'ot, near Titov Veles (Macedonia). They resulted from lateritization of Paleozoic ultrabasic massifs, which took place in the Lower Cretaceous.

Table 12–3 summarizes trace element contents such as Ni, Cr, Mn, and Sr from samples of Vavdos, Greece, and Hodkovece, Czechoslovakia. Magnesite nodules from Servia, Greece are also cryptocrystalline, but of sedimentary origin. Rare earth distribution patterns of magnesites have been determined by neutron activation spectrometry. Compared with metasomatic-crystalline magnesites, the magnesites from ultramafics show low to very low REE contents (Table 12–4). The same results were obtained by Morteani et al. (1983). Most of the REE are below the detection limits of the method (compare Fig. 12–2). I investigated cryptomagnesite deposits from Vavdos and Euboea, Greece.

DISTRIBUTION OF DEPOSITS

Most of the cryptocrystalline magnesite deposits occur in belts of serpentinized ultramafic rocks. Magnesites of this type are found as fillings in cracks, faults, shear zones, and stockworklike structures. Well known as a standard deposit is Kraubath in Austria, a vein–stockwork mineralization of cryptomagnesites in serpentinites. Italy, with small magnesite deposits from the serpentinite mountain zone near Florence and highly impure cryptomagnesites from serpentinite on the island of Elba, counts as one of the less important producers.

Supergene magnesites from Czechoslovakia were described by Martiny and Rojkovič (1977) in ultrabasic bodies near Hodkovece. Magnesite, dolomite, and subordinate calcite and aragonite occur in veins and nests in a weathering crust with a maximum thickness of 40 m. Important deposits of cryptomagnesites are in the belt of ultramafics in southern Europe (Yugoslavia, Albania, Greece, and Turkey). The largest and economically most interesting Balkan

Table 12-3. Trace Element Contents in Different Types of Magnesite Deposits (Ni, Cr, Mn, Sr)

Magnesite Deposit	*Ni*	*Cr*	*Ni/Cr*	*Mn*	*Sr*	*Type*	*Literature*
Servia (magnesite)	19	10	1.9	45	27	limnic-sedimentary	Wetzenstein and Zachmann (1977)
Servia (hydromagnesite)	23	5	4.6	43	100	limnic-sedimentary	Wetzenstein and Zachmann (1977)
Servia (huntite)	12	9	1.3	285	432	limnic-sedimentary	Wetzenstein and Zachmann (1977)
Vavdos (magnesite)	194	38	5.1	59	n.d.	cryptocrystalline	Franz et al. (1982)
Hodkovece (magnesite)	157	97	1.6	211	8	cryptocrystalline	Martiny and Rojkovič (1977)
Hodkovece (magnesite)	245	168	1.5	424	44	cryptocrystalline	Martiny and Rojkovič (1977)
Bankov (magnesite)	9	2	4.5	1243	5	sedimentary (?)	Martiny and Rojkovič (1977)
Ochtina (magnesite)	3	1,5	2.0	2408	53	sedimentary (?)	Martiny and Rojkovič (1977)
Limbo (magnesite)	4	4	1.0	275	55	evaporitic	Franz et al. (1979)
Biele Vody, etc. (magnesite)	8,5	4,5	1.9	2234	93	evaporitic	Martiny and Rojkovič (1977)

Table 12–4. Rare Earth Elements (REE) of Cryptocrystalline Magnesites in Ultrabasic Rocks and in Sedimentary Magnesites

1	*2*	*3*	*4*	*5*	*6*	*7*	*REE*
—	—	0.14	—	0.32	0.09	—	Eu
—	—	—	—	1.54	0.24	—	Hf
—	—	0.80	—	15.0	2.80	—	La
—	—	0.13	—	0.22	0.02	—	Lu
0.02	0.01	0.42	0.01	9.30	1.70	0.14	Sc
—	—	0.58	—	2.13	0.31	0.03	Sm
—	—	—	—	0.50	—	—	Ta
—	—	0.86	—	5.50	—	—	Th
—	—	0.90	—	1.30	0.16	—	Yb
<2.8	<2.4	<7.2	<2.2	33.2	<9.4	<2.8	Rb
<32.0	<32.0	415.0	<30.0	<192.0	105.0	<46.0	Sr

1 = Cryptocrystalline magnesite in ultrabasic rocks from Euboea, Greece.
2 = Cryptocrystalline magnesite in peridotites from Vavdos, Greece.
3 = Huntite (lacustrine-sedimentary) from Servia, Greece.
4 = Magnesite nodules (lacustrine-sedimentary) from Servia, Greece.
5 = Banded magnesite-dolomite ore (evaporitic-sedimentary) from Limbo, Bolivia.
6 = Spongy magnesite (lacustrine-sedimentary) from Bascesme, Turkey.
7 = Magnesite (lacustrine-sedimentary) from Kümbet–Eskisehir, Turkey.

deposits of vein magnesite are those in Yugoslavia (Ilič, 1968). Three districts are mentioned by Harders and Kienow (1960): Maglaj in Bosnia, Sumadia in the surroundings of Užiče, Valjewo and Kraljewo near Čačak, as well as the district between Raška and Skoplje. The most important deposit is situated here: Goleš, with reserves of several million tons. Another vein deposit in the same district is Dubovac with vein fillings between 50 m and 300 m long and from 1 m to 3 m thick, locally 20 m. Ilič explains the formation of these deposits by hydrothermal activity, considering the following parameters: length, thickness, and considerable vertical extent of magnesite veins (e.g., Goleš has some veins with a horizontal extension of about 1.2 km, thickness up to 20 m, and vertical extent of over 300 m); association with fault zones and volcanic activity (thermal springs); and laboratory investigations.

Greece also has large deposits of cryptocrystalline magnesites that are being mined. The deposit located at Vavdos and Ormylia in the Chalkidiki region close to Thessalonika has been well investigated. This ultramafic complex contains serpentinites, dunites, and peridotites with vein–stockworklike magnesite mineralization. In the outcrops of pyroxenites, magnesite could not be observed (Franz et al., 1982). The magnesite deposits of the Vavdos district show the magnesite mineralization like "roots of teeth" in serpentinized and unaltered ultramafics. Quantitative chemical trace element analyses of magnesite ores were performed. Table 12–3 lists average values of 14 samples (magnesites of high purity, Mg/Ca-ratio/molar = 98.5). No significant dependence on trace element distribution could be observed with respect to the depth of sampling. Hodkovece cryptocrystalline magnesites show more or less the same trace element distribution. Table 12–3 includes two types of Hodkovece magnesites: the average value of 8 samples with high magnesia content from the upper column and 25 samples of magnesites with higher contents of dolomite from the lower column. Mn and Sr values are higher in the dolomite-rich type. The low contents of REE from Vavdos and Euboea magnesites are listed in Table 12–4. Petrascheck (1972) and Franz et al. (1982) propose that the vein-stockwork mineralization, controlled by the Tertiary land surface of Chalkidiki, was formed by descendent CO_2-bearing waters. Dabitzias (1980) explains the magnesite formation by ascending CO_2-rich hydrothermal solutions. Geochemical trace element and REE patterns do not confirm the latter genesis.

Some important deposits are exploited in the ultramafic belt on Euboea. Three types of magnesite deposits are present: a vein type, a stockwork type, and a massive type. The massive type is in lenticular concentrations of magnesite with volumes of a few to several thousand cubic feet. This type of a local anomaly of the vein or stockwork magnesite types (Wicken and Duncan, 1983).

Cryptocrystalline magnesites also occur in the provinces of Eskisehir and Kütahya in Turkey (stockwork deposits with magnesite in weblike cracks in serpentine; Harben and Bates, 1984). Magnesite nodules and crusts in serpentinites and surrounding sediments from the Eskisehir district were described by Petrascheck (1963, 1972).

Large deposits of cryptomagnesites are exploited in India around Salem (Chalk Hills). These are a vein type of compact magnesite in serpentinized ultramafics. Reserves range between 5 million tons and 15 million tons (Harders and Kienow, 1960).

Some smaller deposits of cryptocrystalline magnesite are reported from the U.S.S.R. (Orsk, southern Urals), Australia (Marlborough, Queensland), the United States (California, Nevada), and South Africa (Lydenburg area, Kaapmuiden area, area north of the Zoutpansberg, Letaba district). The formation of South African cryptocrystalline magnesite deposits are considered as a product of weathering (Herzberg, 1976).

CRYSTALLINE MAGNESITES IN SEDIMENTARY AND METAMORPHIC ROCKS

The deposits of crystalline, macrocrystalline, and sparry magnesites are widespread and the most important commercially. Most of the large sparry magnesite deposits occur in Paleozoic (e.g., Austria) and Proterozoic (e.g., U.S.S.R.) rocks. There is not a shadow of doubt about the sedimentary formation of magnesites from evaporating ocean waters and the precipitation of magnesite from solutions contacting surface waters after being involved in the hydrolysis of Mg-bearing minerals (olivines, pyroxenes). The difficulty is to recognize the original formation of magnesites, especially in metamorphic complexes. From the physicochemical conditions that lead to the formation of magnesite, many deposits can be explained by metasomatism, especially metasomatic replacement of carbonates by Mg-ion supply.

The sparry magnesites in Austria and Spain were discussed and investigated by the sedimentary and metasomatic schools. A sedimentary to early diagenetic formation of magnesite is proposed by Leitmeier, 1953; Siegl, 1955; Llarena, 1953; Ney, 1956; Lesko, 1972; and Höll and Maucher, 1967. Leitmeier (1953) is of the opinion that $MgCO_3$ is an alteration product of $MgCO_3 \bullet 3H_2O$. According to Lesko (1972, p. 61),

> Contrary to the theory of metasomatism, the chiefly macrocrystalline spar magnesites with high iron content owe their existence to primary enriching of water-containing magnesium compounds in an O_2-poor, H_2S-, CO_2- and NH_4- copius reducing sedimentary environment. Chlorite minerals with high Mg^{2+}-content—equal to the microcrystalline magnesites—are involved as the origin of magnesium.

Sparry magnesites are considered as a metasomatic-hydrothermal replacement product by Angel and Trojer, 1953*a*, 1953*b*; Petrascheck, 1932, 1972; Friedrich, 1953; and Clar 1953. Hercynian or an Alpidian age of replacement in the eastern Alps is discussed. Different sources for Mg^{2+} are supposed: Volcanic activity under submarine conditions (scheelite–magnesite deposit of Lanersbach; Höll and Maucher, 1967, 1976); andesite magmatism (Petrascheck, 1932), ultrabasic magmatism (Angel, 1964); Mg-rich skeletons of echinoderms in the sedimentary host rock and mobilization of magnesium during low-grade metamorphism (Mostler, 1973); generation of Mg-rich solution during regional metamorphism (Clar, 1956; Friedrich, 1963).

More recently, the formation of sparry magnesites from the eastern Alps has been investigated by Morteani et al. (1983). From the study of REE distribution patterns in the magnesite samples, combined with petrological and physicochemical considerations, the formation of sparry magnesites from the northern graywacke zone and the Innsbruck quartz phyllite might be possible by metasomatic replacement of a carbonate precursor by magnesite, most probably due to a local to regional anomalous increase of temperature and/or Mg^{2+} ion supply.

> The simplest model for large-scale magnesite formation in the northern graywacke zone and the Innsbruck quartz phyllite requires an irregular geothermal gradient caused by an overthrusting phenomenon. Along such a perturbated geothermal gradient, ascending solutions, originally in equilibrium with calcite and dolomite at a lower temperature, might have been able to convert a dolomitic country rock into magnesite at a higher temperature. The Mg^{+2} ion might have been supplied by mineral reactions induced by low-grade regional metamorphism. The latter is supported by the rather high level of rare earth element contents in some of the dolomite samples with positive Eu anomalies in coexistence with magnesite. This result based on rare earth element distribution in the carbonates, tectonic aspects, and physicochemical considerations agrees with the suggestions of Clar (1956), Friedrich (1963), and Lesko (1960) on the formation of the sparry magnesite mineralizations in the graywacke zone from metamorphogenic solutions.
>
> With respect to this model it is interesting that many magnesite mineralizations are found within the graywacke zone along the actual borders of the tectonic sheets. It should also be mentioned that not only magnesite mineralizations but also siderite and sulfide occurrences are well known in this zone (Schulz, 1972*a* and 1972*b*). The probably rare interplay between tectonic overthrusting and metamorphic reactions might explain why this type of magnesite formation is not more frequently observed in other areas of similar lithology, such as that of the northern graywacke zone and the Innsbruck quartz phyllite. (Morteani et al., 1982, p. 629)

The formation of magnesite (early diagenesis and diagenesis) under euxinic environmental conditions is an additional possibility. The low-grade metamorphic phyllites and dolomites of the Lanersbach magnesite–scheelite deposit are enriched in graphite. The negative Eu anomaly identified by Morteani et al. (1982) from samples of the Lanersbach deposit is a further argument that euxinic conditions were dominant during carbonate formation.

The question of sedimentary or hydrothermal-metasomatic genesis is discussed in the case of many sparry magnesite deposits. A few examples are mentioned. In the Carboniferous of the Gemerides (Czechoslovakia), magnesite deposits occur near Bankov in metamorphic series of graphitic schists and phyllites, layers of banded limestones, diabase tuffs and tuffites, dolomites and dolomitic limestones. The deposits are in rocks of Upper Carboniferous age and are formed by three layers. These magnesites contain thin beds of graphitic phyllites and sandstones (Ťapák, 1967). Trace elements (45 samples) of magnesite were determined by Martiny and Roskovič (1977; see also Table 12–3).

The deposit of Ochtiná, Czechoslovakia, also in Carboniferous metasediments shows similar features. Magnesite is found with graphite schists, which are gradually changing into dolomite shales, schistose dolomites to compact dolomites (Abonyi, 1971). The trace elements are listed in Table 12–3 (average value of 12 samples).

The trace element distribution pattern of the latter deposits show low Ni and Cr contents but high Mn values compared with other magnesites. Typical trace elements for these deposits are Fe, Mn, Ti, B, and V (Martiny and Rojkovič, 1977). Boron leads to the formation of tourmalines in the deposit of Bankov, due to volcanic activity. Environmental conditions and trace element distribution points to the fact that magnesite formation took place under euxinic conditions (graphite, vanadium anomalies) combined with volcanic activity (tuffites, boron anomalies).

Some of the world's largest crystalline magnesite deposits are located on the western slope of the Urals: the Satka deposits. The Bakal–Satka mining district is associated with the western structural and facies zone of the Bashkirian meganticlinorium. This zone is made up of Upper Proterozoic (dolomites, phyllitic shales, limestones, carbonatic shales, granite, and diabase dikes and sills). Only carbonate rocks of the Satka suite are magnesite-bearing, with two types of magnesite occurrences: (1) fine grained and sparsely dispersed and stratigraphically widespread throughout the carbonate rocks; and (2) crystalline bodies associated with definite stratigraphic horizons. Geologic exploration supports sedimentary formation of the magnesites as the stratigraphic control of magnesite mineralization. A sedimentary hypothesis of the origin of Satka magnesites was propounded by Garan (1954).

Perepelitsyn et al. (1971) described the content of graphite in magnesites from the Satka group. The average concentration of carbon in surrounding dolomites was determined by these authors as 0.3–0.8 percent. More recent investigations of Satka deposits point to a "finely dispersed sedimentary mineralization and large massive deposits, epigenetic with respect to the enclosing formations. The conclusion is that the Satka deposits are metasomatic apodolomitic formations originated in catagenesis of dolomites with finely dispersed magnesite." (Anfimof et al., 1980, p. 115).

The sole natural magnesite producer in the United States is the deposit at

Gabbs, Nevada (Harben and Bates, 1984). Magnesite ores with brucite generally occur in dolomite of the upper member of the late Triassic Luning formation, north and northwest of a granodiorite stock. The Triassic carbonate formation has been divided into lower, middle, and upper members by Vitaliano et al. (1957). The upper member consists of two distinct rock types: (1) a dense, blocky, finely laminated, very fine-grained, dark gray to black, locally cherty dolomite, which contains the magnesite and brucite; and (2) thin-bedded, slaty, black limestone. The gray color of the dolomites and magnesites is due to the presence of finely divided carbonaceous matter. Other impurities of the magnesite are dolomite, talc, tremolite, forsterite, grossular garnet, chalcopyrite, calcite, and clay minerals. Magnesites and recrystallized dolomites are believed to have been formed by hydrothermal replacement of the regionally metamorphosed sedimentary dolomite (Schilling, 1968). Martin (1960) states that the magnesite bodies are sedimentary in origin, having been deposited, with the enclosing dolomite, in shallow, restricted, unstable, marine embayments, similar to some in the Caspian Sea in which magnesite and dolomite are presently being precipitated.

The genesis of sparry magnesites has been discussed in many publications. One of the first to postulate a sedimentary origin for magnesite was Llarena (1950, 1951). In the foothills of the Pyrenees (Spain and France), magnesites occur in Devonian and Carboniferous formations. The well-bedded magnesites, $MgCO_3$, ripple marks, horizontal extension, and the intercalation of fossiliferous (including plants) slates in the magnesite horizons are arguments for a sedimentary origin of Upper Carboniferous age (Llarena, 1951).

In the discussion on the origin of the magnesite deposits in the Kumaun Himalaya (India), the role of algae was presented (Valdiya, 1968). The coarsely crystalline magnesite in a chain of lentiform deposits is restricted to a narrow stratigraphic interval of the Gangolihat dolomites and is almost invariably associated with the stromatolite-bearing dolomite. Formation took place in the back-reef shelf in barred embayments, attributed to barrier-building algae. Later recrystallization during regional metamorphism converted the fine-grained Mg-carbonates in sparry magnesites. Similar sedimentary formation of bedded deposits was described from the Sinian magnesite of Malyghingen (Manchuria and northern China). Those deposits are irregular, though concordant (Nishihara, 1956).

DISTRIBUTION OF DEPOSITS

World production of natural magnesite in 1981 was some 11.4 million tons (see Table 12–1). An estimated 80 percent are crystalline magnesites. Proterozoic to Cambrian carbonatic rock sequences contain the largest magnesite reserves of the world, for example, the Satka/Ufa Province (U.S.S.R.) with more than 250 million tons. Identified reserves of crystalline magnesite in the U.S.S.R. are on the order of 400 million tons in the southern Ural Mountains and 2 trillion tons in the Savan Mountains of eastern Siberia.

The reserves of North Korea are estimated at about 3 trillion short tons (main deposits in Kankyo Province), and those from China at about 5 trillion tons in the famous Manchurian deposits (Wicken and Duncan, 1983).

The deposits of Brazil are another example of recrystalline magnesite in later Precambrian dolomites and quartzites. Known reserves are of the order of 500 million short tons. These deposits, located in the Alencar area of Ceará Province and the Eguas range near Brumado in southwestern Bahia, were described by Bodenlos (1954) as replacement formations of dolomites.

The deposits of Gabbs (U.S.A.) are being worked by the open-pit method. Total magnesite reserves are estimated at more than 25 million short tons. At present, the Chewelah deposit (Washington State) is not worked. In Canada a rather unique deposit is mined near Kilmar (Quebec). The intermixed rock of magnesite–dolomite is thought to be the product of hydrothermal dolomitization of limestone (Bray and Hilchey, 1957). The high-lime magnesite is of vein type.

Most of the famous European spar magnesite deposits occur in Paleozoic rocks (Austria, Czechoslovakia, Spain). Austria, the oldest producer in western Europe, contains three groups of magnesite deposits (Harders and Kienow, 1960):

1. Coarse-grained magnesite with more than 3 percent FeO (in crude magnesites): Veitsch, Breitenau, Lassing, Leogang, Dienten, St. Oswald, and Tragail
2. Coarse-grained magnesite with less than 3 percent FeO: Trieban, Radenthein, and Tux-Lanersbach
3. Fine-grained magnesites with FeO contents between 2 percent and 3 percent: Fieberbrunn.

The reserves of the Austrian magnesite deposits are estimated at approximately 500 million tons, a fifth of which is in the Fieberbrunn deposit. The crystalline magnesites of Czechoslovakia are in carboniferous host rocks (Martiny and Rojkovič, 1977; Kužvart, 1981); reserves are on the order of 100 million short tons (Wicken and Duncan, 1983).

SALINE TO HYPERSALINE AND/OR EVAPORITIC MAGNESITES

Early diagenetic formation of dolomite, huntite, magnesite, gypsum, bassanite, celestite, and halite from a Recent, supratidal evaporitic environment along the Trucial coast (Persian Gulf) was described by Kinsman (1967). From Tuz Gölü (Turkey), a seasonal lake in Central Anatolia, huntite, magnesite, dolomite, polyhalite, and authigenic idiotopic gypsum crystals were investigated (Irion and Müller 1968). The formation of huntite, magnesite, calcite, and gypsum was observed in salt pans of Miocene age at Zarzis, Tunisia (Perthuisot, 1971).

A hypersaline environment with a high Mg/Ca ratio in the solutions are the principal parameters for the formation of sedimentary magnesites, whether primary or early diagenetic (metasomatism of preexisting carbonates).

Paleoevaporitic minerals have been recognized in early Archean metasediments in Australia (Golding and Walter, 1979). Franz et al. (1979) describe paleoevaporitic magnesites (? Proterozoic) undergoing exploitation in Bolivia. The distinction between ancient marine evaporite sequences is problematic, whether in near-shore marine sabkha facies, continental playa lakes, or lacustrine-lagunal basins in tectonically and metamorphically overprinted Paleozoic and Proterozoic complexes.

My co-workers and I (Franz et al., 1979) investigated magnesite deposits of evaporitic and metasomatic origin in Cambrian(?) or Proterozoic(?) rocks of the Bolivian Andes. The magnesites are exploited at the eastern slope of the Andes in the Alto Chapare area (Department of Cochabamba) in a paleocoastal zone between the Arequipa and the Brazilian Platform. The reserves of crude magnesite are limited and do not exceed 20 million tons of industrial quality (Wetzenstein, 1983). Magnesite deposits occur in two stratigraphic units:

1. Evaporitic magnesites with gypsum, dolomite, danburite, boracite, tourmaline, and lower content of talc (lower unit).
2. Bedded metasomatic magnesites with dolomite, calcite, and higher content of talc (dressing and separation of talc is necessary as SiO_2 content exceeds limits). Demagnesitization and secondary formation of dolomite is typical. The clayey-carbonatic series were overprinted by low-grade metamorphism (upper unit).

The overlying Cambrian metasediments show boron anomalies, especially in their sandy facies. Tourmaline-banded metasandstones and tourmaline-bearing metapelites are unique in this part of the Andean metallotect. During the sedimentary stage, dolomite, calcite, aragonite, magnesite, anhydrite, gypsum, boracite, danburite, and tourmaline were formed. During the Hercynian orogenesis, hydrothermal-metasomatic processes produced the following minerals:

1. Schorl, ferridravite (Walenta and Dunn, 1979), danburite (B-metasomatism),
2. Microcline, micas (K-metasomatism),
3. Aegirine (Na-metasomatism),
4. Magnesite and asbestos (riebeckite and Mg-riebeckite; Avila, 1977) (Mg-metasomatism).

The importance of evaporites in ore-forming processes (especially in tin mineralization) in this part of the Andes has been pointed out by Lehmann (1979).

Proterozoic to Cambrian evaporitic magnesites may be more widespread in South America, as recently proposed. Possibilities exist in eastern Bolivia (Brazilian Shield), where the following depositional phase of the Brazilian Orogenic Cycle is represented by the sediments of the western margin of the north-trending Paraguay–Araguaia Belt, one of the main elements of the Brasiliano Cycle in Brazil, along with those of the subsidiary west–northwest-trending Tucavaca basin. Three different lithostratigraphic groups can be distinguished: those of Boqui, Tucavaca, and Murciélago (Upper Proterozoic). The facies of the Boqui group changes to the east in a thicker eugeosynclinal affiliation of conglomerates, graywackes, banded ironstones (Fe-Mn-deposit of Mutun), cherts, limestones, and calcilutites. The basal unit of the Tucavaca group is formed by the 150 m thick Pororó formation composed of dolomitic limestone with local conglomerate bands. A sample checked by me indicated the presence of magnesite. This sequence is followed by an arenaceous unit with variable shale fraction.

The Murciélago group, corresponding in time to the Chapare sequence, is tentatively ascribed to the Cambrian since it represents a marine regression over fault-controlled Upper Proterozoic basins and in neighboring Brazil contains algal structures. The Murciélago group is made up of approximately 300 m of unfolded reef and nonreef limestones (unstratified and locally oolitic) (Berrangé and Litherland, 1982).

Due to their genesis, evaporitic magnesites are characterized by low trace element contents of Ni and Cr. Table 12–3 lists the determinations of Ni, Cr, Mn, and Sr. Magnesites were selected from the Alto Chapare (Limbo, Bolivia, average value of 8 samples) and from Permian–Triassic magnesites in Czechoslovakia (15 samples). Boron contents have been determined, but the presence of B-minerals indicates higher values of this element. Sr contents in magnesites of different origin hardly ever depend on their genetic environment. Sr values in magnesites show a strict relation to the paragenesis with Ca-bearing minerals (Fig. 12–1). REE contents (Table 12–4) of some magnesites from the Alto Chapare show the characteristics of metasomatic and sedimentary formation. In Figure 12–2 the REE distribution patterns of different magnesite types have been correlated with results from Morteani et al. (1983).

Evaporitic magnesites overprinted by regional metamorphism are described from the Rum Jungle Uranium Field in Australia (Crick and Muir, 1980; Morteani et al., 1983). Magnesites occur in dolomitic country rock of Proterozoic age in the closer surroundings of the U deposits. A sabkha model for the deposition of carbonates of the Proterozoic McArthur Group (northern Australia) has been published by Muir (1979).

Occurrences of Proterozoic evaporitic metasediments are specific for parts of Gondwanaland. They can be related to a phase of continental stretching. The influence of palaeoevaporitic rock sequences in post-Proterozoic metallogenesis must be seriously considered. Proterozoic evaporites have been described from Australia, Africa, and South America.

Borates are common in evaporitic settings, and metamorphosed evaporites generally contain tourmaline-rich concentrations (e.g., northern Australia, Muir, 1979; southern Australia, evaporites from the Adelaide Supergroup, Rowlands, 1974; African Copperbelt, Damara orogen, Behr et al., 1981, 1983). According to Sawkins (1976), the Proterozoic continental rifting is connected with elevated subcontinental hot spot activity. Boron-rich environments in evaporites could be explained by this model, whether boron is a product of leaching or of postvolcanic activity.

Mesozoic evaporitic magnesites are known from Czechoslovakia (Martiny and Rojkovič, 1977). Magnesite is mainly present in evaporites with gypsum, anhydrite, halite, dolomite, and calcite. Magnesite from these occurrences is only of academic interest (localities: Biele Vody, Novoveská Huta, Smizany, and Sankovece).

Magnesite occurrences in the Permian and Scythian series of the eastern Alps in Austria were described by Niedermayr et al. (1983, p. 97):

> Magnesite is bound to basin sequences as well as to sabkha sediments. It occurs as cement in sand- and siltstones, as recrystallized magnesite-rich matrix, in concretionary

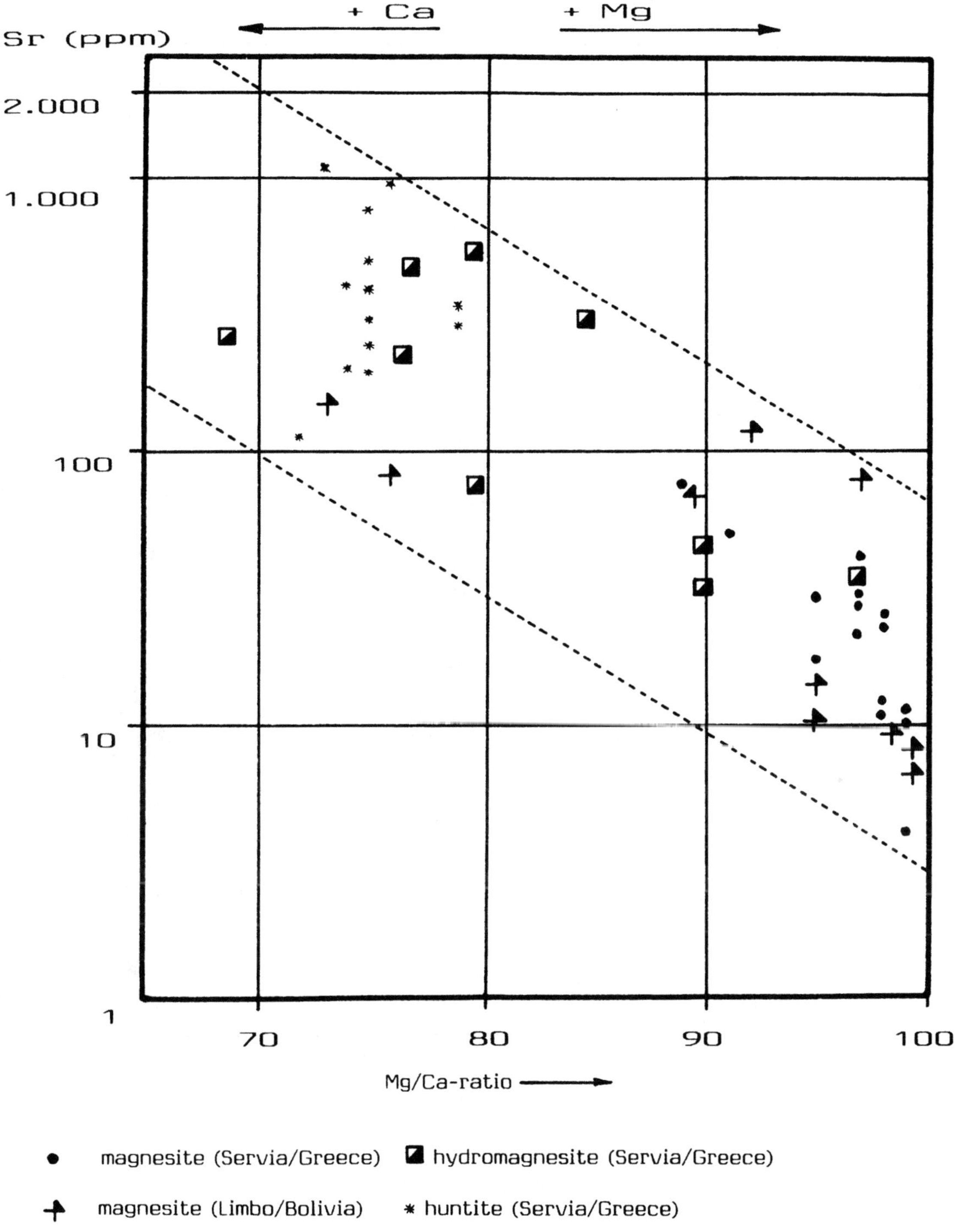

Figure 12-1. Strontium distribution in magnesites, hydromagnesites, and huntites.

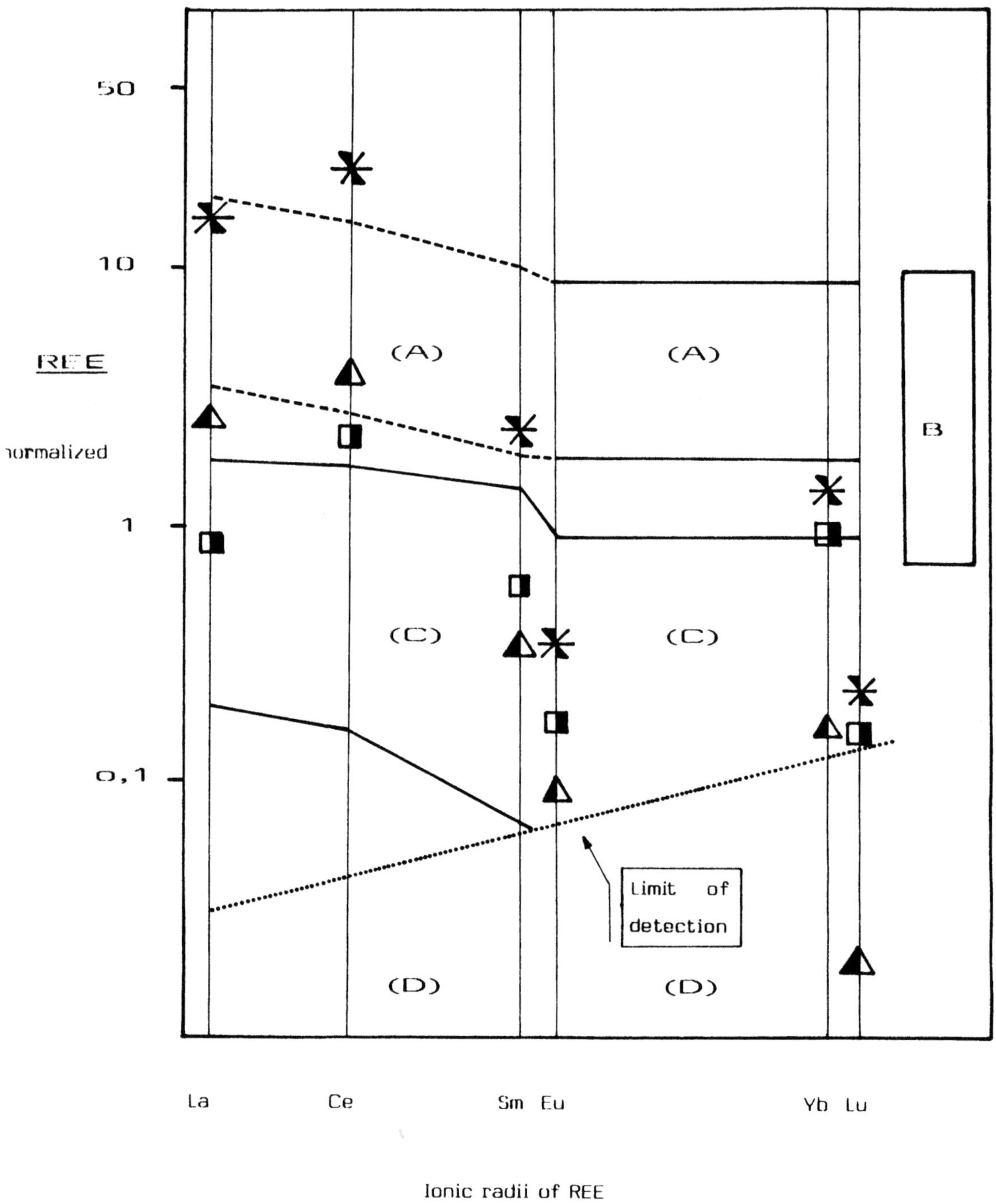

Figure 12–2. REE distribution patterns of magnesites from different occurrences and the detection limit of the method used (INAA; from Morteani et al., 1983, p. 111).
(A): metasomatic magnesites, (B): sparry magnesites, (C): sedimentary magnesites, (D): cryptocrystalline magnesite.

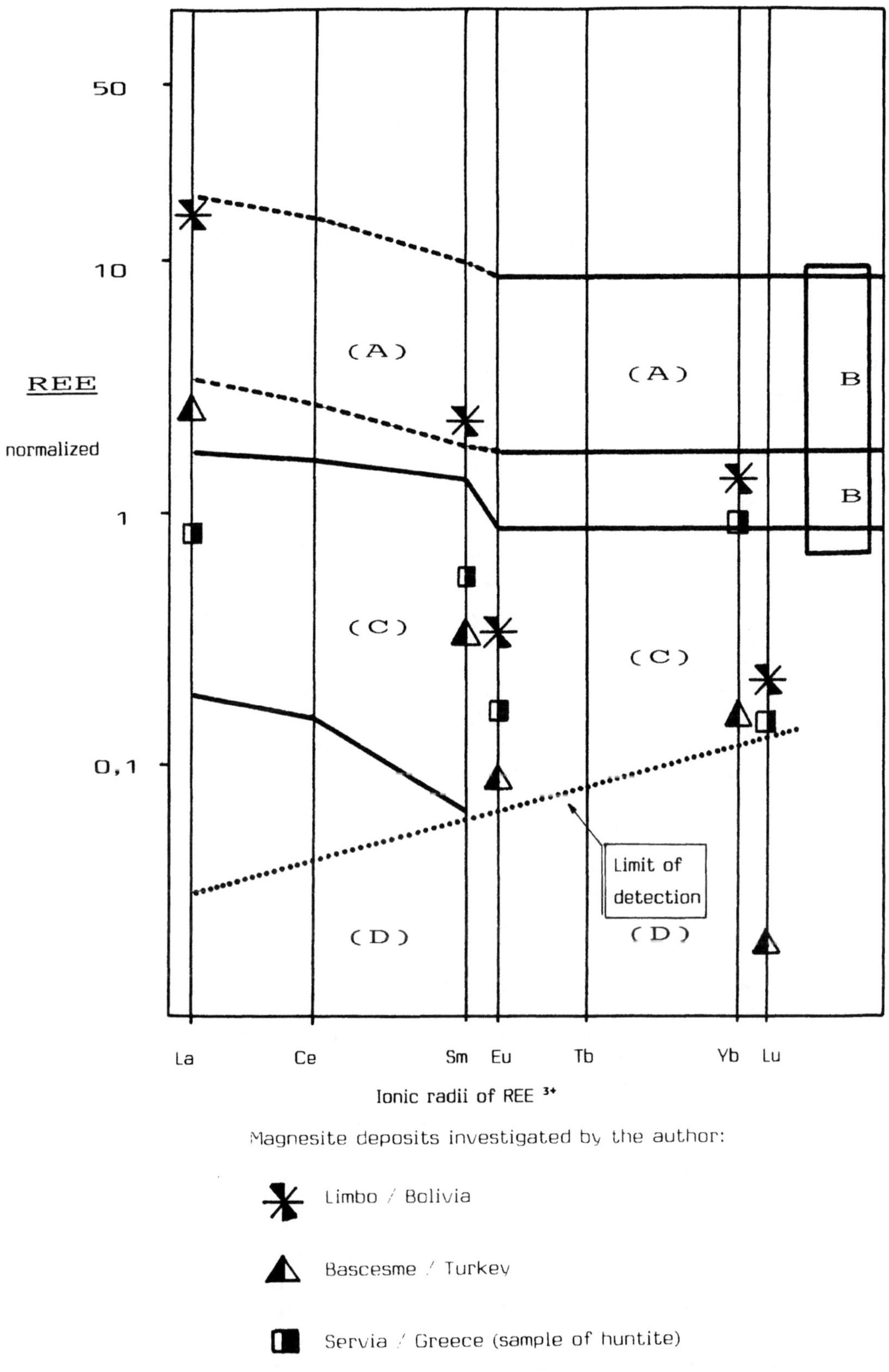

Figure 12–2. (Continued)

forms and as mobilizations in veins crosscutting the primary magnesite-bearing rocks. These primary magnesites are poor in Fe and Mn. The latter exceeds Fe in the Permian magnesites of the Drauzug. In contrast, all Scythian magnesites, as well as the Permian magnesites of the Northern Calcareous Alps, normally contain more Fe than Mn.

The alpidic metamorphism which has involved parts of the Permian and Scythian series of the Eastern Alps caused remobilization of magnesite in places. These metamorphically generated magnesites are characterized by variable to high Fe contents. In conclusion, both the Permian and Scythian magnesites are crypto- to megacrystalline and show evidence for a rather complex genesis influenced by sedimentary environment, diagenesis and Alpidic metamorphism.

Economic magnesite deposits of early diagenetic origin are reported from Kharidunga, Nepal (Anonymous, 1973). The deposits of the Sindhupalchok district are situated in the Lower Himalayas. The epizonal, overprinted, magnesite–talc-bearing host rock is of Precambrian(?) age. The formation of magnesite is considered to be syngenetic-sedimentary in isolated, epicontinental, sea water basins. The origin of talc is thought to be in a saline environment (Braitsch, 1962).

The Jatulian dolomites (Middle Proterozoic) in southern Karelia (U.S.S.R.) contain low-grade magnesites (Aksenov et al., 1975). The formation of dolomite took place at the end of transgressive–regressive sedimentation cycles in the stage when the basin shoreline was stationary, and the salinity relatively high. Aksenov et al. consider the southern Karelian magnesites to be of "protosedimentary" origin.

Magnesite deposits of economic interest, formed in high-saline sea water basins, are reported from Kumhar and Kulhar Khetar in the Sherwan area of Pakistan (Ashraf et al., 1971). Magnesite deposits are embedded in light gray dolomite and dolomitic limestone of Upper Paleozoic age.

MAGNESITES AND RELATED Mg-BEARING MINERALS IN INTRAMONTAIN FRESHWATER BASINS

At very high Mg/Ca ratios and high Mg concentrations, hydrous Mg-carbonate deposits (hydromagnesite, nesquehonite) can be formed. In this case the sources of Mg are normally serpentinized ultramafic rocks and greenstones. Mg is transported by weathering solutions (e.g., fluviatile). Barnes and O'Neill (1969) determined Mg/Ca ratios between 20 and 40 in meteoric waters associated with ultramafic rocks. Intramontain freshwater basins are favored regions for the formation of these types of Mg minerals.

The first magnesite deposit in the Miocene freshwater basin of the Inner Dinarides of Yugoslavia was discovered in the Bela Stena/Jarando Miocene basin of the Kapaonik region (Ilič, 1952). Once this economic type of deposit was detected, Yugoslav geologists began to investigate Miocene freshwater basins of the Inner Dinarides and the results are evident. The following deposits were discovered: Beli Kamen in the Strezovce basin (one of the largest magnesite deposits under exploitation in Yugoslavia); Nevade, Shilopaj, and Lunjevica in the Milanovac basin; the deposits at Rvati, Pranjani, Dobrinje, the Chachal-Kraljevo; Kacher, Kremna, Bioska, Semegnjevo, among others.

Freshwater deposits of the same type have been discovered in northern Greece in the Servia basin near Kozani (Donath, 1955; Wetzenstein, 1974*a*, 1974*b*; Wetzenstein and Zachmann, 1977) and in Turkey in the Neogene basin of Denizili at Bozkurt (Petrascheck, 1962). Almost certainly, this type of magnesite–hydromagnesite–huntite–dolomite deposit has been overlooked because of its insignificant exterior, but it is by no means unique. I recently discovered these Mg-carbonate formations on the Island of Margarita (Venezuela).

Since mineralogical composition, trace element, and REE distribution of the Servia/Aiani basin carbonates are available, this deposit is discussed in more detail, based on the investigations by Wetzenstein and Zachmann (1977). Tectogenetically, the area studied is an intramontain basin in the Pelagonian massif. In contrast to the Plio-Pleistocene of Italy, the backdeep in Greece never led to marine development. There was, however, an expansion of the Pannonian-euxinic-limnetic system, with lignite deposits typical of intramontain basins, of which there is, however, very little in the Servia basin, quite in contrast to the well-known basin of Ptolemais. It is to be supposed that because of the nearly total lack of lignite, the terrigenous-detrital, limnetic-fluviatile, sandy, marly to clayey, and dolomitic to limy series have never received close attention. Relics of karstic marbles in the basin show the old land surface. Serpentinized ultramafic rocks are situated to the west of the basin (Vourinos Mountains).

The mineralization is seamlike in different horizons, and from an economic point of view, two ore types can be differentiated: magnesite–huntite ore and hydromagnesite–aragonite ore. Carbonate-bearing minerals are: hydromagnesite, magnesite, huntite, dolomite, aragonite, and very low contents of calcite.

Layers of huntite–magnesite ore with a thickness from 5 cm to 3 m were determined in different places of the basin. Nodules of magnesite (diameter up to 1 m) layers with huntite nodules, and rarely aragonitic sinters with huntite are typical. Normally, huntite forms nodules in a clayey, sandy, or dolomitic matrix. The grain size of huntite varies between 0.5 μm and 1.5 μm. Fault zones with magnesite sinters are rare, whereas locally the thickness reaches 10 m. Magnesite nodules can be selectively mined. Dressing is necessary for huntite–magnesite ores (CaO-rich huntite of small grain size). Hydromagnesite is widely distributed in the Servia basin. At Aiani, hydromagnesite layers, mainly accompanied by aragonite, can be observed over an extension of 2 km with a thickness up to 5 m. Sintering tests of this material run extremely good. Dressing of hydromagnesite ores is without problems, and CaO limits between 1 percent and 3 percent can be reached.

The geological reserves of Mg-bearing minerals in the Servia basin were estimated at 400 million tons. The probable reserves of hydromagnesite ores in an area of 46 km^2 were calculated as 95 million tons. An area 10 km^2 can be mined by open-pit methods. Our investigations led to a quintuple increase of Greek magnesite reserves.

Magnesites derived from weathering solutions by precipitation should logically be low in trace element and REE contents. The average values of 14 magnesite samples from the Servia deposit are listed in Table 12-3. As demonstrated in Figure 12-1, Sr content increases with growing Ca content (mainly huntite). Pure magnesites contain ± 10 ppm Sr. The same phenomenon is observed in hydromagnesites with increasing aragonite content (Fig. 12-1). The average Sr content of 9 hydromagnesite samples was calculated between 100 ppm and 432 ppm in huntite (13 samples), with Ni and Cr distribution in magnesite, hydromagnesite, and huntite of the same dimension. REE distribution patterns (Fig. 12–2) point to sedimentary formation in the case of huntite. REE contents of Servia magnesite nodules are below the detection limit of the method (compare Morteani et al., 1982) for most of the elements. Servia magnesite demonstrates the same REE distribution pattern as cryptocrystalline magnesites from Euboea and Vavdos (Table 12–4). The same results were recorded by Morteani et al. (1982) on magnesites from Bela Stena and Aiani.

Weathering solutions (fluviatile as well as in situ) are considered sources of Mg). CO_2 may be produced from lignitic series in the depth and/or transported by CO_2-bearing juvenile waters. The boron content of the magnesites point in this direction. The large and extensive hydromagnesite mineralization took place under superficial conditions, related to the formation of aragonitic sinters. No transformation from hydromagnesite into magnesite could be observed.

Sedimentary freshwater deposits of magnesite in Tertiary clayey, marly, and sandy rock sequences occur 60 km southwest of Denizili in Western Anatolia, Turkey. From Bascesme, Petrascheck (1962) described two seams of high-grade magnesite (respective thicknesses of 2.8 m and 1.5 m). The magnesite is extremely fine grained (5 μm to 10 μm) and of high porosity. REE data on spongy magnesite from Bascesme are shown in Figure 12–2 (sedimentary field) and Table 12–4. Additional analysis of trace elements show the following values: Sr, 50 ppm (dolomite content); Mn, 135 ppm; Fe, 555 ppm; and Ni, 40 ppm for the spongy magnesite, and Sr, 65 ppm; Mn, 220 ppm; Fe, 8000 ppm; and Ni, 375 ppm for the dolomitic series in the hanging wall of the magnesites. The trace element distribution points to a juvenile CO_2 supply (boron anomalies were determined).

Lacustrine sedimentary magnesite occurrences were found in the Eskisehir basin at Kümbet (for REE data, see Table 12–4). The cryptocrystalline magnesite is extremely pure: CaO, 0.01 percent; Sr, 2 ppm; Mn, 5 ppm; Fe, 30 ppm; and Ni, 22 ppm.

Going ahead with sedimentary formation of Mg-minerals, thick layers of tuffaceous clays from Miocene–Pliocene saline alkaline and lacustrine deposits from the intramontain basin in the north of Mihaliççik (between Ankara and Eskisehir), Turkey, mainly consist of sepiolite, loughlinite, and analcime. They result from syngenetic and early diagenetic reactions between pyroclastics, clay minerals, and magnesium–sodium-rich saline weathering solutions in the Neogene basin. Silica, arising from the devitrification of volcanic glass, was completely consumed when the silicates were formed. Quartz and any other SiO_2 phase are lacking. The present composition, that is, the high content of MgO and Na_2O, reflects the allochemical diagenetic alteration. The formation of magnesite was observed in analcime-bearing formerly vitric tuffs (Echle, 1974).

ACKNOWLEDGMENTS

I am indebted to Dr. P. Möller (Hahn-Meitner-Institut für Kernforschung, Berlin) for the determination of REE by neutron activation spectrometry.

REFERENCES

Abonyi, A. 1971, Magnezitové loziská Slovenska. *Mineralia Slovaca, Bratislava* **3**(12-13):319-342.

Aksenov, Y. E., S. V. Ignatyev, V. Z. Negrutsa, and V. Z. Solontsov. 1975. Magnesite in Jatulian Carbonate Rocks of southern Karelia. *Doklady Akad. Nauk SSSR* **227**:59-61.

Anfimov, L. V., B. D. Busygin, and L. E. Demina. 1980. Formation conditions of magnesites in the Satka group of ore deposits, South Urals. All-Union Geological Institute, Leningrad. *Litologiya Poleznye Iskopaemye*, Leningrad **1**:115-128.

Angel, F. 1964. Petrographische Studien der Ultramafit-Masse von Kraubath (Steiermark). *Johanneum Mineralog. Mitt.* **2**:1-125.

Angel, F., and F. Trojer. 1953*a*. Der Ablauf der Spatmagnesit-Metasomatose. *Radex-Rundschau*, Radenthein, pp. 315-334.

Angel, F., and F. Trojer. 1953*b*. Zur Frage des Alters und der Genesis alpiner Spatmagnesite. *Radex-Rundschau*, Radenthein, p. 374-392.

Anonymous. 1973. Magnesit Kharidunga, Nepal. Deutsche Gesellschaft für Technische Zusammenarbeit (GTZ) GmbH, internal information, Essen, vol. 1.

Anonymous. 1974. World magnesia shortage—particularly in high-purity grades. *Industrial Minerals* (May 1974):23-34.

Anonymous. 1975. Synthetic magnesia will be produced from dolomite and seawater in major Greek project. *Min. Mag.* (December 1975):449-451.

Ashraf, M., R. Haq, M. H. Qureshi, and F. A. Faruqi. 1971. Magnesite Occurrence in Hazara district. *Pakistan. Jour. Sci.* **14**(6):538-541.

Avila, W. 1977. La crocidolita magnesiana del Chapare. *Rev. Banco Min. Bolivia*, La Paz **92**:25-43.

Bain, G. W. 1924. Types of magnesite deposits and their origin. *Econ. Geol.* **19**:412-433.

Barnes, I., and J. R. O'Neill. 1969. The relationship between fluids in some alpine type ultramafics and possible modern serpentinization, western United States. *Geol. Soc. America Bull.* **80**:1947-1960.

Behr, H. J., H. Ahrendt, A. Schmidt, and K. Weber. 1981. Saline horizons acting as thrust planes along the southern margin of the Damara orogen (Namibia, SW Africa). In *Thrust and Nappe Tectonics*, Geological Society of London, pp. 167-172.

Behr, H. J., H. Ahrendt, H. Martin, H. Porada, J. Röhrs, and K. Weber. 1983. Sedimentology and mineralogy of Upper Proterozoic Playa-lake deposits in the Damara Orogen. In *Intercontinental Fold Belts,* eds. A. Martin and K. Weber, pp. 577-610. Berlin: Springer-Verlag.

Berrangé, J. P., and M. Litherland. 1982. Synopsis of the geology and mineral potential of the Proyecto Precambrico area. Inst. Geol. Sciences, Overseas Division, United Kingdom and Servicio Geológico de Bolivia, Regional Santa Cruz, Report No. 21 (unpublished), 106p.

Bodenlos, A. J. 1954. Magnesite deposits in the Serra das Eguas, Brumado, Bahia, Brazil. *U.S. Geol. Surv. Bull.* **975C**:87-170.

Braitsch, O. 1962. *Entstehung und Stoffbestand der Salzlagerstätten*. Berlin: Springer-Verlag.

Bray, W. T., and G. R. Hilchey, 1957. Magnesite deposits of Kilmar, Quebec. In *Geology of Canadian Industrial Mineral Deposits*, pp. 164-166. Montreal: Canadian Institute of Mining and Metallurgy.

Clar, E. 1953. Geologisches Gefüge und Formentwicklung von metasomatischen Karbonatlagerstätten (Siderit und Magnesit) der Ostalpen. In *19th International Geological Congress*, Algiers, 1952, Comptes Rendus, Section 10, pp. 83-96.

Clar, E. 1956. Zur Entstehungsfrage der ostalpinen Spatmagnesite. *Carinthia II* **20**:22-32.

Crick, I. H., and M. D. Muir. 1980. Evaporites and uranium mineralization in the Pine Creek geosyncline. In *Uranium in the Pine Creek Geosyncline*, eds. J. Ferguson and A. B. Goleby. Vienna: International Atomic Energy Agency.

Dabitzias, S. G. 1980. Petrology and genesis of the Vavdos cryptocrystalline magnesite deposits, Chalkidike Peninsula, Northern Greece. *Econ. Geology* **75**(8): 1138–1151.

Donath, M. 1955. Die jugoslawischen Magnesitlagerstätten. *Tonindustrie Zeitung*, Goslar **79**:267–274.

Echle, W. 1974. Zur Mineralogie und Petrogenese jungtertiärer tuffitischer Sedimente im Neogen-Becken nördlich Mihaliççik (Westanatolien, Türkei). *Neues Jahrb. Mineral. Abh.* **121**(1):43–84.

Eisbacher, G. H. 1969. Neue Beobachtungen zur Deutung der knolligen Magnesite im Tonschieferkomplex des alpinen Buntsandsteines der nördlichen Kalkalpen. *Mineralium Deposita* **4**:219–224.

Franz, E. D., D. Marazidis, and W. Wetzenstein. 1982. Geochemie magnesiumkarbonatischer Feuerfestrohstoffe von Vavdos—Chalkidike, Griechenland. *Radex-Rundschau*, Radenthein, pp. 699–706.

Franz, E. D., J. Ponce, and W. Wetzenstein. 1979. Geochemie und Petrographie der Magnesitlagerstätten des Alto Chapare, Bolivien. *Radex-Rundschau*, Radenthein, pp. 1105–1119.

Friedrich, O. M. 1963. Zur Genesis des Magnesites vom Kaswassergraben und über ein ähnliches Vorkommen (Diegrub) im Lammertel. *Radex-Rundschau*, Radenthein, pp. 421–432.

Friedrich, O. M. 1953. Erzlagerstättenkarte der Ostalpen. *Radex-Rundschau*, Radenthein, pp. 371–407.

Garan, M. J. 1954. Deposits of crystalline magnesite in the Satka area, South Urals. *Mineralogy of the Urals*, vol. 1, Zzd. Akad. Nauk SSSR.

Golding, L. Y., and M. R. Walter, 1979. Evidence of evaporite minerals in the Archean Black Flay beds, Kalgoorlie, Western Australia. *BMR Jour. Australian Geol. Geophys.* **4**:67–71.

Harben, P. W., and R. L. Bates. 1984. *Geology of the Nonmetallics*. New York: Metal Bulletin, Inc., 392p.

Harders, F., and S. Kienow. 1960. *Feuerfestkunde*. Berlin: Springer-Verlag.

Herzberg, W. 1976. Magnesite. In *Mineral Resources of the Republic of South Africa*, ed. C. B. Coetzee, pp. 385–388. Pretoria.

Höll, R., and A. Maucher, 1967. Genese und Alter der ScheelitMagnesit-Lagerstätte Tux. *Sitzungsber. Bayerische Akad. Wiss. Math. Nat. Kl.* **1**:1–11.

Höll, R., and A. Maucher. 1976. The strata-bound ore deposits in the eastern Alps. In *Handbook of Strata-bound and Stratiform Ore Deposits*, ed. K. H. Wolf, pp. 1–37. Amsterdam: Elsevier.

Ilič, M. 1952. *The Bela Stena Magnesite Deposit*. Beograd: Zbornik radova Rudarskog i geološkog fakulteta, vol. 1.

Ilič, M. 1968. Problems of the genesis and genetic classification of magnesite deposits. *Geol. Carpathica*, Bratislava **19**(1):149–160.

Irion, G., and G. Müller. 1968. Huntite, dolomite, magnesite and polyhalite of recent age from Tuz Gölü, Turkey. *Nature* **220**:1310–1311.

Johannes, W. 1970. Zur Entstehung von Magnesitvorkommen. *Neues Jahrb. Mineral. Abh.* **113**(3):274–325.

Kinsman, D. J. J. 1967. Huntite from a carbonate-evaporite environment. *Am. Mineral.* **52**:1332–1341.

Kužvart, M. 1981. Industrial minerals and rocks of Czechoslovakia. *Industrial Minerals* (March 1981):19–35.

Lehmann, B. 1979. *Schichtgebundene Sn-Lagerstätten in der Cordillera Real, Bolivien*. Berliner Geowissenschaftliche Abhandlungen, Berlin: Verlag D. Reimer, 136p.

Leitmeier, H. 1953. Die Entstehung der Spatmagnesite in den Ostalpen. *Tschermaks Mineralog. Petrog. Mitt.* **3**:305–331.

Lesko, I. 1960. Geologische und lagerstättenkundliche Untersuchungen im Raume Oberdorf a.d. Laming, Steiermark. *Ges. Geol. Bergbau Stud. Mitt.* **12**:3–65.

Lesko, I. 1972. Über die Bildung von Magnesit-Lagerstätten. *Mineralium Deposita* **7**:61–72.

Llarena, J. G. de. 1950. La magnesite (giobertite) de la haute vallée de l'Arga (Pyrénées de la Navarre espagnole et d'Urepel (Basses-Pyrénées). *Acad. Sci. Comptes Rendus* **18**(12):231.

Llarena, J. G. de. 1951. Die Lagerstätten des sedimentären Magnesites im Paläozoikum der Westpyrenäen. *Berg-und hüttenmänn. Monatsh.* **96**(11):221–227.

Llarena, J. G. de. 1953. Über die sedimentäre Entstehung des ostalpinen Magnesits "Typus Veitsch." *Montan-Zeitung* **69**:55–62.

Martin, C. 1960. Origin of crystalline magnesite deposits (abs.). *Geol. Soc. America Bull.* **71**:1921–1922.

Martiny, E., and I. Rojkovič. 1977. Trace elements in magnesites of Slovakia (Central West Carpathians). *Geol. Carpathica* **28**(2):311–322.

Morteani, G., P. Möller, and F. Schley. 1982. The rare earth content and the origin of sparry magnesite mineralizations of Tux-Lanersbach, Entachen Alm, Spiessnägel, and Hochfilzen, Austria, and the lacustrine magnesite deposits of Aiani-Kozani, Greece, and Bela Stena, Yugoslavia. *Econ. Geology* **77**:617–631.

Morteani, G., F. Schley, and P. Möller. 1983. On the formation of magnesite. In *Mineral Deposits of the Alps and of the Alpine Epoch in Europe*, ed. H. J. Schneider, pp. 105–116. Berlin: Springer-Verlag.

Mostler, H. 1973. Alter und Genese ostalpiner Spatmagnesite unter besonderer Berücksichtigung der Magnesitlagerstätten im Westabschnitt der nördlichen Grauwackenzone (Tirol, Salzburg). *Veröffentlichungen Univ. Innsbruck* **86**:237–266.

Muir, M. D. 1979. A sabkha model for the deposition of part of the Proterozoic McArthur Group of the Northern Territory, and its implications for mineralization. *BMR Jour. Australian Geol. Geophys.* **4**:149–162.

Ney, P. 1956. Zum gegenwärtigen Stand des Magnesitproblems. *Zeitschr. Deutsche Geol. Gesell.* **108**:203–220.

Niedermayr, G., A. Beran, and E. Scheriau-Niedermayr. 1983. Magnesites in Permian and Skythian Series of the Eastern Alps, Austria, and its petrogenetic significance. In *Mineral Deposits of the Alps and of the Alpine Epoch in Europe*. ed. H. J. Schneider, pp. 97–104. Berlin: Springer-Verlag.

Nishihara, H. 1956. Origin of the bedded magnesite deposits of Manchuria. *Econ. Geology* **51**(7):698–711.

Perepelitsyn, V. A., S. M. Perepelitsyna, K. V. Simonov, and L. D. Bocharov. 1971. *Magnesite and Surrounding Rocks in the New Deposits of the Satinsk Group*. East Institute of Refractories, Magnezit Factory, Ogneupory, No. 6, June 1971, pp. 36–41.

Perthuisot, J. P. 1971. Présence de magnésite et de huntite dans la sabka el Melah de Zarziz. *Acad. Sci. Comptes Rendus* **272D**:185–188.

Petrascheck, W. E. 1932. Die Magnesite und Siderite der Alpen. Vergleichende Lagerstättenstudien. *Sitzungsber. Wien Akad. Wiss. Math. Nat. Kl.* **141**(1):195–242.

Petrascheck, W. E. 1963. Die Meerschaum-Lagerstätten bei Eskisehir. *Mineral Res. Expl. Inst. Turkey Bull.* (MTA), Ankara **61**:11–14.

Petrascheck, W. E. 1972. Beziehungen zwischen krypto-kristallinem und spätigem Magnesit. *Radex-Rundschau*, Radenthein **5**:339–350.

Petrascheck, W. E. 1962. Neue Beobachtungen an griechischen und türkischen Magnesitlagerstätten. *Radex-Rundschau*, Radenthein **6**:303–310.

Redlich, K. A. 1909. Die Typen der Magnesitlagerstätten. *Zeitschr. prakt. Geol.* **17**:300–310.

Rowlands, N. J. 1974. The gitology of some Adelaidean stratiform copper occurrences. In *Gisements Stratiformes et Provinces Cuproiferes*, ed. P. Bartholomey, pp. 419–427. Geological Society of Belgium.
Sawkins, E. J. 1976. Metal deposits related to intracontinental hotspot and rifting environments. *Jour. Geology* **84**:653–671.
Schilling, J. H. 1968. The Gabbs magnesite-brucite deposit, Nye County, Nevada. *Ore Deposits of the United States, 1933–1967*, vol. 2, cited in J. D. Ridge, pp. 1607–1622. New York: American Institute of Mining and Metallurgical Engineers.
Schulz, O. 1972*a*. Horizontgebundene Altpaläozoische Kupferkiesvererzung in der Nordtiroler Grauwackenzone, Österreich. *Tschermaks Mineralog. Petrog. Mitt.* **17**:1–18.
Schultz, O. 1972*b*. Neue Ergebnisse über die Entstehung paläozoischer Erzlagerstätten am Beispiel der Nordtiroler Grauwackenzone. *International Symposium on Mineral Deposits of the Alps*, Ljubljana, vol. 15, pp. 125–140.
Siegl, W. 1955. Zur Entstehung schichtiger und strahliger Spatmagnesite. *Berg- und Hüttenmänn Monatsh.* **100**:79–84.
Ťapák, M. 1967, cit. in Slávik, J., 1967.
Valdiya, K. S. 1968. Origin of magnesite deposits of South Pithoragarh, Kumaun Himalaya, India. *Econ. Geology* **63**(8):924–934.
Vitaliano, C. J. et al. 1957. *Geology of Gabbs and Vicinity, Nye County, Nevada*. U.S. Geol. Surv. Mineral. Inv. Field Study, MF-52, 1:24.000.
Walenta, K., and P. J. Dunn. 1979. Ferridravite, a new mineral of the tourmaline group from Bolivia. *Am. Mineral.* **64**:945–948.
Wetzenstein, W. 1974*a*. Sedimentpetrographische Untersuchungen an limnischen Magnesit-Huntitlagerstätten im Plio-Pleistozän des Serviabeckens, Nordgriechenland. *Neues Jahrb. Geol. Paläontol. Monatsh.* **10**:625–642.
Wetzenstein, W. 1974*b*. Limnische Huntit-Hydromagnesit-Magnesit-Lagerstätten in Mazedonien, Nordgriechenland. *Mineralium Deposita* **10**:129–140.
Wetzenstein, W. 1983. *Estado actuàl de los trabajos ejecutados en los yacimientos de asbesto y magnesita del Alto Chapare, Departamento de Cochabamba, Bolivia*. La Paz: Int. Inf. Grupo de Cooperación Minera (GCM-GTZ), 27p.
Wetzenstein, W., and D. Zachmann. 1977. Sedimentäre magnesium-karbonatische Bildungen im Servia-Becken, Nordgriechenland. *Radex-Rundschau*, Radenthein **1**:29–49.
Wicken, O. M., and L. R. Duncan. 1983. Magnesite and related minerals. In *Industrial Minerals and Rocks*, vol. 2, cited in S. J. Lefond, pp. 723–1446. New York: American Institute of Mining, Metallurgical, and Petroleum Engineers.

13

Some Stratabound–Stratiform Fluorite Ore Fields

Horacio Magliola-Mundet

In this review of stratabound deposits, which are fundamentally emplaced in sedimentary environments, a relatively neutral interpretation of the possible source of fluorspar mineralization has been followed.

Since Blondel's initial concept of ore deposit classification by types (1942, 1951), geologists have tried to construct approaches on a descriptive basis that, free of genetic dogmatism, allow them to evaluate the importance of the geologic context where deposits are located as a key to interpretation and prediction.

In the case of fluorite, a positive advance is evident in an integral study of the fluorite geology and resources in the United States (Shawe, 1976). The study concentrates on the wide variety of geologic environments, the diversity of its presentation, and the adoption as classification of deposits that in which they are associated with different rock types. In this way, in the sedimentary and volcanic rocks, mantos* are included, leaving aside the inveterate expression "bedding replacement," of deep epigenetic connotation. As pointed out by King (1967), stratiform mineral emplacement of this nature obtains a spatial and chemical omnipotence, which in an a priori way would even eliminate the additional material and the synchroneity of mineralization with host

*_Manto_ is a Spanish term that passed to English and French with the meaning of tabular mineralized body, usually in near horizontal attitude and confined to a determined stratigraphical horizon. According to Amstutz (1971) in some Latin American countries it is also applied to any flattened deposit, stratum, or mineral in an irregular plain mass.

rock. An unquestionable fact of observation is the frequent coincidence of the similar fluorite concentrations in veins, mantos, pipes, and breccias. This has usually been interpreted as a metallogenic bond, in spite of injuring the condition of mineralized stratigraphic confinement. Consequently, it would be unreasonable to make morphological discrimination but rather to look for the integration of heterotypism of the concentrations within the same fluoritic domain.

On the other hand, as a reflex of the tendency of fluorine to remain within a certain region (Peters, 1958), although subjected to different crustal contingencies (Lindgren pointed out the character of "persistent" element), by successive heritages, fluorite appears in deposits of varying morphology at various times.

If this elemental approach is followed for the sedimentary environments (Selley, 1970), one should observe first, the natural context, geometry of materials, and geologic age, and then, with the structural paleogeographic and sedimentologic interpretation, establish a frame of reference for the environmental evolution (diagenesis, volcanic-plutonism, metamorphism, pedogenesis, etc.). From here one recognizes the preferential position of stratabound fluorite at the margins of sedimentary basins and over epicontinental platforms and its talus, both in the base of transgressive series and within the carbonatic series.

Routhier (1980) proposed a unifying regional scheme that recognizes (from the mining potential) a zone of preconcentration or deep connections—the so-called primordial metallotect—the superposition or intersection of other metallotects that act as "revealers" of the potential to form accumulations.

In the case of fluorine in which geochemical behavior marks two poles in its geologic relationships, by magmatic differentiation or biochemical accumulation (phosphates contain important amounts) in the convergence point of a lineamentist–platist balance, abundance is noted near and within taphrogenic environments (taphro = trench). Routhier developed this idea for Europe, giving a submeridian control originally linked to the late Hercynian tension of the Atlantic opening. Similar treatment of the evolution of the African continent can link it to a South American area, apparently unconnected in an exclusively continental context. Perhaps a great South African region covered by the Karoo and Cape subgroups (Anhaeusser and Button, 1976) acts as a screen that separates the Transvaal from the North Patagonia region in South America.

DEPOSITS OF LACUSTRINE ENVIRONMENT

Fluorite has long been known to occur in old sedimentary rocks, but since the late 1960s it has been recognized as an important rock-forming constituent in certain Cenozoic lacustrine rocks (Sheppard and Mumpton, 1984). At all the following localities the fluorite has been interpreted to be either a primary precipitate or a diagenetic product in rocks that were not subjected to hydrothermal alteration.

LATIUM REGION (ITALY)

Since the beginning of this century (Clerici, 1920), the presence of fluorspar and barite in Quaternary sediments has been known surrounding the city of Rome (toward the north following the via Cassia). However, many decades passed before this mineralogical observation inspired mining investigations. From the 1970s, the extractive activity found this type of fluorite accumulation in the following places (Fig. 13–1): (1) south of Lake Bracciano, Monte Loreto, and Pianciano in Castel Giuliano; (2) between Lake Bracciano and Civitavecchia in La Tolfa; and (3) northwest of Viterbo and west of Lake Bolsena in Valentano.

The geologic setting is extremely complex. Conspiring against its comprehension are the soft geomorphology, the dense vegetation-covered surfaces (the Mediterranean ''maccia''), and the modern volcanic cover. On one side, the regional map delimited these formations and others of sedimentary origin (Sheets Civitavecchia and Bracciano, published in 1888), but up to 1949 the Jurassic limestone near Sasso had not been described (Segre, 1950) nor the nappe structures recognized (Merla, 1951).

In the Monte della Tolfa area (Fazzini et al., 1972) are various units of an

Figure 13–1. Latium region location (Italy).

autochthonous basal complex, mainly formed by marine limestones (Toscana group), overlain by other allochthonous formations (1st outer: Tolfetano Flysch series and 2nd inner: Pietraforte series). Finally, the neoallochthonous is superposed, generally through an unconformity, represented by two Pliocene transgressive series where the first volcanic manifestation appears. During the Quaternary, cinerites and lavas (Tolfetano–Cerite complex) and the fluviolacustrine, palustrine, and beach sediments were deposited. In the basal complex, only outcropping over a few square kilometers at Monte delle Fate, fluorspar was recognized for the first time and mined (Vighi, 1951).

In general, the Quaternary fluorspar manto is one individual layer or horizon separated by barren intercalations (diatomites, paleosoils including carbonaceous matter, or pyroclastic sediments). The pyroclastics, where present, form the footwall or the cover of the deposits. Therefore many of the first signs of mineralization were produced in a casual way when roads and waterwells were made. Barren cover thickness is important since it seems impossible to mine this material except by open pit. The undisturbed structure and the manto thickness (up to 30 m at Valentano) give high tonnages in the surface unit. Consequently, a hundred million tons of ore with grades over 15 percent CaF_2 exist in this region.

The fluorspar ore is soft and friable. It is present in two different varieties: The first, known as *sabbioso* (sandy) is a fine-grained clastic; the other, grayish-brownish pellets, known as *argilloso* ore (argillaceous) may be confused at first glance with the diatomites also present in the zone. At Pianciano (Spada, 1969), the sandy ore forms discontinuous lenses of some meters thickness at the mineralized manto base, and the contact with the overlying argillaceous ore is sharp. When the sandy ore is absent, the argillaceous ore regularly and extensively coincides with the pyroclastic footwall. In the footwall, the ore presents irregular erosion surfaces over which the cineritic products were deposited.

Changes in the nature and relationships among sediments give rise to other problems about the evolution of the lacustrine basin itself, its climatic conditions, and contemporary volcanic episodes. Nevertheless, there is an evident contrast between the environment in which the important quantities of fluorite could have accumulated and the environment of the organic formations. The ecological requirements of diatoms, thriving in fresh and clean waters, is positively correlated to the silica availability coming from the volcanic activity. The lack of detrital minerals suggests that the mineral contribution may come from springs within the basin.

Other major mineralogic components are barite, calcite, and apatite. Other accessories are present as chalcedony-type silica, ferruginous ochres, and some clays, mainly halloysite. According to a mineralogic study (Carta et al., 1975), fluorite appears predominantly in colloidal state, less commonly cryptocrystalline and microcrystalline. Calcite is always crystalline. In the Pianciano ore, Carta et al. (1975) observed that all the varieties classified have about 80 percent weight in grains of less than 5 μm diameter. Table 13–1 lists percentages of common ores corresponding to different horizons.

According to other investigations (Spada, 1969), small proportions of biotite, augite, chlorite, and feldspar always exist with crystalline forms in the major granulometric portion. Spada, analyzing the physical properties of the

Table 13–1. Mineralogical Percentages in the Pianciano Fluorspar Ores

Ore	*White*	*Gray*	*Brown*	*Sandy*
Fluorite	70.02	66.92	28.14	15.70
Barite (partly baritocelestite)	16.88	11.82	9.40	3.71
Apatite	3.94	6.46	1.87	1.39
Clays	5.16	8.95	49.95	2.55
Calcite	3.85	3.11	8.99	76.20

various minerals, classified them as products of chemical deposition (fluorspar, barite, apatite, opal, and chalcedony), whereas others would be clastic even pyroclastic, rejecting Cavinato's opinion (1968), which recognized only one detrital accumulation of all minerals. Consequently, the first hypothesis postulated an exhalative-sedimentary lacustrine emplacement (Matteucci, in Zucchetti, 1975).

Interest in this fluorspar was neutralized because of the extremely fine granulometry of its ores: due to the lack of an appropriate treatment method, they could not be economically mined by conventional techniques. Neither flotation nor other tried procedures have the expected results. A preconcentration system was invented by the French Bureau de Recherches Géologiques et Minière (B.R.G.M.), based on extraction with hydrocarbons followed by other metallurgical refinements applied in the United States and Italy, resulting in concentrates (85 percent CaF_2 pellets) with a competitive cost.

PIANCIANO

The deposit is owned by Soricom SpA, a subsidiary of the Australian company Southland Mining Ltd. Despite many delays, the processing problems have been overcome, but the project is inactive at present. The clayey ore is extremely fine grained with 70 percent less than 2 μm in size and an average fluorite content of 55.69 percent. The sandy ore consists predominantly of calcite in the 40–100 μm size range and the fluorite content of 20.74 percent is present in fine form. The reserves in the mining property's 2042 hectares are 9.43 million tons of sandy ore and 5.83 of clayey ore. Restricting operations to only 272 hectares with a predominance of clayey ore, it is estimated that 7.78 million tons with a grade of 43.88 percent CaF_2 can be mined, which means 16 years of exploitation at 200,000 tons per annum of metallurgical-grade concentrates (metaspar).

MONTE LORETO

In a sampled zone of approximately 200 hectares, about 4 million tons were estimated in 1972. The seam has a 1.8 m average thickness and the cover a maximum of 5 m. The grades are between 45 percent and 50 percent CaF_2, 10 percent $BaSO_4$, between 0.5 percent and 1 percent phosphorous, with strong variations in the silica content (6 percent to 38 percent).

VALENTANO

At the Acquaforte mining concession (800 hectares), the tonnage estimated by drilling, up to a depth of 65 m and pits up to 14 m, totals about 80 million tons as follows:

18.0 million tons with less than 10 percent CaF_2,
57.0 million tons with between 10 percent and 30 percent CaF_2,
1.0 million tons with between 30 percent and 60 percent CaF_2,
0.9 million tons with between 60 percent and 70 percent CaF_2, and
1.0 million tons with more than 70 percent CaF_2.

ROME, OREGON (U.S.A.)

Fluorite was first identified near Rome, Oregon, in 1967 (Studer, 1967) when Tertiary zeolitic tuffs were being studied. According to Sheppard and Gude's description (1969), the mineralized layer forms part of the top of a lacustrine fluviatile sedimentary sequence of Pliocene age "not subjected to hydrothermal activity." The chemical analysis and X-ray on 67 samples from 7 profiles indicate that practically all the fluorine present is contained as fluorite, which in its great part is built up of small granules of less than 1 μm in diameter, irregularly distributed. The key horizon, the indicator of the mineralization, is a zeolitic tuff that can be subdivided into three units: the upper gray and friable, with about 5 percent CaF_2; the middle yellow, practically barren; and the lower gray, about 5 m thick with grades of 2.5–5 percent up to 16.4 percent CaF_2. The integrated thickness is 7.5 m. At the bottom is a tuffaceous mudstone, slightly mineralized. According to Sheppard and Gude (1969), zeolites, potassic feldspar, and at least some clay minerals, including the fluorite itself, were formed during diagenesis. The fluorspar abundance in this manto could be a consequence of an exceptional concentration of fluorine in the lake water, which constitutes a strongly saline depositional environment with alkaline content, judging by the present molds of dissolved salts and silicate association. In a 7–8 km elongate north-trending zone, a conservative estimate is 12 million tons of fluorspar. Fluorspar along with associated zeolites are only a potential resource, but if the zeolites could be extracted the fluorspar could perhaps be recovered.

OTHER LOCATIONS

In recent years findings of fluorite have been mainly associated with zeolitic lacustrine sediments, generally as a consequence of X-ray studies. Such are the cases of the Eastgate (Nevada) zeolitic deposits; the Gila (New Mexico) conglomerate; and also the Lake Magadi (Kenya) Quaternary sediments (Sheppard and Gude, 1980). The latter has an important stratiform concentration of trona (Na_2CO_3), which locally contains up to 22 percent of villiaumite (NaF) and 6 percent fluorite. In the other cases, there are close analogies of the mineralized setting with that described at Rome, Oregon.

A detail emphasized by Sheppard and Gude (1980) as indicative of an alkaline environment deposit is the chert nodule layer. In Lake Magadi, the silica is dense and homogeneous, light gray in lobulated form, and with a thin reticulated white crust surface.

When the water and sediments of the Lake Kivu (2100 km^2 in Ruanda and Zaire) were investigated, they proved an exceptional environment (Degens and Kulbicki, 1973). This lake is in the western branch of the African Rift System. With respect to fluorine, it represents 1.3 million tons accumulation in the water and, for the sediments deposited during the last 5000 years, about 2.1 million tons. In this case, a coincidence was established between the volcanic and hydrothermal activity periods and the climatic epochs of greatest rainfall. The lacustrine mineralization depended on the load carried by the meteoric water through the surrounding volcanic terrain. Other saline African lakes, including Magadi, have very high fluorine values, a fact that leads one to suspect (Van Alstine and Schruben, 1980) that new fluorspar concentrations could be found, especially in environments of Cenozoic lacustrine sediments near the rift zones.

DEPOSITS AT THE BASE OF SERIES TRANSGRESSIVE OVER THE BASEMENT: CASES OF THE MORVAN AND THE VOSGES (FRANCE)

Widespread stratabound fluorite with accessory barite mineralization conforms to the Mesozoic sedimentary series transgressive over the Hercynian basement of France. In the Morvan region, the Liassic or Infraliassic cover is superposed on the basement, at the southeastern edge of the Paris basin. Here the basement was uplifted as a tectonic horst now forming mountains.

This stratabound fluorite concentration was the first described in this country (Soule de Lafont, 1967). Fluorite of the so-called silicified levels was known in the old lead and zinc mines, as a mineralogic curiosity, and its great economic importance was not even suspected.

The five deposits demonstrated to date (Soule de Lafont and Lhegu, 1980), have the following estimates (CaF_2 tonnages):

Pierre Perthuis, 1.3 million tons over 50 hectares (Fig. 13–2);
Pontaubert, 0.5 million tons over 30 hectares (Fig. 13–2);
Marigny-sur-Yonne, 0.5 million tons over 20 hectares;
Courcelles-Fremoy, 1.0 million tons over 50 hectares; and
Antully, 0.9 million tons.

Other areas are being investigated, therefore, the mineralized areas contain more than 5 million tons of fluorite in ores with grades between 32–40 percent CaF_2, which can easily be mined by open-pit methods. At the present, these layers constitute the main French reserves, which in 1980 were estimated to total 7 million tons (Lhegu and Touray, 1980).

From the paleogeographic point of view, at the beginning of the Mesozoic

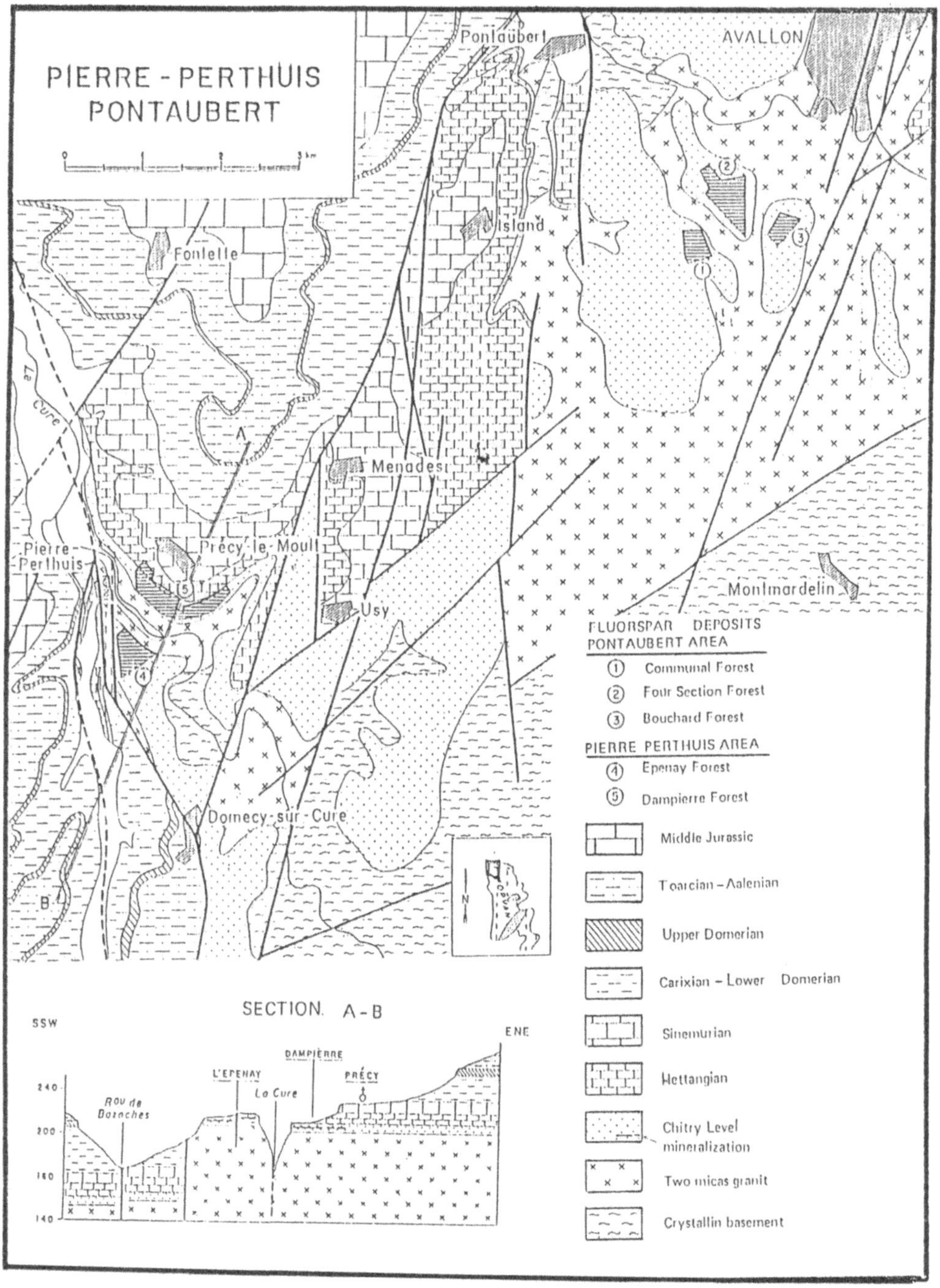

Figure 13–2. Geologic map of Pierre-Perthuis and Pontaubert areas (Morvan, France; after Soule de Lafont and Lhegu, 1980).

the Morvan was an emerged mass progressively invaded by the sea. When the basal dolomitic rocks at the northern and western edges were investigated (Bois et al., 1980), their relationship with the basement was clarified by more than 200 drill holes. The materials present, partially fossiliferous (algae), correlate with the zones of main subsidence; its lower passage is progressive and its geographic distribution appears related to the limits of transgression. Frequent subsidence structures are related to the disappearance of old evaporites, similar to the rare dolomites that have replaced gypsum in the Peruvian Gulf coast sabkhas.

The fluorite concentrations are always associated with silification of host horizons. This phenomena, more or less pronounced, appears to be developed through phases of cementation and corrosion, especially of feldspars, always near the contact of basement and cover.

Besides fluorite and barite, galena and sphalerite are present, sometimes abundantly and, as accessory minerals, pyrite and chalcopyrite. Three main types of mineralized layers can be distinguished, first directly over the basement:

1. A limestone–dolomite horizon, frequently breccia type, developed in the western part of the region, designated as the *Chitry* layer;
2. Some fine sandstone, lightly feldspathic, cemented by carbonates, in the eastern part; and
3. Some calcarenites, to the north.

The deposit mineralogy is the same in the three cases; likewise, there is not appreciable change with the variation of the basement composition. Nevertheless the origin of mineralization must be looked for in the basement and in certain specific horizons, perhaps detrital of the transgressive series. Mobilization could be related to the paleohydrography of the basin edges, without taking into account the agency of the juvenile waters. The age of the mineralization process could have an upper limit, so in the Chitry layer overlying formations have been recognized with boulders of the first, including fluorite and silification.

As prospecting guides, Soule de Lafont (1967) indicates that the outcrop alteration has caused the disappearance of the fluorite up to some centimeters deep. The best indication is given by siliceous boxworks, sometimes with vugs having quartz, fluorite, and barite crystals. At the surface the dissolution of small lenses of carbonaceous concentrations have created karstic openings.

The paleotopographic study of the basement and calcic fluoride accumulations (product of width times grade) shows preferential distribution of ore in certain paleochannels or paleobasins. In detail, the mineralization is peneconcordant.

Soule de Lafont (1967) states that the doubling of the number of drill holes, passing from a grid of 50 m to 35 m does not represent an increase in the accuracy of the significant mineral quantification.

In conclusion, the geologic environmental features would be the following (Soule de Lafont and Lhégu, 1980):

- Location at the paleogeographic basin–continent limit forming the basal part of a transgressive series over the basement;

- The mineralization and silicification are included in a small stratigraphic width along some 200 km around Morvan;
- The bearing levels show different lithologies—sandstones and carbonatic rocks—but clearly distinctive from the intercalated and overlying barren formations, which are mainly marl. Consequently, those levels are located between two impervious entities, the basement and the barren formations.

As appendix to this case it is convenient to remember that in the Vosges, Lougnon (1974, 1980) described a metallogenic anomaly with an estimated fluorite content of more than 3 million tons. There is a big tonnage of low-grade content in this stratabound deposit (70 million tons with 4–5 percent CaF_2), nearly 2 m thick, spread over an area of about 14 km^2 within the Saint-Dié basin. The deposit was discovered by alluvial prospecting, followed by hydrogeochemistry.

The Upper Permian geologic context of this northern part of the Vosges is entirely continental although with lacustrine episodes. The sedimentary environments are located over a relatively irregular pre-Permian surface, modified by various infra- and intra-Permian volcanic episodes.

In this case, the metallotect is of paleogeographic type, a paleorelief product of a basement within which the modification of the sedimentary sequence is given by detritic transgressive accumulations cemented by carbonates where dolomitic lenses with red siliceous nodules are individualized in particular horizons.

The fluorite, generally purple, is in a multilenticular position, basal to the deleterious syndiagenetic dolomitic rocks and also to the microfissured chert crossed by stylolites. In detail, the mineral distribution is irregular and its grain has a great variation (average 50 μm). Some barite, iron oxides, calcite, and so forth are also present.

The circumstances that fluorite is confined vertically to a unique carbonatic horizon, although there are others of analogous facies in the vicinity, and the complete lack of vein mineralization led Lougnon (1974) to suggest that these concentrations are responses to a late diagenetic process and that the fluorine was introduced by thermomineral water linked to the Permian acid volcanism.

NORTHERN MEXICO FLUORSPAR MANTOS

North of the twenty-eighth parallel in Coahuila state are a dozen fluorite districts. Those that stand out are Pico Etéreo, La Encantada–Buena Vista, and El Tule. Of lesser rank are La Mariposa, San Manuel, San Vicente, and Santo Domingo. In all except the first, stratabound concentrations prevail, but without exception they have an analogous stratigraphic position (Fig. 13-3C).

This region is constituted of moderately folded Mesozoic sedimentary formations. It extends over 200 km from the U.S. boundary to Melchor Múzquiz City and is limited in the east by the Serranía del Burro and in the west by the

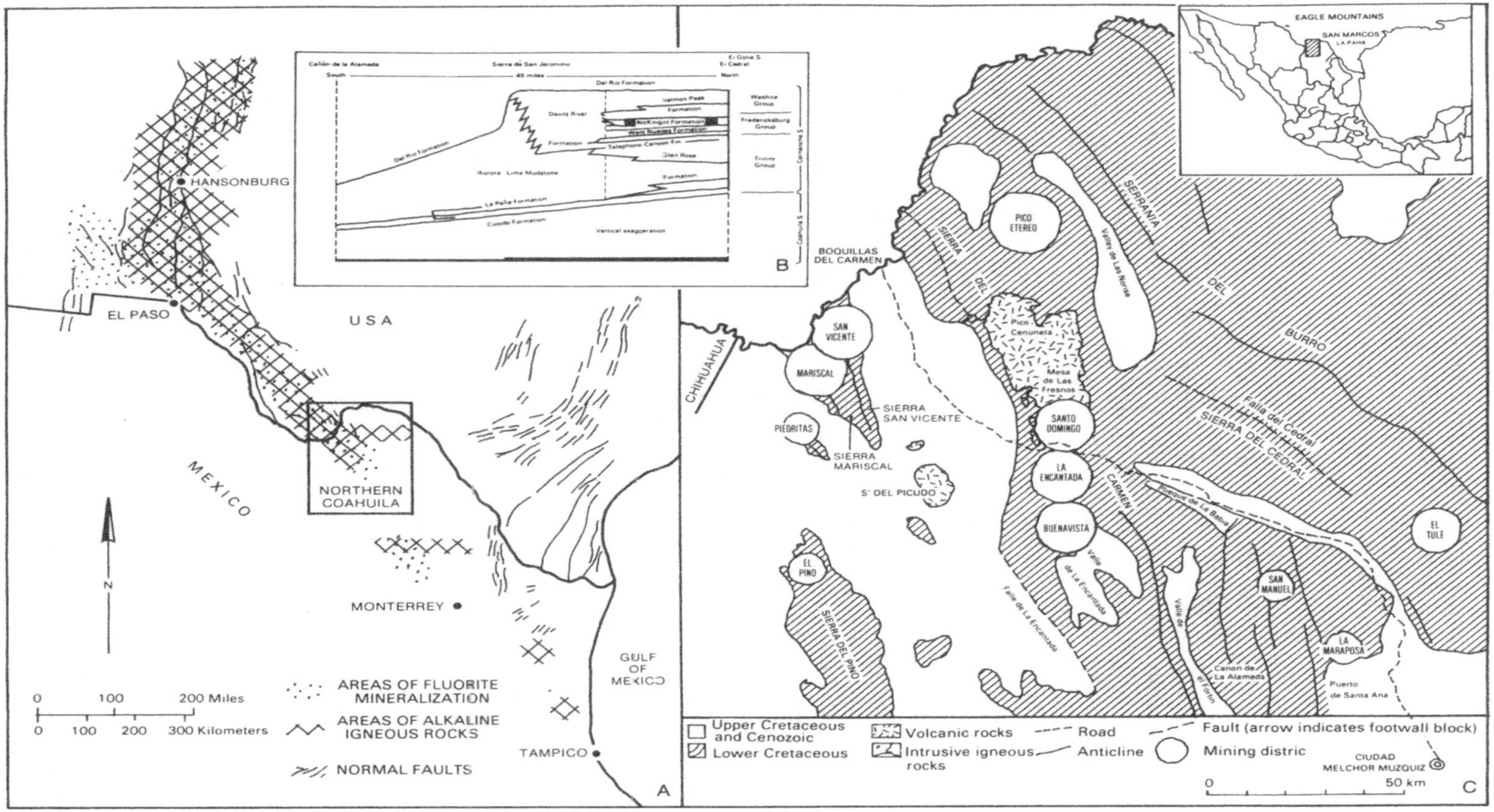

Figure 13–3. Northern Mexico fluorspar mantos. (A) Distribution of fluorite deposits, alkaline igneous rocks and extensional faults (after Kesler, 1977, p. 217). (B) Stratigraphic column of Lower Cretaceous (after Smith, 1966). (C) Fluorite districts of Coahuila State (after Magliola-Muncet, 1972*a*).

desert plains of Chihuahua state. In the middle of the area, over the Mesa de los Fresnos, effusive rocks cover the Cretaceous sediments. Between the Serranía del Burro and the Sierra del Carmen, in the Pico Etéreo district, there are several asymmetrical domes attributed to laccolith intrusions with a group of ring dikes.

The stratigraphic column (Fig. 13–3B) includes over 1000 m of sedimentary rocks, which can be grouped into three series (Smith, 1970): Coahuila and Comanche from the Lower Cretaceous and El Golfo from the Upper Cretaceous. The El Golfo does not appear related to the fluoritic mantos and for this reason is not detailed here. The Coahuila series has sediments from the Mexican geosyncline, when the Coahuila Peninsula existed as emerged ground. First, they were originated by erosion of fine terrigenous basement sediments (shales). Then a marine transgression deposited carbonates—limestones and dolomites—in littoral waters.

The Comanche series started afterwards, also with terrigenous clastic materials (La Pena formation), presumably brought from an emerged region situated to the north. Immediate subsidence, caused by the differences between the speed of the submersion and the consequent formation of barrier reefs, gave a varied sedimentary complex collectively called Limestone Aurora formation in this region and south Texas. In it, there is a lateral equivalence between the McKnight formation from the eastern area, which contains evaporites because of the restriction in the water's renewal caused by the marginal reefs, and the Devils River formation formed by organisms.

Tectonic uplift followed a period of subaerial erosion marked just below the limestone's top, called generically by the miners "floor limestones." On this contact, terrigenous clays, which sometimes have pyrite nodules, fossils, and some undulate beds of clean fine sands, were deposited in disconformity, particularly in the basins. The unconformity surface is irregular, reddish, and perforated by organisms. The sediments, supposedly coming from the north and west, put an end to the favorable conditions for the development of the building organisms. In the same way, they were infiltrated into the dissolution cavities and channels (paleokarst). These materials, designated by the miners as "lamajo," belong to the Del Rio formation that has a wide regional distribution and is closely related to the fluoritic mineralization. This formation is constant for 350 km in the northwest (Eagle Mountains and Trans Pecos, Texas) and for 200 km in the south (Sierras de San Marcos and La Paila, in the meridional part of Coahuila; Fig. 13–3A).

Finally, the Comanche series finishes with massive limestones of porcelainous aspect belonging to the Buda formation (because of the deformation of the Spanish term *viuda*). This last is the result of abrupt subsidence and delineates the end of Lower Cretaceous sedimentation.

The igneous activity is represented by lava flows, necks, dikes, sills, laccoliths, and so forth that cut the Upper Cretaceous sediments. Thus, their age is Tertiary but relative dating of the several representatives is difficult. In any event, the intrusives and the volcanic manifestations, including the pyroclastic ones, took place after the Laramide orogeny, which flooded the sediments. Older igneous rocks exist in the basement where the sedimentary basins were formed at the end of the Jurassic and Cretaceous, particularly, in the old Coahuila Peninsula. From the petrologic point of view, the Tertiary domes of the Pico Etéreo district and the less spectacular ones from the Sierra del Carmen, in the south, are intrusive complexes of essentially sodic rocks that extend the alkaline rock belt to the south, east of the Rocky Mountains, as Lindgren (1933) described them almost a half century ago.

Summarizing, all the fluoritic stratabound deposits are confined to the terminal Lower Cretaceous. This same location is also known along a belt of about 700 km from western Texas to the south of Coahuila State. Credibly, the mineralizations are related with the intra-Cretaceous erosional unconformity that, essentially, forms a regional metallotect (Magliola-Mundet, 1972*a*, 1972*b*). In detail, other subordinated factors, such as those dependent on paleogeography, diagenetic evolution, Tertiary igneous activity, paleohydrological system, played different roles in mineral concentration. When beds are present, where the fluorite is included in the Del Río formation, the corresponding stratigraphic thickness in the districts of La Encantada-Buena Vista, El Tule, La Mariposa, and so forth has 20 m as a maximum and is frequently less than 12 m.

The Pico Etéreo district deposits are spatially related to Tertiary igneous rocks, particularly rhyolites. It presents the only known case where some great mineralized bodies of inverted conical or tabular form are discordant. Mining at Aguachile, Cuatro Palmas, Malabrigo, and so forth has demonstrated that they do not continue more than a hundred meters below the contact of the Del Río formation and the infralayer limestones.

In all these deposits, the dominant mineral is fluorite, generally with calcite and less quartz and chalcedony. For this reason it is an excellent ore for acid concentrates. In the El Tule district where the beds overlie the McKnight formation, which bears evaporites, the first bed has an important quantity of celestite. This strontium mineral is interpreted as a product derived from saline materials. Other components present in small proportions are: iron oxides, gypsum, barite, clays, organic matter, and dolomite. The Aguachile deposit also has a beryllium mineral, bertrandite (Levinson, 1962).

The fluorite presents great color and textural variations. It is of high grade and crystallized, also impure and fine or microcrystalline, of earthy appearance and brown or purple in color. Also, it is frequently present in mineral concretions and ribbon banding, mammillary and botryoidal and other forms. Small stalactite concentrations in Cuatro Palmas mine are formed in open spaces. Fluid inclusions with hydrocarbons have had interesting geochemical and thermometric studies (Kesler, 1977; Kesler et al., 1983) suggesting mineralization transitional with that in New Mexico (Roedder et al., 1968).

Van Alstine (1976) noted morphostructural relations that may associate high-angle deep faults with the upward leakage of large volumes of fluorine in the carbonatic rocks of this region, and thus renews the early opinion that attributes this fluoritic accumulation directly and exclusively to the Tertiary magmatism as the cause of the metasomatic substitution of favorable beds. The adjustment of this point of view to the lithostratigraphic peculiarities is inadequate and the regional and local paleogeographic knowledge is insufficient to obtain definite conclusions.

The most pragmatic application of the geologic evidence, shallow pitting

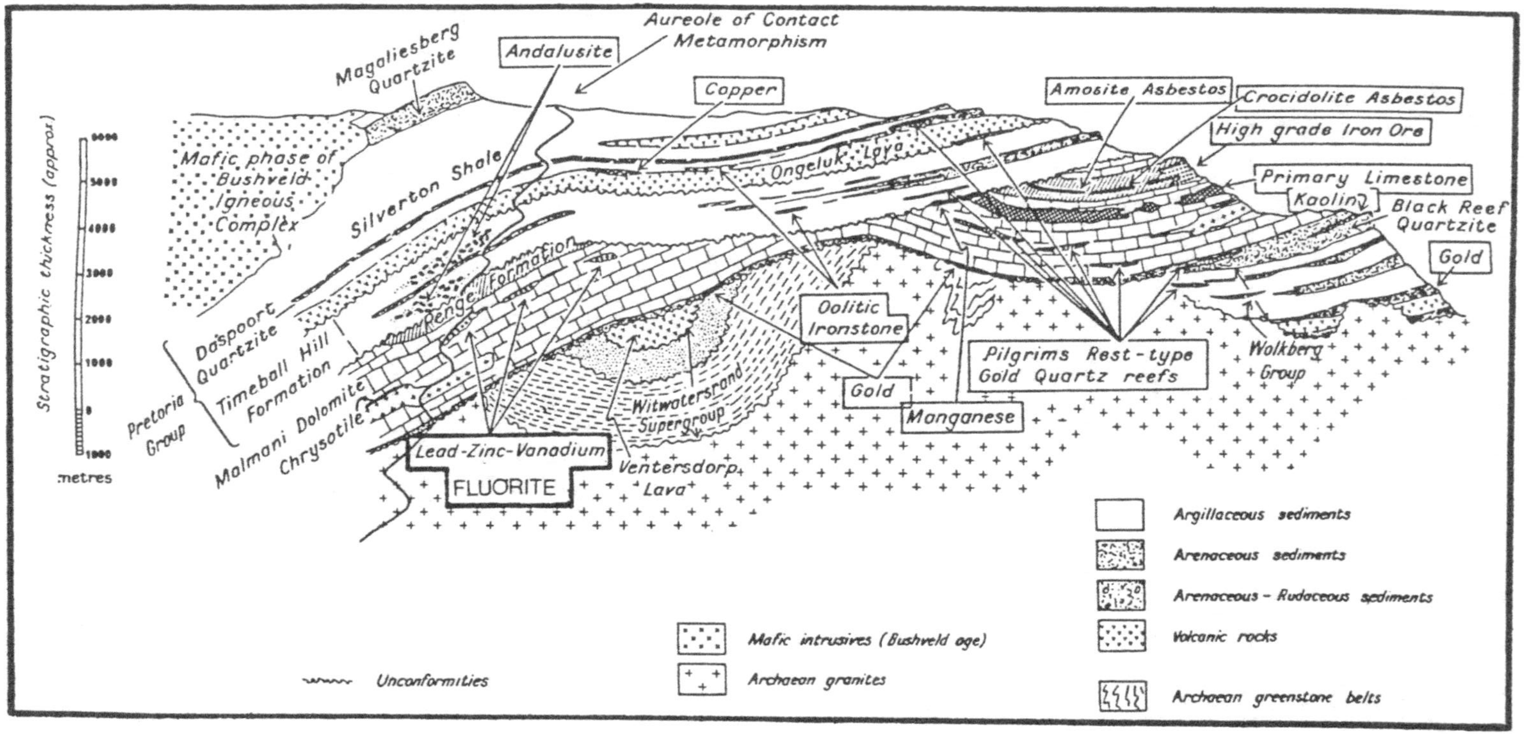

Figure 13–4. Stratigraphic position of fluorite deposits in the Transvaal basin (after Anhaeusser and Button, 1976, p. 276).

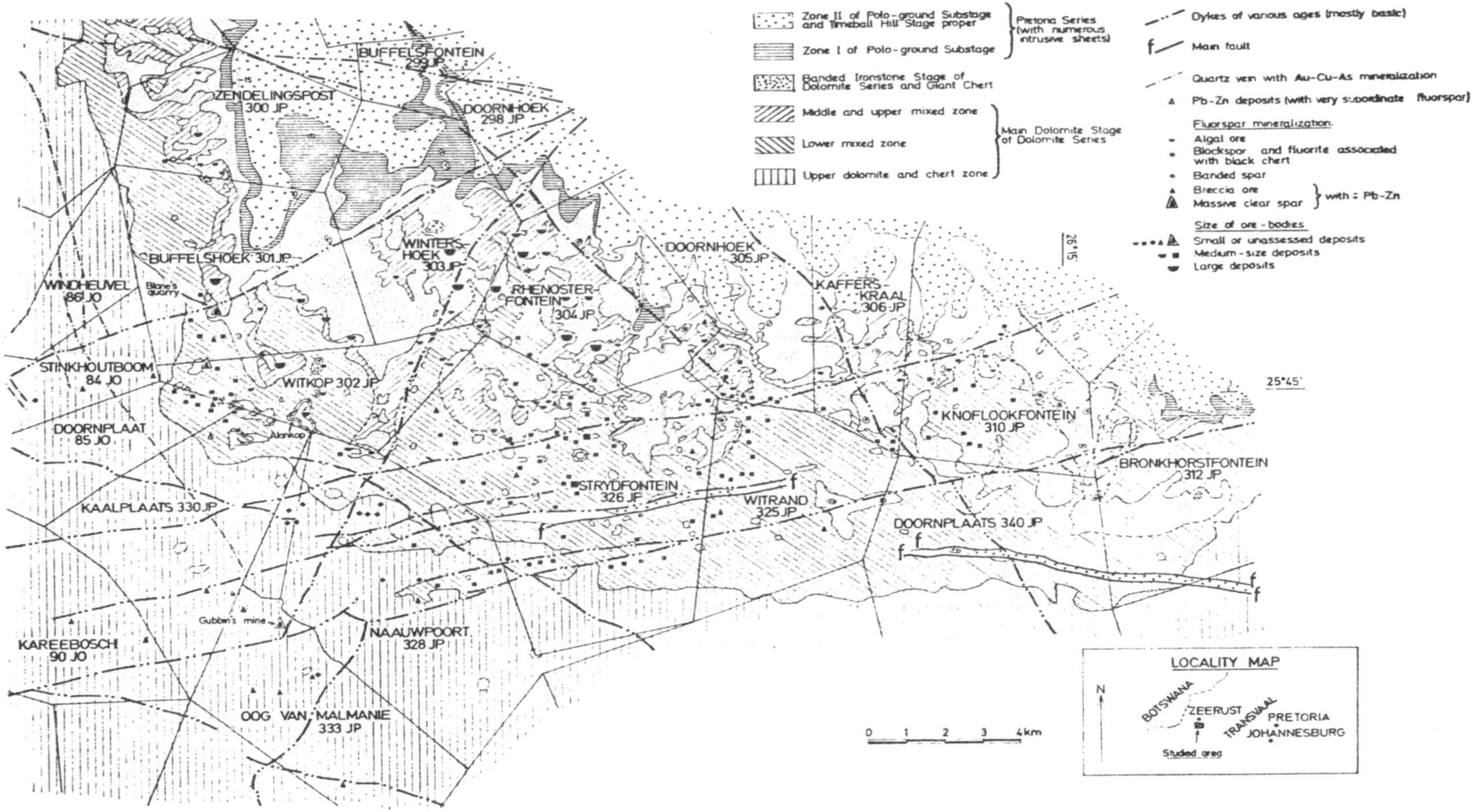

Figure 13–5. Geologic map of the Marico district (Transvaal, South Africa; from Martini, 1976, p. 627).

and trenching guided by the contrasting outcrops of Del Río formation over the base limestones, led to the discovery and development of the greatest part of these rich concentrations of fluorspar (Williamson, 1961; Pickard, 1974).

DEPOSITS IN THE DOLOMITE SERIES OF THE TRANSVAAL

As a complement to the idea of the basement relationships with the cover, Chazan and Fauvelet (1958) introduced the distinction of the intermediate basement. In this situation, sedimentary rocks, volcanic or metamorphic, folded or not, might be incorporated into a cratonic zone. For the metallogenic interpretation, this means a discontinuity with respect to the crystalline basement. Looking at the general stratigraphic column of the mineral deposits of the Transvaal (Fig. 13–4), South Africa (Anhaeusser and Button, 1976), one can distinguish the Precambrian Dolomite Series (now in the Transvaal sequence) as belonging to the intermediate basement. In its upper part there are fluoritic concentrations of the Marico District, which for its morphology could be considered, according to Anhaeusser and Button, as mineral deposits of older Mississippi Valley type. Naturally they have suffered metamorphism caused by the Bushveld igneous complex.

This district is 20 km south of Zeerust and has been known since the nineteenth century. According to the latest information (Van Alstine and Schruben, 1980), it may contain reserves of 80–100 million tons with 15 percent CaF_2 in a zone 60 km long. On a regional scale, the mineralization appears located in a paleorelief (Martini, 1976). In detail, the ore locations are mostly controlled by paleoporosity of the dolomite such as vuggy horizons and paleokarst developed during the elevation of the area before the deposition of the Pretoria series (Fig. 13–5).

Several types of mineralization, such as algal ore, blockspar, banded spar, and breccia spar can be distinguished. The first is the most important because it represents about 85 percent of the reserves. Some of the microalgal fluorite varieties and the black fine crystalized fluorite are difficult to recognize in the field.

The fine rhythmic stratification of the ore guides one to think of synsedimentary accumulation. From other interpretations, the most attractive explanation would be to attribute the concentration to a diagenetic process (Martini, 1976), although it cannot explain the fluorine's origin (magmatic or shale of the Pretoria series, or even the dolomite).

REFERENCES

Amstutz, G. C. 1971. *Glossary of Mining Geology.* Stuttgart: Ferdinand Enke Verlag, 196p.

Anhaeusser, C. R., and A. Button. 1976. A review of Southern African stratiform ore deposits—their position in time and space. In *Handbook of Strata-bound and Stratiform Ore Deposits,* vol. 5, ed. K. H. Wolf, pp. 257–319. Amsterdam: Elsevier.

Blondel, F. 1942. La recherche scientifique des gisements minéraux. *Bureau d'études Geol. Min. Coloniales,* Paris, **18:** 1–83.

Blondel, F. 1951. La classification des gisements minéraux. *Chron. Min. Coloniales,* Paris, **179:** 126–129.

Bois, M., L. Courel, and R. Coustau. 1980. Des lagunes paraliques à la base de la transgression mésozoique: cas de la dolomie de base de la bordure du Morvan. *Cent. Rech. Expl. Prod. Elf Aquitaine Bull.* **4**(1):81–97.

Carta, M., G. B. Alfano, C. Del Fa, M. Ghiani, P. Massacci, and F. Satta. 1975. Investigation on beneficiation of ultrafine fluorite from Latium. *11th Mineral Processing Congress,* Cagliari, pp. 1–26.

Cavinato, A. 1968. La fluorita nel Lazio. *L'Industria Mineraria,* Roma **19**(1):1–10.

Chazan, W., and E. Fauvelet. 1958. Facteurs de répartition de certains indices téléthermaux de plomb-cinc-cuivre au Maroc. *Chron. Mines Rech. Min.* **269**:311–320.

Clerici, E. 1920. Nuova giacitura di minerali presso di Roma. *R. Acc. Lincei,* Roma **29**(10):318.

Degens, E. T., and G. Kulbicki. 1973. Hydrothermal origin of metals in some East African Rift lakes. *Mineralium Deposita* **8**:388–404.

Fazzini, P., R. Gelmini, M. P. Mantovani, and M. Pellegrini. 1972. Geologia dei Monte della Tolfa. *Soc. Geol. Ital. Mem.* **11**(1):65–144.

Kesler, S. E. 1977. Geochemistry of fluorite deposits, Northern Coahuila, Mexico. *Econ. Geology* **72**:204–218.

Kesler, S. E., J. Ruiz, and L. M. Jones. 1983. Strontium-isotopic geochemistry of fluorite mineralisation (Coahuila, Mexico). *Isotope Geosci.* **1**(1):65–75.

King, H. F. 1967. Geological significance of stratiform deposits. In *Sedimentary Ores,* ed. C. H. James, pp. 97–104. Leicester: Department of Geology, University of Leicester, Special Pub. 1.

Levinson, A. A. 1962. Beryllium-fluorine mineralization at Aguachile Mountain, Coah., Mexico. *Am. Mineral.* **47**:67–74.

Lhegu, J., and Touray, J. C., 1980, Les gisements francais filoniens et stratiforms de fluorine et barytine. *Cent. Rech. Expl. Prod. Elf Aquitaine Bull.* **3**:277–334.

Lindgren, W. 1933. Differentiation and ore deposition, Cordilleran of United States. In *Ore deposits of the Western States* (Lindgren volume AIME), pp. 152–180. New York: American Institute of Mining, Metallurgical, and Petroleum Engineers.

Lougnon, J. 1974. Contribution à l'étude d'un gisement stratiforme de fluorine encaissé dans le Permien supérieur de la dépression de Saint-Die (Vosges, France). *Soc. Geol. Strasburg Bull.* **27**(3):195–217.

Lougnon, J. 1980. Le massif des Vosges, in Les Ressources Minières Françaises. *Annules des Mines,* Saint-Etienne **186**(7–8):42–45.

Magliola-Mundet, H. 1972*a*. Contexto geológico de los yacimientos de fluorita de Múzquiz, Estado de Coahuila, México. *Geol. Min. Bol.,* Madrid, **82**(1):48–67.

Magliola-Mundet, H. 1972*b*. Généralisation d'un contrôle stratigraphique des gisements de fluorine du Nord de l'Etat de Coahuila, Mexique. *Acad. Sci. Comptes Rendus* **274D**:3164–3167.

Martini, J. E. J. 1976. The fluorite deposits in the Dolomite Series of the Marico district, Transvaal, South Africa. *Econ. Geology* **71**(3):625–635.

Merla, G. 1951. Geologia dell'Appennino Settentrionale. *Soc. Geol. Ital. Boll.* **70**(1):95–382.

Peters, W. C. 1958. Geologic characteristics of fluorspar deposits in the Western United States. *Econ. Geology* **53**:663–688.

Pickard, G. 1974. Geology of Mexican fluorspar deposits. In *Symposium on the Geology of Fluorspar,* Special Publication No. 22, ed. D. W. Hutcheson, pp. 23–30. Lexington, Ky.: Kentucky Geological Survey.

Roedder, E., A. V. Heyl, and J. P. Creel. 1968. Environment of ore deposition at the Mex-Tex deposits, Hansonburg district, New Mexico. *New Mexico Bur. Min. Bull.* **22:**1–245.

Routhier, P. 1980. Où sont les métaux pour l'avenir? *BRGM Orléans Mém.* **104**:1–410.

Segre, A. G. 1950. Relaxione sommaria dei rilevamenti ai Monti del Sasso e Ceriti. *Soc. Geol. Ital. Boll.* **69**:593–594.

Selley, R. C. 1970. *Ancient Sedimentary Environments.* Ithaca, N.Y.: Cornell University Press, 237p.

Shawe, D. R., ed. 1976. Geology and resources of fluorine in the United States. *U.S. Geol. Surv. Prof. Paper 933,* pp. 1–99.

Sheppard, R. A., and A. J. Gude. 1969. Authigenic fluorite in Pliocene lacustrine rocks near Rome, Malheur County, Oregon. *U.S. Geol. Surv. Prof. Paper 650-D,* pp. 69–74.

Sheppard, R. A., and A. J. Gude. 1980. Diagenetic fluorite in the Eastgate zeolite deposit, Churchill County, Nevada, *U.S. Geol. Surv. Open File Report 80-506,* pp. 1–8.

Sheppard, R. A., and F. A. Mumpton. 1984. Sedimentary fluorite in a lacustrine zeolitic tuff of the Gila conglomerate near Buckhorn, Grant County, New Mexico, *Jour. Sed. Petrology* **54**(3):853–860.

Smith, C. I. 1970. *Lower Cretaceous Stratigraphy, Northern Coahuila, Mexico,* Report of Investigation 65. Austin, Texas: Bureau of Economic Geology, University of Texas.

Soule de Lafont, D. 1967. Les gîtes de fluorine stratiformes de la bordure du Morvan. *Chron. Mines Rech. Min.* **361**:85–108.

Soule de Lafont, D., and J. Lhegu. 1980. Les gisements de fluorine du Morvan, *26th International Geological Congress,* Paris, pp. 1–39.

Spada, A. 1969. Il giacimento di Fluorite e baritina esalativo-sedimentario in "facies" lacustres, intercalato nei sedimenti piroclastici della zona di Castel Giuliano, in prov. di Roma. *L'Industria Mineraria,* Roma **20**(10):501–518.

Studer, H. P. 1967. Mineral analysis of natural zeolite deposits. In *Clays and Clay Minerals,* ed. S. W. Bailey, p. 187. (Abs.). New York: Pergamon.

Van Alstine, R. E. 1976. Continental rifts and lineaments associated with major fluorspar deposits. *Econ. Geology* **71**(6):977–987.

Van Alstine, R. E., and P. C. Schruben. 1980. Fluorspar resources of Africa. *U.S. Geol. Surv. Bull.* **B1487**:1–25.

Vighi, L. 1951. Segnalazione di un'interessante formazione filoniana di fluorita in comune de Cerveteri (Roma). *Soc. Geol. Ital. Boll.* **70**:632.

Williamson, D. R. 1961. Mexican fluorite deposits, *Min. Ind. Bull. Colorado Sch. Min.* **4**(2):1–15.

Zucchetti, S., ed. 1975. Le fluoriti sedimentarie laziali, *29th Congress Società Italiana di Mineralogia e Petrologia,* Roma, pp. 1–53.

14

Celestite, Barite, Magnesite, and Fluorspar: Stratabound Settings Through Time and Space

Victor A. Ramos and
Milka K. de Brodtkorb

The stratabound settings of the main nonmetallic ore fields such as celestite, barite, magnesite, and fluorspar show a wide variety of tectonic environments through the geologic record. Even so, there are defined habitats for some specific ore-mineral assemblages, which indicate an active tectonic control for the mineralization.

Unfortunately, the paleogeographic and tectonic knowledge of these stratabound ore fields is not enough in most cases to define the precise settings. Therefore, the different ore fields are described based on the available information, and concluding remarks are only made where possible.

Within each ore mineral, the different ore fields are described in age order, beginning with the pre-Paleozoic ore fields whenever such deposits are known.

Celestite Ore Field Settings

Mode of Primary and Diagenetic Precipitation

Although in recent years some doubts have arisen about a simple evaporitic origin of huge celestite ore fields, present knowledge shows a complex interaction of different processes that have a common habitat: thick and extensive accumulation of evaporitic deposits. The environment where these are deposited controls the sequence of sedimentation, the early and late diagenetic processes including different secondary transformations.

Therefore, the celestite ores show a variety of textures and occurrences that testify to their complexity, but a definite lithologic assemblage is common to most ore fields: the close association with limestones and gypsum (or anhydrite). This is produced due to the close solubilities of celestite and gypsum, and even barite, which favor successive replacements during sea level fluctuation in tidal flats. In addition, the interaction of ground water recharged with different sea waters may rapidly change the precipitation conditions, contributing to the saturation or undersaturation of a definite ion.

As a result of such processes, most of the celestite ores are characterized by extensive layers, sometimes up to several meters thick, which alternate with minor lenses, associated with various types of limestone and gypsum or anhydrite.

The most prominent environments where celestite ore fields are found are those related to littoral or intracontinental settings, and this is one striking difference from the barite occurrences, in which significant concentrations are known in deep ocean basins. One favorable environment where large and thick layers of celestite are formed through direct evaporation is coastal lagoons; other occurrences are concentrated during early to late diagenetic processes.

No celestite ore deposits older than Late Paleozoic are as yet known, although the genesis of possible ancient deposits may be obliterated by postdiagenetic and metamorphic phase changes of the original strontium evaporitic concentrations. Therefore, the genesis is somewhat obscured by the age of the mineralization, which is why the most complete knowledge is derived from the most recent deposits, mainly those of Late Mesozoic age and those belonging to the Andean Chain.

There is no doubt that sea water bears a significant concentration of strontium, which averages 8 gr/t at the present time. This is testified to by the large amounts of Sr that were introduced during the precipitation of celestite of the large carbonate platform occurrences. The primary precipitation of celestite in those environments is well defined. It required a sea water reduction between one-third and one-fifth of the original volume to begin and to produce the main precipitation of celestite. This reduction can be obtained only in strong evaporitic settings.

However, it is necessary to emphasize that although evaporitic deposits are widespread in most of the Mesozoic sequences related to the opening of the South Atlantic, for example, the lagoon facies and evaporites of continental margins of Angola and Brazil, no important deposits have as yet been recorded in those environments. Present knowledge indicates that in the early phases of the opening of an ocean, as in the Red Sea evaporites, although celestite occurrences are known, no significant or exploitable ore field has been found.

On the other hand, most of the celestite deposits presently known in South America, as well as many others from Mexico and Europe, are closely related to basins belonging to orogenic environments. This indirect relationship with active continental margins, and the relative insignificance of celestite from passive margin sequences, may indicate some kind of control between the tectonic setting and the development of celestite ore fields.

TECTONIC SETTINGS

Based on the previous premise, we analyze the typical tectonic setting of the presently known occurrences in order to elucidate the regional control that crustal evolution exerts on mineralization.

ANDEAN CELESTITES

The main celestite ore fields of the Andes are presently known in the Neuquén Basin, developed along the retroarc side of the Andean Chain (see Fig. 14–1).

As previously discussed in Chapter 4, the main celestite ore fields are linked to the climax of the regressions of the Cuyean (Early Jurassic), Lotenian (Late Jurassic), and Mendocian (Early Cretaceous) cycles of sedimentation. These cycles are controlled by tectonoeustatic sea level changes as described by Mitchum and Uliana (1985). The precipitation of celestite shows a striking coincidence with those periods of important lowstands.

It is important to bear in mind that during this long period no shortening or tectonic unlift took place on the Cordilleran side of the basin. The subsidence was controlled by thermotectonic mechanisms in a tensional regime. This regime was controlled by the early break-up phases of western Gondwanaland fragmentation. As a result, a sedimentary pile up to 6 km thick has filled the basin, deposited during a high heat flow period.

This tectonic evolution of the basins rules out some feasible mechanisms for the concentration of the celestite, such as those proposed for the Betic Cordilleran ore fields. No significant relief existed during the accumulation of the celestite. The provenance of the material was mainly from the foreland, where a stable craton was only slightly eroded. Source rocks consist mainly of medium-grade metamorphic and plutonic rocks. The influx of volcanic debris to the basic was limited to the westernmost border, as testified to by the absence of ash falls, tuffs, and other volcanic materials.

Therefore, the accumulation of celestite in that tectonic environment was mainly controlled by restrictions of the sea water influx to the basin, related to tectonoeustatic sea level changes.

The increased volcanism noted in the active margin during the rift and drift stages has contributed to the isolation of the Pacific transgression and the development of restricted circulation in the basin. Such an environment is depicted in Figure 14–2.

This environment controlled the deposition of the primary beds of celestite in typical stratiform ores. Examples of this type of occurrence are the Cuchillo Cura, Naunauco, and Continental ore fields in the central Neuquén Basin (see Chapter 4).

Important tectonism took place soon after the deposition of the Huitrín formation occurrences. Strong deformation folded and thrust the sedimentary sequences. This stage was responsible for important migration and relocation in fractures and folds of some of the previous stratiform celestite beds, as in the stratabound San José and Julieta prospects, of the Pichi Neuquén River ore field (Brodtkorb et al., 1975).

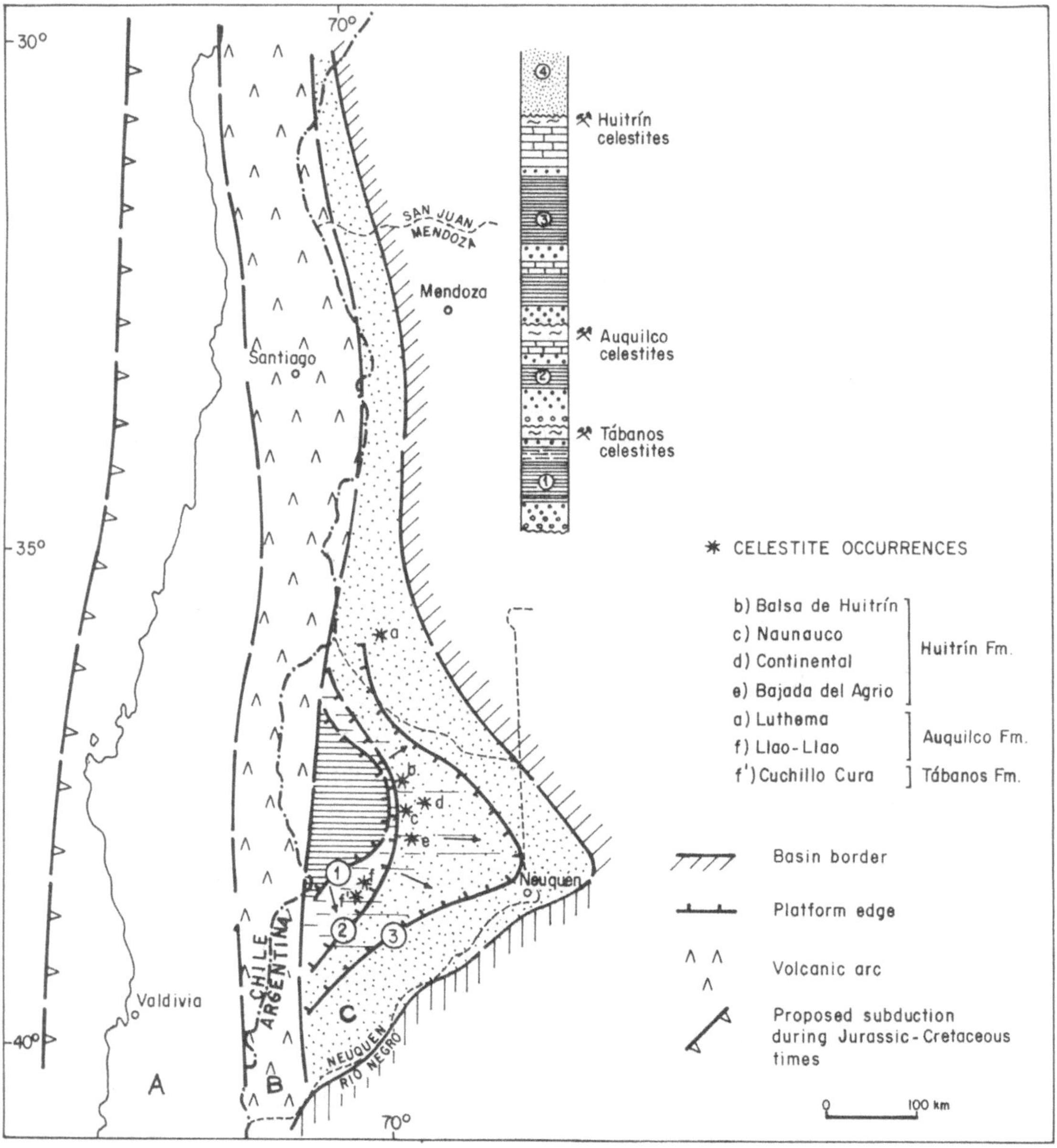

Figure 14–1. Tectonic setting of the celestite deposits of the Neuquén basin, with indication of the principal tectonic units (after Ramos, 1985; Mpodozis and Ramos, 1986).

(A) Forearc area; (B) magmatic arc; (C) outline of the retroarc Neuquén basin with the migration of the platform edges during the deposition of the three phases of celestite mineralization (Middle Jurassic to Early Cretaceous) and dominant provenance of the clastic material.

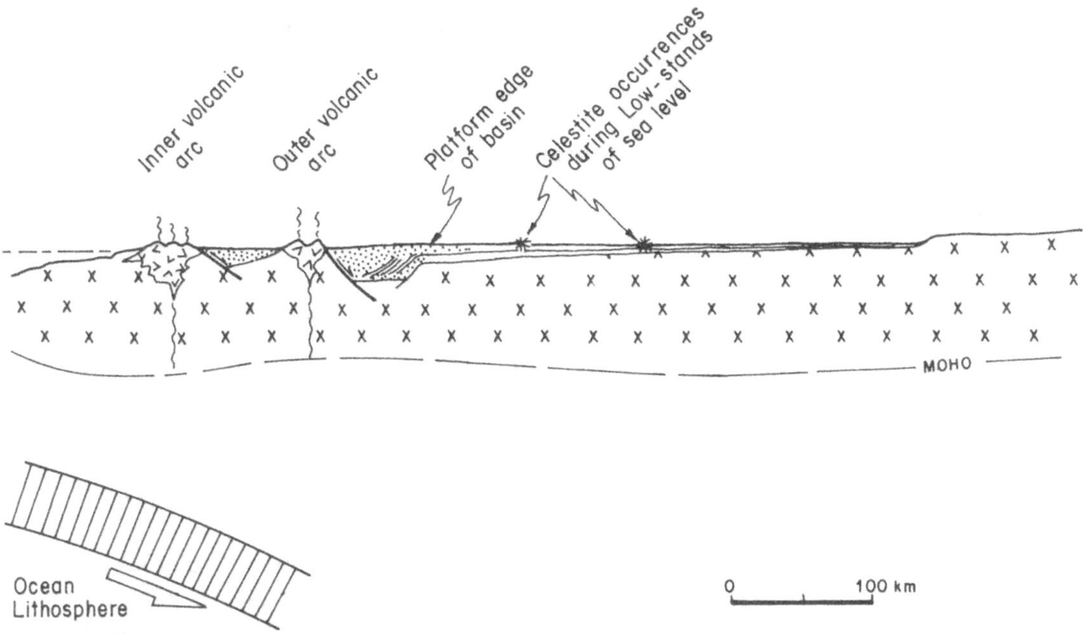

Figure 14–2. Schematic cross section of the Middle Jurassic to Early Cretaceous tectonic setting of the Neuquén basin with indication of the celestite occurrences.

MEXICAN ORE FIELDS

The Mexican ore fields located in the northeastern part of the country, along the Gulf of Mexico, although of Early Cretaceous age like the Andean occurrences, show a strikingly different genesis. As a common feature they are closely related to evaporitic deposits, but in this case are associated with a back-reef setting. A series of sea level changes produced the submersion of the Late Neocomian–Aptian sequences, which were covered by pelagic lime mudstones (see Chap. 3).

These changes in sea level favor the interaction of continental fresh water with sea water, producing important replacements of Sr and Ba in the limestones. The strontium has been derived from either the diagenetic release of the limestones or some immature arkoses, as indicated by the isotopic composition. The required sulfur has been derived from the nearby evaporites. These diagenetic replacements need the interaction of two different local solutions for the precipitation of the celestite.

EARLY TO DIAGENETIC CONCENTRATIONS

Several studies during recent decades have shown the significant role of diagenetic processes in the secondary concentration of celestite ores, for example, diagenetic transformation of aragonite into calcite or the dolomitization of the aragonite, due to the considerable amount of strontium released during those transformations. There is also some strontium released during the transformation of gypsum to anhydrite. The strontium released may precipitate as celestite when proper conditions of salinity are obtained.

Even though such diagenetic processes could be locally important, the secondary concentration is generally as significant a concentration agent as the primary evaporation seen in the huge celestite ore fields.

BRITISH ORE FIELDS

The deposits of the Bristol area, where huge deposits at Yate are well known, are good examples of the interaction of primary and diagenetic processes. The celestite ores are of Late Triassic age, when most of this region was under desertic and arid climates. The paleogeography of the celestite occurrences are characterized by widespread sabkha conditions, where, in a supratidal environment, halite, celestite, and gypsum were deposited.

The different settings of the Late Triassic celestite deposits can be grouped in three habitats: celestite beds in a stratiform evaporitic sequence, nodular and disseminated celestite above the unconformity with Paleozoic rocks, and celestite veins and infilling hosted in the Paleozoic rocks.

The genesis of these deposits has been interpreted as a primary precipitation of celestite, in arid desert basins with inland drainage, where percolating ground water remobilized the Sr, probably during dolomitization of Carboniferous limestones.

CONCLUDING REMARKS

The main celestite ore fields of the world are depicted in Figure 14–3, with indication of their relative importance.

Each ore field has its own complex history, controlled by a series of phenomena superposed during evaporation and diagenesis, in littoral and intracontinental settings. Common ore textures are widely represented by geodes, veins, stalactites, and several other primary and secondary structures. Therefore, in a single deposit syngenetic celestite layers, together with early, late, or even epidiagenetic celestite infillings are commonly found.

BARITE ORE FIELD SETTINGS

Although the precise setting of the older ore fields are sometimes obliterated by transformations and changes of the original facies, the younger deposits show a remarkable control of the different environments. The stratabound barite deposits are not good indicators of a precise geologic environment, as were, for example, the previously described celestite ores. The barite deposits are associated with a broad variety of host rocks, which is a clear indication of the different paleoenvironments where these ores are formed.

Even so, they may be tentatively subdivided into two distinctive settings, one dominated by coastal environments and the other belonging to the oceanic

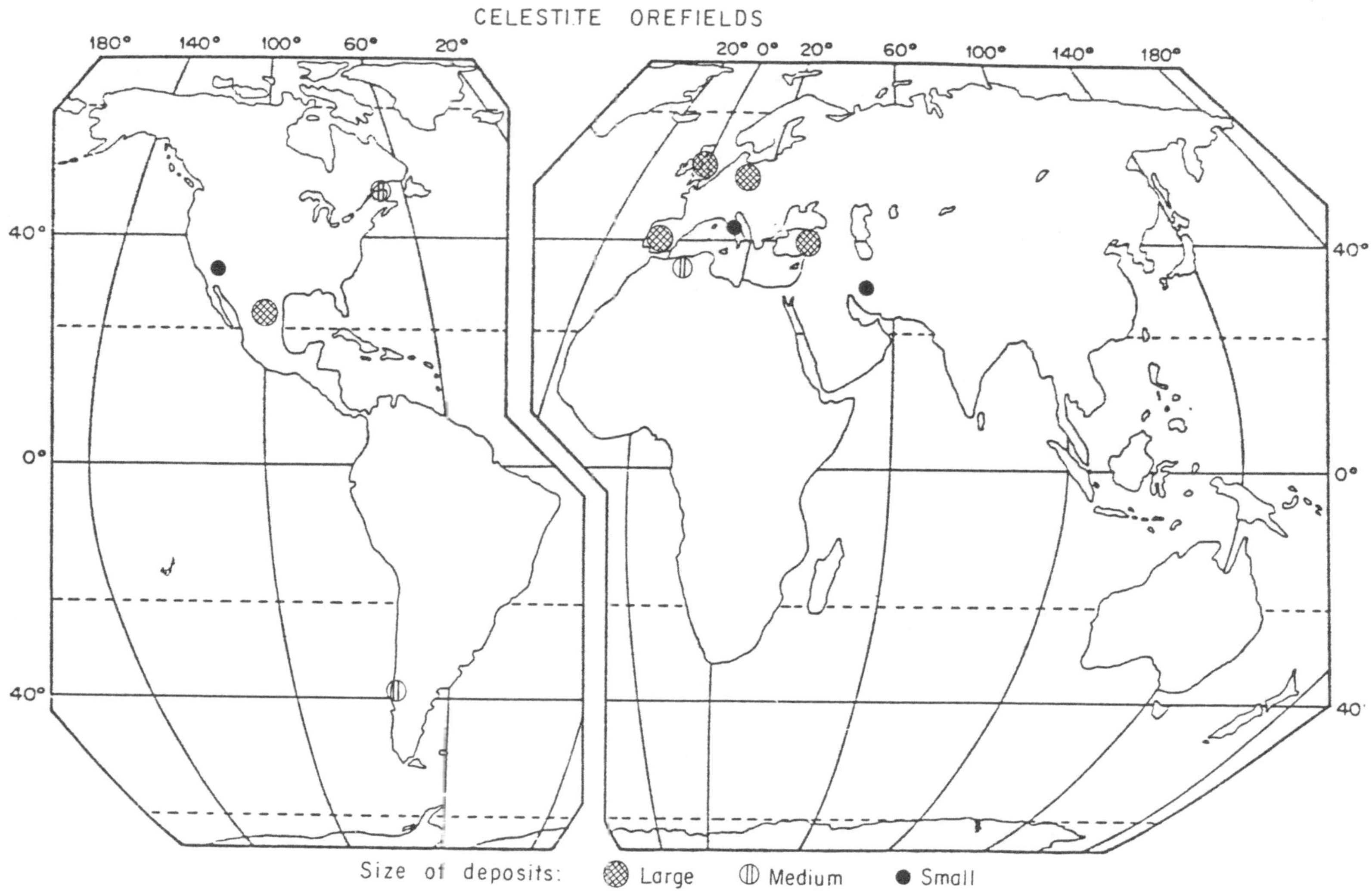

Figure 14-3. Main celestite ore fields of the world weighted by their relative sizes.

realm. Both ore types are primary settings and could be related or not to the occurrence of sulfides in the main ore. There is also no time restriction in the different settings, because both primary types are known in different epochs. Another important setting is represented by the karstic ore fields.

GENERAL REMARKS

It is interesting to note that, although the average sea water content of barium is approximately of 0.06 ppm, several authors have reported larger concentrations, such as Chow and Goldberg (1960), who reported Ba concentrations of up to 10 ppm in surface waters of the Pacific Ocean. Analyses of pelagic sediments in the ocean also suggest a barium concentration in the range of 0.05 percent to 0.3 percent (Shawe et al., 1969). On the other hand, Arrhenius and Bonatti (1965) mentioned that some radiolarian ooze in the equatorial regions contains as much as 10 percent or more of barite. At that time, these abnormal concentrations of barite in the southeastern Pacific Ocean were interpreted in terms of a giant system of counter-clockwise currents as the main factor that controlled the deposition of barite (see Fig. 14-4).

More recently, Chase et al. (1985) demonstrated the active precipitation of barium as well as several sulfides of Zn, Cu, Fe, Pb, and so forth in connection with hydrothermal vents on an axis seamount of the Juan de Fuca Ridge. Chase et al. localized some chimneys where rock samples have up to 15 percent barium in locations closely related to some sort of calderas that developed in the vicinity of an axial seamount, as depicted in Figure 14-5. There are also

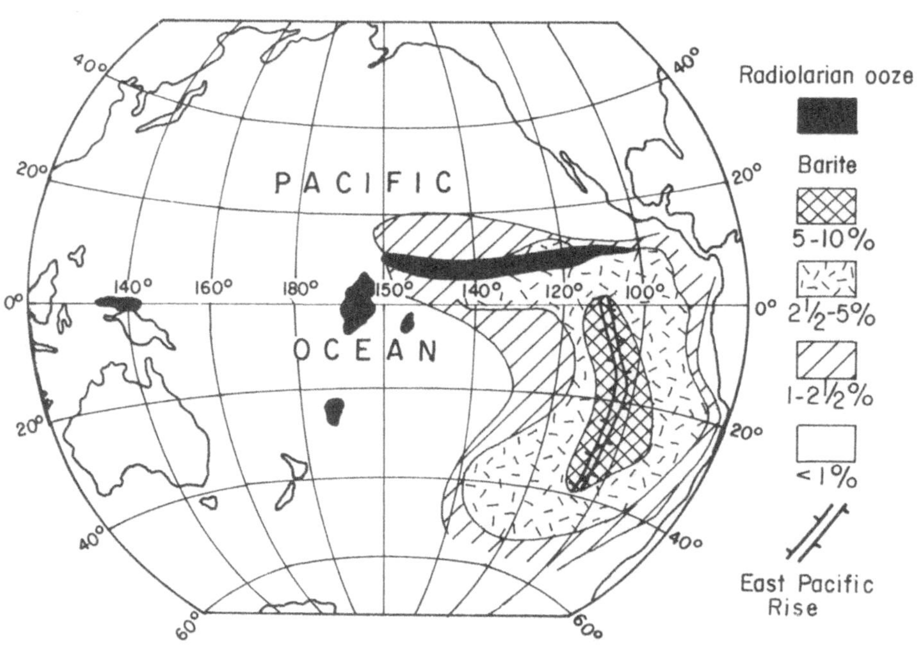

Figure 14-4. Abnormal concentration of barium in radiolarian ooze and pelagic sediments of the Pacific Ocean (after Arrhenius and Bonatti, 1965).

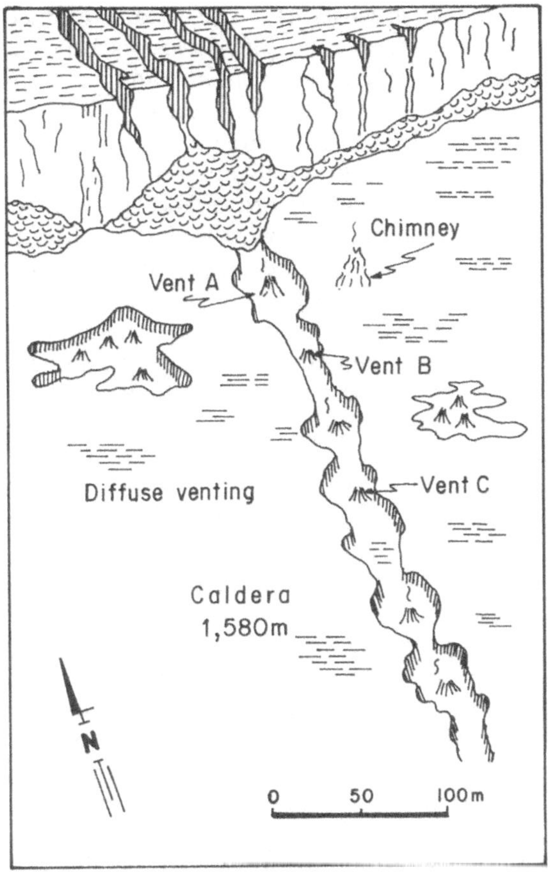

Figure 14–5. The northern caldera wall and floor on axial seamount, with indication of a chimney where an abnormal concentration of Ba was found (after Chase et al., 1985).

recent sulfide deposits in the Tyrrhenian Sea that have barium concentrations as high as 48 percent (Puchelt, 1986).

The result obtained from the combination of biogenic ooze with hydrothermal activity is shown by Marchig et al. (1985). On both flanks of the eastern end of the Galapagos rift, between 10 km and 40 km from the spreading axis and at approximately 2500 m depth, five cores were obtained to investigate the relationships between sea-floor spreading, tectonics, volcanism, hydrothermal activity, and metallogenesis. Although active surface volcanism appears to be limited to the axial zone, hydrothermal activity also occurs at the flanks, as shown by the presence of hydrothermal alteration in the rocks. The analyzed sediment is an ooze with a large content of biogenic material (mainly foraminiferal tests, but also radiolarians, and diatoms cemented by clay minerals), ranging from marls to calcareous oozes with high silica content. The hydrothermal impact produced chemical changes during diagenesis, when Mn, Co, and Cu, and to a lesser extent Ba, Zn, and Ni, were mobilized in the reduction zone and precipitated in the oxidation zone.

Crystal nodules and small lenses of barite are also reported in the Atlantic Ocean near the northwestern coast of Africa as reported by Dean and Schreiber (1977), who described the barite as totally or partially replaced by calcite. The barite seems to have been originated by diagenetic processes, under oxidant conditions in materials located in the sediment/water interphase during intervals of no sedimentation. The barium originated from oxidation of organic matter, whereas the sulfate provenance is from the sea water, which imbibes the sediments, or by oxidation of pyrite and filtration and rising of the sulfate content of some deeper and preexisting evaporitic sequences.

Barite nodules in Recent marine sediments were found by Jones (1887) when studying the sediments off Colombo (Sri Lanka) at a depth of 1235 m (Revelle and Emery, 1951). The nodules were spherical to cylindrical in shape with some concentric growths and lengths up to 10 cm. They were composed of 75 percent barium sulfate and 25 percent other minerals, foraminifera, radiolaria, and so forth.

The dredging performed in 1938 in the ocean floor near California showed that between 650 m and 800 m depth, in a 100 kg of vadose ooze, 250 nodules of barite were found as well as some fragments of andesite (Revelle and Emery, 1951). The barium sulfate content of these nodules varied between 62 percent and 67 percent.

Nodules formed in a marine environment of Miocene age have been recorded on the coast of Japan (Kim et al., 1970).

Present knowledge of such occurrences shows that they are confined to specific tectonic settings. They are associated with large seismic ridges and leaky fault zones, which allow the rise of barium-enriched solutions that interact with the interstitial water of the sediments, which may contain a high sulfate ratio (Moro Benito and Arribas Moreno, 1981).

On the other hand, several authors proposed the liberation of barium cations from feldspar during weathering of crystalline massifs as the main source of barium for the coastal examples. The cations adsorbed in clays are carried toward the sea by the fluvial systems (Puchelt, 1967; Hanor and Chan, 1977). Barium is liberated when the salinity increases and reacts with the sulfates present in sea water.

For the specific case of the Arkansas deposits, Hanor and Baria (1977) suggest that barite precipitated as the interphase between formation water and overlying sulfate-rich sea water.

Recent understanding of the different transformations that occurred in the tidal flat environment, as well as the interrelations of sea water and ground water recharges, together with the hydrodynamic potential of evaporite pumping as proposed by Hsü (1984), are perhaps the most important processes that lead to the development of a barite ore field.

Ancient Stratabound Deposits

For decades most of the important ore fields were restricted to Phanerozoic or Late Proterozoic settings, and several authors have argued about the different conditions that prevailed during early Precambrian times that hindered the formation of stratabound barite ores. However, in recent years important discoveries of stratabound ore fields deposited as far back as Swazian times have been made.

SWAZIAN ORE FIELDS

Swazian ore fields are known from the Swaziland supergroup of South Africa (Reimer, 1980), East Pilbara block from Western Australia (Hickman, 1973), and the Peninsular Gneissic complex of India (Radhakrishna and Vasudev, 1977). The ages of the host rocks are as old as 3.0 Ga and even 3.5 Ga (Lambert and Groves, 1981).

The ancient bedded barite deposits, according to Lambert and Groves (1981), were the result of: (1) the existence of a stable shallow water evaporitic environment; (2) the influx of juvenile sulfur exhalations, and (3) superficial oxidation, probably involving microbial photosynthesis, which gave rise to, at least locally, high sulfate/sulfide ratios in the hydrosphere.

The largest of these three ore fields is the North Pole ore in the Pilbara. The barite beds occur within a sequence consisting of laminated chert, silicified arenite, and conglomerate. Various structures and textures suggest that barite and chert replaced the original gypsum–carbonate sequence (Dunlop, 1978; Lambert and Groves, 1981). For instance, the barite beds are composed of bottom-nucleated crystal groups, radiating across bedding. In some undeformed single crystals it was possible to measure interfacial angles indicating what were typical gypsum replacements. This fact was also found by Walker et al. (1977), who described pseudomorphs in sulfate evaporite deposits in the Proterozoic McArthur group of Australia.

The finding of a sedimentary barite deposit in the Fig Tree group of South Africa is mentioned by Heinrichs and Reimer (1977). A few years later, five major sedimentary barite occurrences were described by Reimer (1980) from the Swaziland supergroup, in the Barbertown Mountain Land of South Africa. The Swaziland supergroup is subdivided into the Onverwacht, Fig Tree, and Moodies groups, and barite ores were found in the first two groups.

The Onverwacht group consists of about 12,000 m of ultramafic, mafic, and felsic lavas and tuffs, interlayered with chert bands, and so far three barite deposits are known. The Fig Tree group is composed of a sequence of cherts, shales, and impure arenaceous sediments up to 2100–2500 m thick. Within the middle part of the southern facies a barite deposit is known, while another but smaller occurrence is known from the lower part of the sequence. The overlying Moodies group, about 3600 m thick, consists of conglomerates, sandstones, quartzites, shales, banded ironstones, and subordinate lava horizons. The metamorphic grade of the sediments and part of the volcanics is remarkably low, even though the minimum age of the sequence is about 3.35 m.y. (Reimer, 1980).

The barite deposits found in the metavolcanics of the Lower Onverwacht group comprise up to 12 m thick zones alternating with shales. The barite is fine to medium grained and contains finely disseminated pyrite and other sulfides, for instance, sphalerite. A lens-shaped barite deposit is found within the chert band, which further east shows evidence of deposition as a carbonate band within an evaporitic sequence. These deposits are considered to be of hydrothermal-exhalative origin.

The larger of the Fig Tree group deposits occurs above a pronounced zone of intraformational erosion. The barite beds contain detrital barite grains, together with detritical pyrite, chromite, zircon, magnetite, ilmenite, molybdenite, and gold. Diagenetic recrystallization textures as well as the so-called cauliflowers are known over large areas at the sediment/water interphase.

The occurrence of barite deposits in northeastern Botswana is also mentioned by Reimer (1980). There are thin bands within greenish quartzites and calcareous metasediments in a metavolcanic sequence from the Matsitama Schist Belt. Two other occurrences are described in the Bulawayan greenstone of Zimbabwe (Rhodesia), for which a sedimentary origin seems feasible. In the Dodge mine, the barite occurs as disconnected lenses in a metasedimentary sequence, consisting of phyllites, banded ironstones, quartzites, and limestones.

Thus far Swazian sedimentary barite deposits have been observed only in eastern Gondwanaland (Australia, India, and Africa). Although the reasons for this situation are beyond the present knowledge of the area, one cannot disregard possible occurrences of bedded barite ores in the Archean cover of the Guyana and Amazon cratons.

KARELIAN ORE FIELDS

Barites are also known in Europe at the Skellefte ore field of Sweden. There the Asen deposit has barite in the transition between volcanic and sedimentary rocks that have been affected by the orogenic Sveco–Karelian cycle (1800 m.y.; Rickard et al., 1979). The barite beds are interbedded with sandstones, mainly quartzites and quartz-sericite schists, up to 8 m thick and developed along 7 km. Besides barite and pyrite, there is chalcopyrite, sphalerite, galena and tetrahedrite. The isotopic study of the sulfur of the barite and pyrite suggest a common volcanic origin, with a later biogenic reduction of the sulfate together with a combination of metal-rich solutions of Kuroko-type (see Chapter 10).

Two ore fields are known in Brazil from the Middle Proterozoic: the Água Clara prospect and some occurrences in the Quadrilatero Ferrifero (see Chap. 5). The Água Clara prospect belongs to a narrow mineralized bed with north and northeast trends. It is located in the southern flank of the Ribeirao da Barra anticline. Host rocks are represented by the Setuva group, composed of a chemical sedimentary sequence where the barite layer lies between quartz–micaschist beds, partially calcareous.

In the Quadrilatero Ferrifero, stratabound metamorphic barite and fluorite occurrences are known in the carbonate sequence of the Minas supergroup.

GRENVILLIAN ORE FIELDS

Younger occurrences are known from the Chapada Diamantina group, which belongs to the Espinazo supergroup (1.7 to 1.1 Ga). The barite-bearing sequence overlies older gneisses in angular unconformity, and it is covered with another angular unconformity of limestones of the Bambui group. The Chapada Diamantina group, which is several thousand meters thick, consists of extrusive volcanic rocks in the lower member; quartzites, schists, conglomerates, and phyllites in the middle units, and limestones in the upper section. In several areas the volcanic sequence hosts stratabound barite deposits.

KATANGAN ORE FIELD

Deposits are known in northeastern Brazil, where quartzites overlying scheelite-bearing skarns are associated with a belt of stratabound barite deposits. Younger ore fields are present in the Bambui group (900–570 m.y.). The Bambui group includes thin conglomerates and thick quartzites and siltstones as well as important carbonatic formations with abundant stromatolites, where Pb–Zn–fluorspar and barite ore fields are hosted. The barite always occurs in the lower part of the carbonate sequences, whereas the fluorspar appears in the upper sections related to regressive facies of the sedimentary cycle.

PALEOZOIC ORE FIELDS

The first occurrence of this era is the Aberfeldy ore field of Scotland, where barite and sulfides are present in the Late Precambrian–Early Cambrian sequence (Coats et al., 1980). This is a recent finding in which the barite is contained in a graphitic schist with a 7 km run, which is up to 100 m thick. This ore field has become of importance due to the large barite and sulfide reserves. The mineralized zone is found in the Ben Eagach formation (Easdale subgroup) of Middle Dalradian (Argyll group)* age. The barite levels are interbedded with quartz–celsian rocks, sulfide-bearing carbonate rocks, muscovite, and graphitic micaschists. Individual barite bands are up to 15.5 m thick and the metal contents, as for example in borehole 4, is up to 8 percent Zn and 4 percent Pb in the carbonate rocks, over a 4 m thickness.

The lowest Paleozoic sequence of Sardinia contains three main barite occurrences: (1) stratiform barite interbedded with dolomites of Lower-Middle Cambrian age; (2) stratabound bodies of quartz and chalcedony with barite and argentiferous galena, located along the Cambrian–Ordovician unconfor-

Series Editor's note:* The tectonic relationship of the Ben Eagach Schist is more readily defined than the stratigraphic classification, and is of some importance in the recognition of ore field type. According to Harris and Pitcher (1975, "The Dalradian Supergroup," in ***Precambrian,* Geological Society of London Spec. Rep. No. 6, pp. 52–75), "The base of the Easdale Subgroup is very well defined over much of the outcrop because the change from the orthoquartzites of the Islay Subgroup is one of the most marked in the whole succession. . . . [O]rthoquartzites give way abruptly to pelites which are commonly carbonaceous." The Islay is the youngest subgroup of the Appin group and the Easdale the oldest subgroup of the Argyll group. Harris and Pitcher continue: "The long period of quiet deposition represented by the Appin Group ended abruptly with the advent of the important Varanger glaciation c. 668 m.y. ago. . . . A very vigorous environment was established following the decline of the ice age."

Thus, though stratigraphic definition of the Precambrian–Paleozoic boundary in the Dalradian is difficult, the tectonic relationship provides excellent definition of the change from the stable platformal regime that marks the end of the Katangan Era to the vigorous environment that marks the onset of the Paleozoic Era in this part of Scotland. The Ben Eagach Schist, host to the Aberfeldy ore field, is thus regarded here as Paleozoic.

mity; and (3) funnel-shaped bodies in Permo-Triassic karst cavities developed in Cambrian limestones.

In South America, barite deposits of Ordovician age are known in the Canota ore field in the Precordillera of Mendoza (Argentina). These occurrences are characterized by nodules, lenses, and crystals, all associated with primary sedimentation of Ordovician shales (Brodtkorb et al., 1987). Stratigraphy and composition of the barite-bearing beds indicate a basinal environment with a heavy organic deposition. The high barium concentrations in this facies cannot be derived from only sea water and might be supplied to the ocean by hot springs such as the present hydrothermal activity of oceanic rift systems. The marine sulfate was bacterially reduced and formed the cores of the nodules, meanwhile the external layers have isotopic sulphur values that correspond to Ordovician sea water sulfate.

Barite deposits are also formed further north in the Precordillera of La Rioja, associated with the San Juan Limestone of Ordovician age. There, in the Helvecia district, the ore is composed of barite and galena located along the angular unconformity that separates the Carboniferous sandstones from the Ordovician limestones. These platform facies limestones also host Pb and Zn primary stratabound deposits (see Chap. 4).

Important barite deposits are also found in New South Wales (Australia). Middle to Late Silurian sedimentary rocks of the Lachlan Fold Belt of the Tasman orogen bear barite deposits (Stevens, 1976, in Moro Benito and Arribas Moreno, 1981). The ore appears as lenses of variable size concordant with quartzites, chlorite, and bituminous schists, cherts, and pyroclastic rocks. The sequence was affected by two deformational phases and low-grade metamorphism. The barite is massive, fine grained, and dark gray, with small amounts of pyrite and other disseminated sulfides. Stevens (1976, in Moro Benito and Arribas Moreno, 1981) proposed a volcano-sedimentary origin for this ore.

Significant stratabound barite deposits are present in the Zamora region of Spain at the Alcañices–Carbajales de Alba synclinorium (Moro Benito and Arribas Moreno, 1981). An Early Paleozoic sequence overlies a gneissic basement of probable Precambrian–Cambrian age. The barite is located in a volcano-sedimentary section at the Siluro–Devonian boundary, associated with a distinctive sedimentary facies.

In Western Europe the stratabound barite deposits have been subdivided by Fuchs (Chap. 8) into three different types:

1. Stratiform lenses associated (or not) with massive sulfide ores, located in pre-Hercynian strata. These ore fields have been interpreted as belonging to the exhalative-sedimentary type like the Rammelsberg, Meggen, and Eisen of Middle Devonian age in West Germany, and the Magcobar ore of Early Carboniferous age in Ireland. These deposits have a distinctive tectonic setting and are located at particular stratigraphic levels, situated at the boundary between platform and basinal facies, similar to the previously described setting of the barite deposits of the Precordillera of Mendoza. The stratigraphic level is characterized by tectonic instability and volcanic activity, although this volcanism is not directly linked as in the Kuroko-type barites. In the German ore fields an open basinal model is preferred, because

isotopic data indicate that the source of the sulfur of the barite is the sea water. Restricted conditions were only obtained at the end of the depositional cycle, as indicated by the isotopic composition of the Grauerzkörper and Liegendes Grauerz ore fields, where sulfur isotopic ratios were higher than sea water for that time (see Chap. 8).

2. Stratiform lenses hosted in carbonate rocks or sandstones unconformably lying on the Hercynian crystalline basement (see barite Mesozoic ore fields for discussion).
3. Barite ore fields with or without fluorspar and sulfides in orebodies located below unconformities with some karst affinities. Examples of this type are found in Sardinia (see Chap. 9), in the northern Pyrenees of France, and at Chaudfontaines, Belgium. In these cases, mineralization took place in karst cavities during emersion of the platform facies and therefore are closely related to an unconformity surface. These deposits represent the last remobilization stage in the continental plate evolution in areas characterized by high barium potential.

The stratabound barite ore fields of North America have been subdivided into four genetic types by Clark and Poole (Chap. 6). Type 1 is a syngenetic–diagenetic stratiform barite that is related to submarine volcanism. The deposits generally formed in island-arc and other sea floor volcanic environments, and were later transported and accreted to the continental margin. Type 2 is a syngenetic–diagenetic stratiform barite that precipitated in fine-grained Paleozoic sedimentary rocks in deep water marine settings adjacent to the continental shelf. Type 3 is epigenetic stratabound barite that is generally hosted by Paleozoic carbonate rocks that were deposited in stable cratonic-platform and continental-shelf environments. Type 4 are residual barite concentrations in colluvium and soil and commonly are associated with type 3 deposits. Type 1 deposits are adjacent to polymetallic massive sulfide ore deposits. Examples are the Lynx and Britannia in British Columbia and Buchans in Newfoundland. Type 2 are the most important ore fields and mainly sedimentary-exhalative, extending along the Cordillera from Alaska to Mexico. Type 3 are similar to the Mississippi Valley-type lead–zinc ore fields, and are found in southeastern Missouri, Sweetwater (Tennessee), and Cartersville (Georgia).

An excellent example of barite stratabound deposits of Late Mississippian age is found in the Ouachita Mountains of Arkansas (see Chap. 7). The tectonic setting involves a trough, developed as a remnant ocean prior to the closure and collision of the Pangaean plates during the Late Paleozoic. Host rocks comprise facies of the submarine continental rise as well as turbiditic fan or flysch sediments. Barite ores are related to shales with slump structures, thin beds of limestone with algal-like layering, and massive barite with mudflow breccias. There is much evidence for syngenetic deposition of the barite ores, such as primary lamination, diagenetic compaction, slumping, and mudflows. These ores present strong similarities with the Meggen ore, pointing out a probable common paleogeographic setting.

Barite of the Arkansas ore fields is different from the barite of Mississippi Valley type. The former are located in siliceous shales of a mobile belt and are not associated with lead and zinc sulfides or fluorspar. Barite occurs in commercial quantities within or near some lead–zinc–fluorspar Mississippi Valley-type deposits, mainly developed in carbonate rocks of different depositional environments with distinctive country rocks and mineral assemblages.

The Sulphur Creek and Mile 472 ore fields of British Columbia (Canada) are associated with a series of coastal platform limestones of Devonian age (Morrow et al., 1978). In this region the barite was deposited in the interphase between sea water under reducing conditions and ground water from a coastal aquifer. Mixing of both kinds of water produced limestone dissolution and precipitation of barite in the previously generated cavities.

MESOZOIC AND TERTIARY ORE FIELDS

Stratiform lenses of barite are known from the Pessens and Chaillac ore fields (France), hosted in Lower Jurassic carbonates (type 2 of Fuchs, Chapter 8). The barite is low grade (5 percent to 20 percent), and isotopic data indicate a probable mixing of continental fresh water and sea water, although these data cannot rule out successive oxireduction processes.

Some stratabound barite deposits are also known in the marine Mesozoic basin of Mendoza and Neuquén (central western Argentina). The sequence encompasses three sedimentary cycles, and barite deposits are known from the littoral and near-shore facies of two older cycles. Over a pre-Jurassic basement, an initial phase of syn-rift deposits preceded two important regression and transgression cycles, which invade the basin from the west and northwest, developing the Neuquén Embayment, which has a large extension toward the east over the foreland region. The first Jurassic cycle is characterized by continuous subsidence, which produced up to 2500 m of sediments. The barite beds of Arroyo Nuevo mine are intercalated in shales underlying evaporites. Related to the second cycle is the Mallin Quemado ore field, as well as the Compadrito, La Salida, and Isidora mines. The Mallin Quemado ore field has barite beds lying conformably within the stratigraphic sequence, although several unconformable veins are emplaced in the overlying sediments.

A small Upper Cretaceous deposit of barite is found in the Judean desert, some 12 km southeast of Jerusalem (Israel), associated with carbonates, chert, and other sedimentary rocks (Bogech and Shirav, 1978). The barite is located in two units: an upper section approximately 3 m thick, separated by thin layers of marly chalks, chalks, and cherts of the lower unit, also nearly 2 m to 3 m thick. Ubiquitous gypsum and anhydrite have been found in the cherts. A 1 m thick breccia body, consisting of chert, calcite, and minor barite, transects the beds on the north flank of a syncline. The barite is interlayered with chert, chalk, and clay lenses and has a conspicuous coarse bedding. A marine or a marine organic origin of the barite was suggested.

A different type of deposit forms the barite ore field associated with the well-known Kuroko-type of Japan (see Chap. 10). This is the result of a subsequent violent submarine volcanism of Miocene age over thick volcanic and sedimentary sequences. The ores can be defined as stratabound polymetallic sulfide–sulfate deposits with stratiform massive bodies and underlying stockwork bodies. They can be subdivided on the basis of total sulfate to sulfide ratios and size of the stratiform bodies in the following subtypes: (1) typical Kuroko, (2) the Kuroko-type network deposit, and (3) the Kuroko-type gypsum deposit and the Kuroko-type barite deposit. The Hokuroku and Hokkaido ore fields are the most important barite deposits. Isotopic data indicate that sulfates are mainly derived from sea water. Barium is provided from neither the sea water nor rhyolitic volcanism. Based on some lead isotopic values of the associated sulfides, a leaching of the older Miocene volcanics and basement rocks is assumed for the barium supply. A convection system, where hydrothermal solutions leached the barium of the older rocks and the interaction of these solutions with sea water, is the model proposed to explain these ore fields.

CONCLUDING REMARKS

This brief worldwide analysis of the different stratabound ore fields indicates that barite has three dominant habitats: (1) coastal or littoral environments, (2) nonlittoral or deep ocean facies related to oceanic ridge volcanic activity, and (3) karst deposits associated with unconformities. The main ore fields classified on that basis are indicated in Figure 14–6.

MAGNESITE ORE FIELDS

Magnesite is found in different geologic settings and, according to Wetzenstein (Chap. 12), occurrences may be divided for practical purposes into four different types:

1. Cryptocrystalline magnesite in or affiliated with ultramafics,
2. Crystalline and sparry magnesites in sedimentary or metamorphic rocks,
3. Saline to hypersaline and/or evaporitic magnesites,
4. Magnesite and related Mg-bearing minerals in intramontain freshwater basins.

The first type (''a'' in Fig. 14–7) constitutes major deposits, occurs in belts of ultramafic rocks, and results from the alteration of serpentine. This type of magnesite deposit is found near the surface and has limited extent in depth. Such deposits are not time-bounded, since the alteration and weathering processes may occur at several times. Ore fields of this type are known in the ultramafic belts of Yugoslavia, Albania, Greece, and Turkey. The largest known Balkan deposit is from Yugoslavia. Some other deposits are known from the United States and the U.S.S.R., related to ultramafic belts. Some interesting occurrences of magnesite have been found in the ultramafic belt of the Frontal Cordillera in Mendoza, Argentina.

The second type of deposit is widespread (''b'' in Fig. 14–7). These have the most important commercial magnesite ores. These ores are formed by evaporation of ocean water or by precipitation of magnesite at the interphase with surface waters from solutions derived from hydrolysis of Mg-bearing minerals. In the metamorphic complexes of Precambrian or Paleozoic age, the genesis is somewhat uncertain, although large deposits are known from the

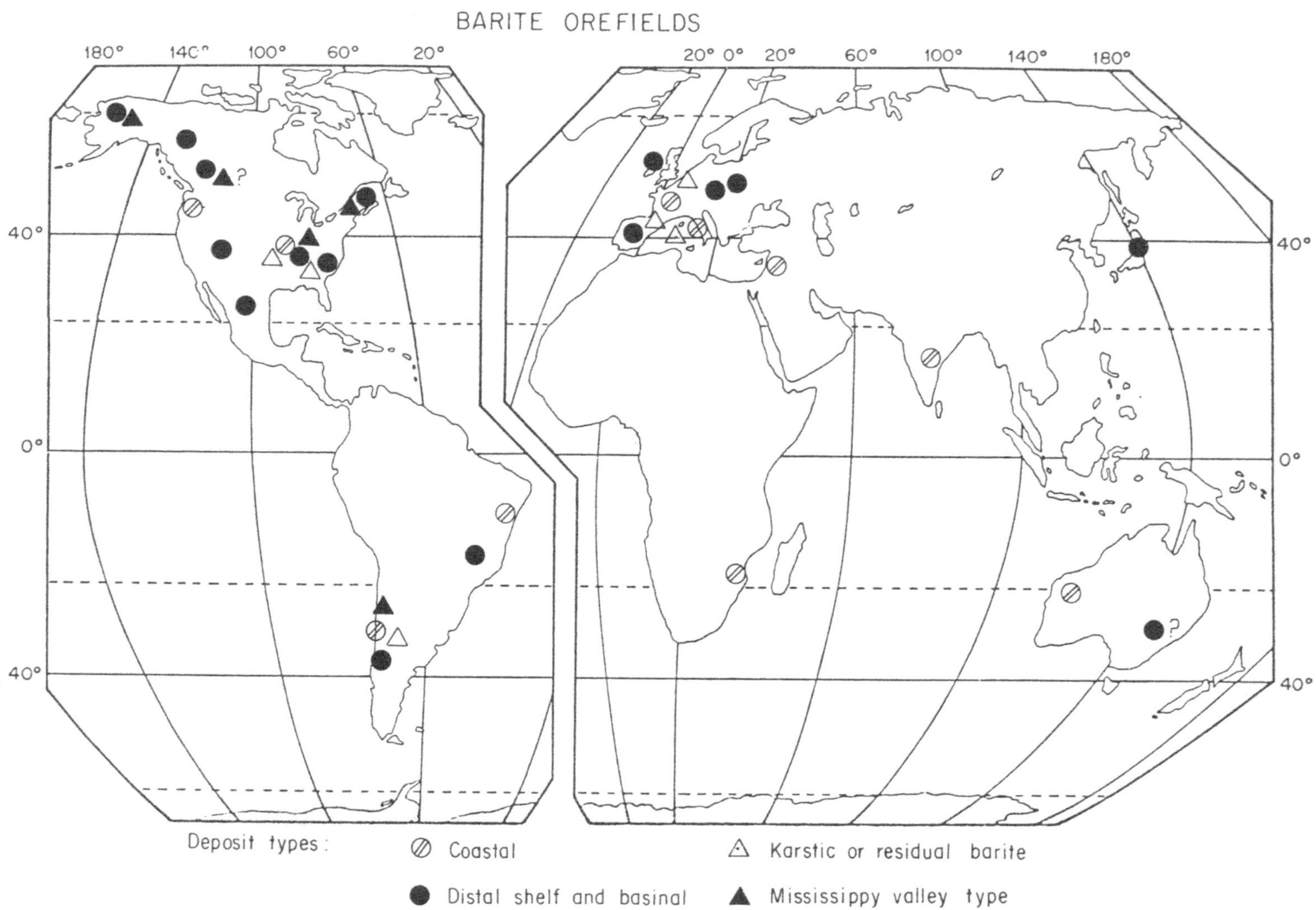

Figure 14–6. Main barite stratabound ore fields classified by their different settings.

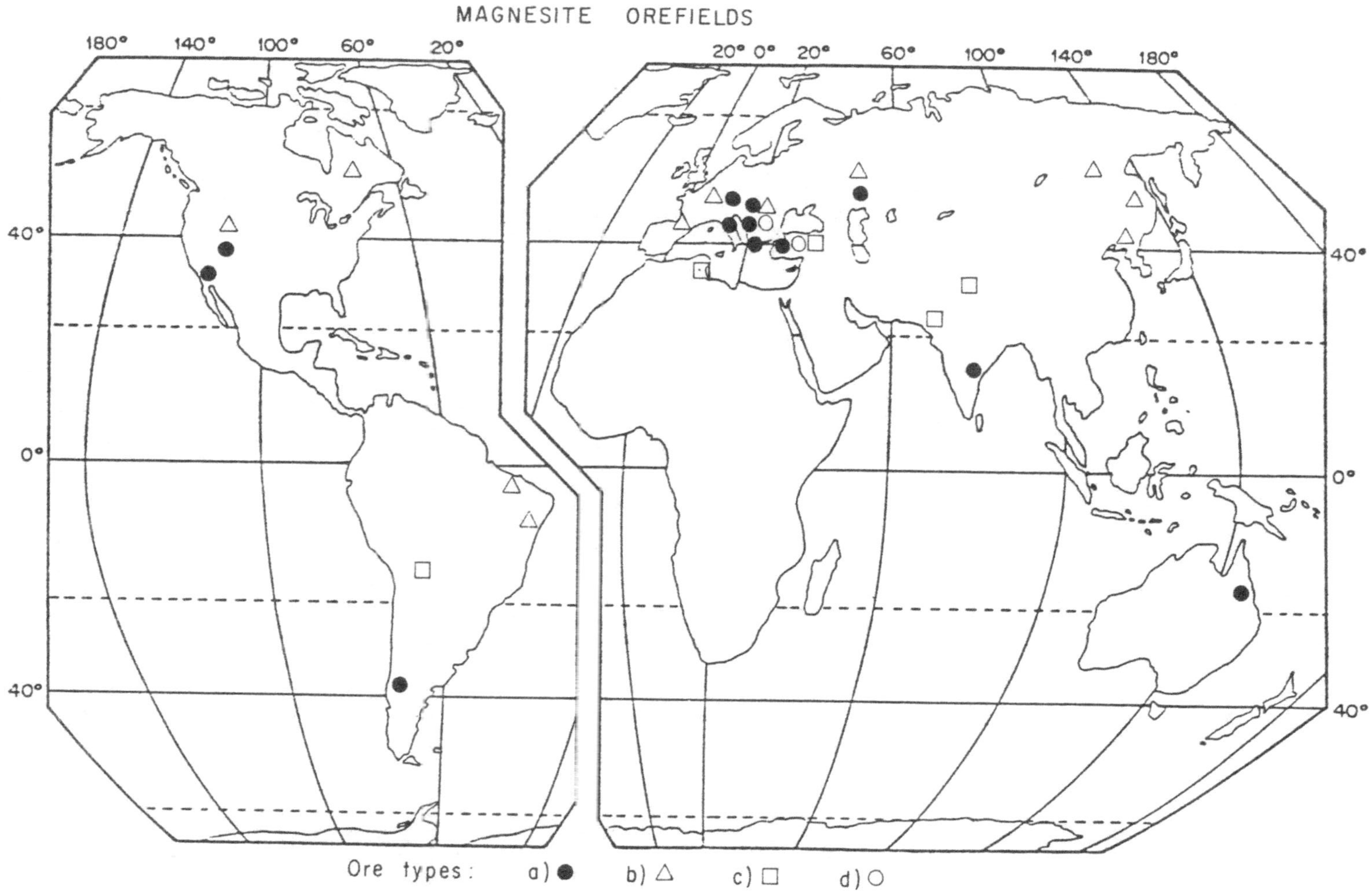

Figure 14–7. Main types of stratabound magnesite ore fields of the world.

a: Cryptocrystalline associated to ultramafics; *b*: crystalline in sedimentary and metamorphic rocks; *c*: saline to hypersaline; *d*: intramontain freshwater basins.

latest Precambrian–earliest Cambrian in the western slope of the Urals (the Satka ore field). Large deposits are also known from the latest Precambrian quartzites and dolostones of the Brazilian craton. Magnesite ores are found in Europe at the famous Paleozoic ore fields of Austria, Spain (Devonian to Carboniferous), and Czechoslovakia (Carboniferous). Younger deposits are known in North America at the Gabbs ore field (Triassic) of Nevada.

The third type ("c" in Fig. 14–7) of ore is well known in a wide time span that covers the Precambrian up to the recently formed deposits of the Trucial Coast of the Persian Gulf. Paleoevaporitic deposits of Proterozoic age, known in Bolivia and Nepal, are under exploitation. The widespread occurrence of evaporitic deposits of latest Proterozoic age, not only in Gondwanaland, but also in other continents, is a clear indication of the break-up of an ancient Pangaea and development of many potential settings for magnesite precipitation. The influence of these paleoevaporitic sequences in the post-Proterozoic mineralization must be taken into consideration as pointed out by Wetzenstein (Chap. 12). Younger deposits have been found in the Late Paleozoic sequences of Pakistan, and in the Permian–Early Triassic of the eastern Alps (Austria).

The magnesite ore fields of the fourth type ("d" in Fig. 14–7) are known in the intramontain basins of Yugoslavia, where they occur in a continental setting related to freshwater Miocene sediments. Similar deposits have been discovered in northern Greece and in Turkish Neogene basins.

Fluorspar Ore Fields

Stratabound fluorspar deposits may be divided into several types based on the different environments of formation (see Chap. 13).

An important type of deposit is that related to a cover over a dolomitic intermediate basement. The cover might be sedimentary, volcanic, or metamorphic, folded or not, but even if incorporated into a cratonic zone, an important discontinuity with the underlying basement plays an important role in the metallogenesis. Examples of these ore fields are found in the Transvaal (South Africa) where the Dolomite series (part of the Transvaal Sequence, 2.0 to 2.3 Ga) have fluorspar concentrations. The Marico ore field, with a distinctive stratabound morphology, has been considered an ancient Mississippi Valley-type deposit. South of Zeerust a large ore field contains reserves of 80–100 million tons with 15 percent CaF_2 along 60 km. On a regional basis, the mineralization is controlled by the paleorelief, but in detail, the ore is located along vuggy horizons and paleokarsts developed in the dolomite during subaerial exposure. The fine rhythmic lamination has been interpreted as a synsedimentary feature, whereas the fluorspar was concentrated during diagenetic processes.

Some other ore fields are also characterized by emplacement at the base of transgressive series, over the contact with the basement. The paleogeography of these settings indicates the close proximity with the basin-continent boundary. Examples of such settings are the Mesozoic transgressive series of France, where the sedimentary rocks deposited over the basement during the Lias produced important ore fields such as Morvan and Vosges. During the Mesozoic the Morvan region was a positive area progressively invaded by the sea. The

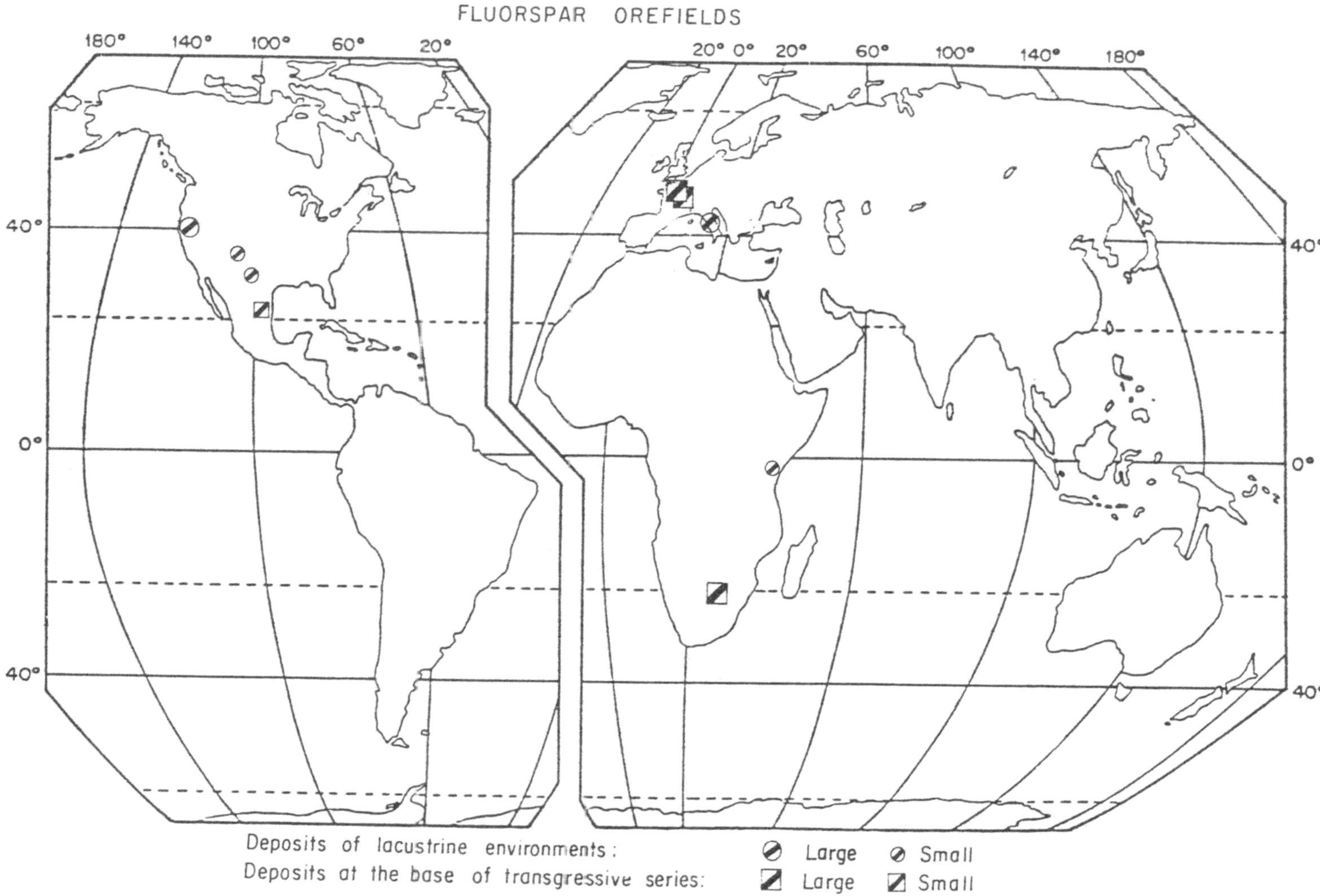

Figure 14-8. Main stratabound fluorspar ore fields.

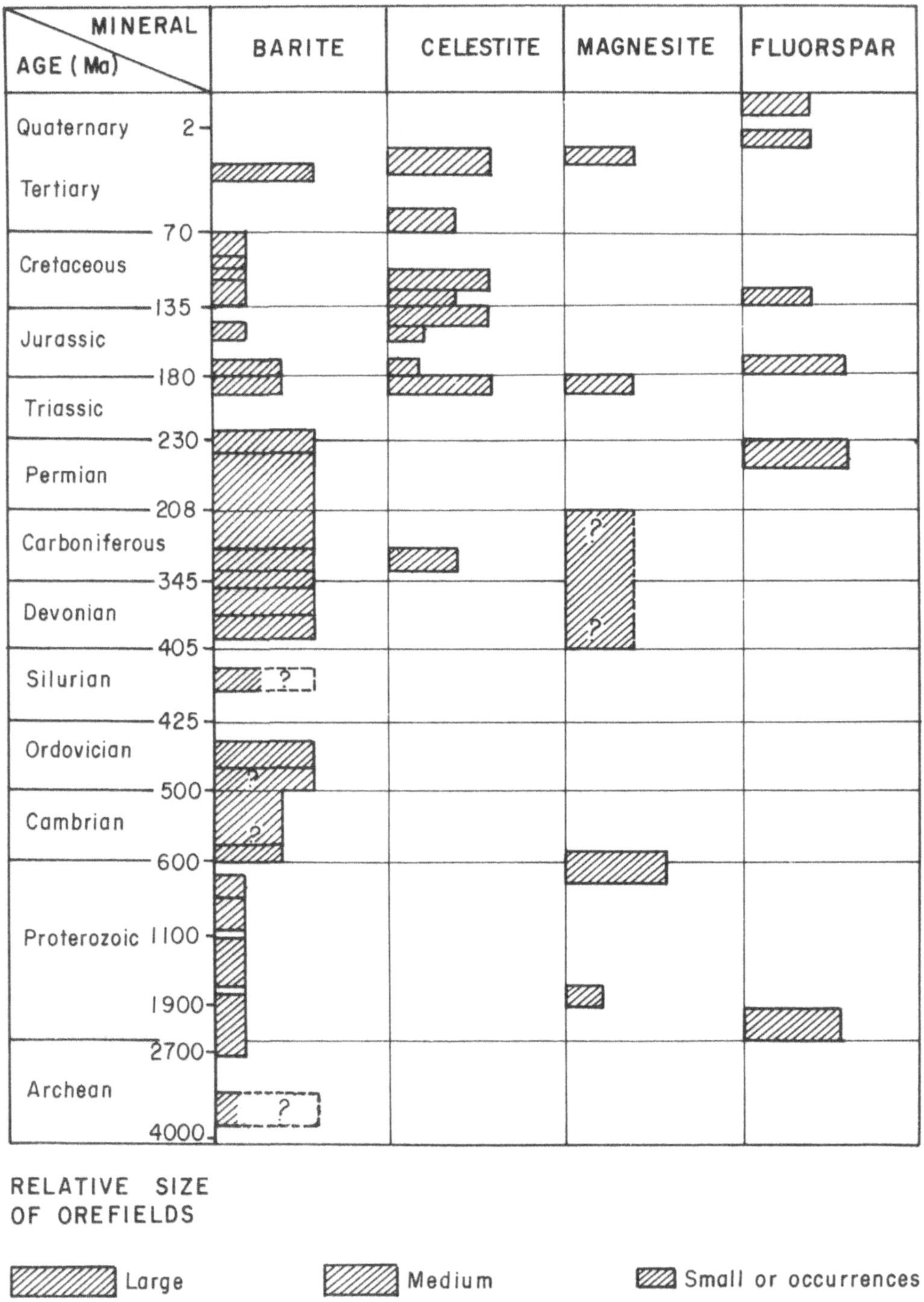

Figure 14–9. Time distribution of stratabound ore fields.

fluorspar is associated with silicified horizons, probably developed during late diagenetic phases derived from the leaching of feldspar. The mineral assemblage is characterized by fluorspar, barite, galena, and sphalerite, with minor amounts of pyrite and chalcopyrite. Mineralization is peneconcordant with the sedimentary sequence, showing a preferential distribution along channels and negative reliefs.

Other famous ore fields are the mantle type deposits of Mexico associated with Lower Cretaceous sedimentary rocks. The ore is located at the contact between limestones and unconformably lying limestones and shales. The mineralization is related to an erosional unconformity that controlled the precipitation of fluorspar. In addition to this paleogeographic control, other factors such as the diagenetic evolution, the interrelation with Tertiary igneous activity, and the paleohydrologic systems active after sedimentation were responsible for remobilization of the fluorspar precipitated earlier.

Although stratabound fluorspar deposits have a worldwide distribution (Fig. 14–8), they are sparse, probably due to the unusual color of the ore and the diagenetic remobilization that obscures the primary features.

STRATABOUND ORE FIELDS THROUGH TIME AND SPACE

The present review shows that stratabound ore fields have a wide variety of tectonic settings and paleogeographic controls that depend on the mineral assemblage and in most cases are independent of time. Figure 14–9 summarizes the time setting of the main ore fields.

The understanding and appraisal of the various and distinctive settings with a common stratabound control allows us to look for new exploration hypotheses that may be used as important tools for prospecting new celestite, barite, magnesite, and fluorspar ore fields in nontraditional environments.

REFERENCES

Arrhenius, G., and E. Bonatti. 1965. Neptunism and vulcanism in the ocean. In *Progress in Oceanography,* vol 3, ed. M. Sears. New York: Pergamon.

Bogoch, R., and M. Shirav. 1978. Petrogenesis of a Senonian barite deposit, Judean Desert, Israel. *Mineralium Deposita* **13**:383–389.

Brodtkorb, M. K. de, V. Ramos, and S. Ametrano. 1975. Los yacimientos estratoligados de celestina-baritina de la F. Huitrín y su origen evaporítico, prov. del Neuquén, Argentina. *2nd Cong. Iberoamericano de Geol. Ec. Bs. As.,* 2, pp. 143–168.

Brodtkorb, M. K. de, I. B. A. Schalamuk, M. Barbieri, H. Puchelt, S. Ametrano, R. Fernandez, and R. Etcheverry. 1987. The nodular barite deposits of Canota, Argentina. VII IAGOD Congress, in press.

Chase, R. L., J. R. Delaney, J. L. Karsten, H. P. Johnson, S. K. Junipe, J. E. Lupton, S. D. Scott, V. Tunnicliffe, S. R. Hammond, and R. E. Duff. 1985. Canadian American Seamount Expedition. 1985: Hydrothermal vents on an axis seamount of the Juan de Fuca ridge. *Nature* **313**:1–3.

Chow, T. J., and E. D. Goldberg. 1960. On the marine geochemistry of barium. *Geochim. Cosmoch. Acta* **20**(3–4):192–198.

Coats, J. S., C. G. Smith, N. J. Forten, M. J. Gallagher, F. May, and W. J. Mc Court.

1980. Stratabound barium-zinc mineralization in Dalradian schist near Aberfeldy, Scotland. *Inst. Min. Metal. Trans.* **89B**:110–122.
Dean, W. E., and B. C. Schreiber. 1977. 33 Authigenic barite, leg 41 Deep Sea Drilling Project. *Initial report of the Deep Sea Drilling Project,* V. XLI, Washington, D.C.: Government Printing Office.
Dunlop, J. S. R. 1978. Shallow water sedimentation at North Pole, Pilbara Block, Western Australia. In *Archean Cherty Metasediments: Their Sedimentology, Micropaleontology, Biochemistry and Significance to Mineralization,* eds. J. E. Glover and D. I. Groves, vol. 2, pp. 30–38. University of Western Australia.
Hanor, J. S., and L. R. Baria. 1977. Controls on the distribution of barite deposits in Arkansas. *Arkansas Geol. Comm. Bull.* **2**:42–49.
Hanor, J. S., and L. H. Chan. 1977. Non conservative behavior of barium during mixing Mississippi River and Gulf of Mexico waters. *Earth and Planetary Sci. Letters* **37**:242–250.
Heinrichs, T. K., and T. O. Reimer. 1977. A sedimentary barite deposit from the Archean Fig Tree Group of the Barberton Mountain Land (South Africa). *Econ. Geology* **72**:1426–1441.
Hickman, A. H. 1973. The North Pole barite deposit, Pilbara gold field. *Western Australia Geological Survey Annual Report, 1972,* pp. 57–60.
Hsü, K. J. 1984. A nonsteady state model for dolomite, evaporite and ore genesis. In *Syngenesis and Epigenesis in the Formation of Mineral Deposits,* eds. A. Wanschkuhn, C. Kluth, R. Zimmermann, pp. 275–286. Berlin: Springer-Verlag.
Jones, E. J. 1987. On some nodular stones obtained by trawling off Colombo in 675 fathoms of water. *Asiatic Soc. Bengal Jour.* **56**:209–212.
Kim, Y., M. Hoshino, H. Aokí, and H. Honma. 1970. Barite nodules collected from the Shiguri–Bank. *Chikyu Kagaku* **24**(2):49–52.
Lambert, I. B., and D. J. Groves. 1981. Early earth evolution and metallogeny. In *Handbook of Stratabound and Stratiform Ore Deposits,* vol. 8, ed. K. H. Wolf, pp. 339–447. Amsterdam: Elsevier.
Marchig, V., P. Möller, H. Backer, and P. Dulski. 1985. Foraminiferal ooze from the Galapagos rift area—hydrothermal impact and diagenetic mobilization of elements. *Marine Geology* **62**:85–104.
Mitchum, R. M., and M. A. Uliana. 1985. Seismic stratigraphy of carbonate depositional sequences, Upper Jurassic–Lower Cretaceous, Neuquén Basin, Argentina. In *Seismic Stratigraphy II,* eds. O. R. Berg and D. G. Woolverton, American Association of Petroleum Geologists Memoir 39, pp. 255–274.
Moro Benito, C., and A. Arribas Moreno. 1981. Los yacimientos españoles de barita estratiforme y su significado metalogenético en el contexto mundial. Tecniterrae. *Rev. Española Geol. Min.* **7**(42):18–45.
Morrow, D. W., H. R. Krouse, E. D. Ghent, G. C. Taylor, and K. R. Danson. 1978. A hypothesis concerning the origin of barite in Devonian carbonate rocks of northeastern British Columbia. *Canadian Jour. Earth Sci.* **15(9)**.
Mpodozis, C. and V. A. Ramos, 1985: The Andes of Chile and Argentina. Symposium on Geology of the Andes and its relation to Hydrocarbon and Mineral Resources, (in press), Santiago de Chile.
Puchelt, H. 1967. Zur Geochemie des Bariums im exogenen Zyklus. *Sitzunsber. Heidelb. Akad. Wiss. Mat. Nat. Kl.* **4**:85–205.
Puchelt, H. 1986. *Sulfid und Oxyderze in der Tyrrhenis in SONNE 41 Cruise.* Karlsruhe, Germany: Institut für Petrographie und Geochemie.
Radhakrishna, B. P., and V. N. Vasudev. 1977. The early Precambrian of the South Indian Shield. *Jour. Geol. Soc. India.* **18**:525–554.
Ramos, V. A., 1985. The Tectonics of the Central Andes: 30°–33° S lat. Geological Society of America. Spec. Pub. (in press).
Reimer, T. O. 1980. Archean sedimentary baryte deposits of the Swaziland Supergroup (Barberton Mountain Land, South Africa). *Precambrian Research* **12**:393–410.

Revelle, R., and K. O. Emery. 1951. Barite concretions from the ocean floor. *Geol. Soc. America Bull.* **62**:707–724.

Rickard, D. T., H. Zweifel, and T. H. Donnelly. 1979. Sulfur isotope systematics in the Asen pyrite-barite deposits, Skellefe District, Sweden. *Econ. Geology* **74**:1060–1068.

Shawe, D. R., F. G. Poole, and D. A. Brobst. 1969. Newly discovered bedded barite deposits in East Northumberland Canyon, Nye Country, Nevada. *Econ. Geology* **4**(3):245–254.

Walker, R. N., M. D. Muir, W. L. Diver, N. Williams, and N. Wilkins. 1977. Evidence of major sulphate evaporite deposits in the Proterozoic Mc Arthur Group, Northern Territory, Australia. *Nature* **265**:526–529.

INDEX

GPSR Compliance

The European Union's (EU) General Product Safety Regulation (GPSR) is a set of rules that requires consumer products to be safe and our obligations to ensure this.

If you have any concerns about our products, you can contact us on ProductSafety@springernature.com

In case Publisher is established outside the EU, the EU authorized representative is:

Springer Nature Customer Service Center GmbH
Europaplatz 3
69115 Heidelberg, Germany

Batch number: 10692442

Printed by Printforce, the Netherlands